Contents

3/9/84

PRINCIPLES OF ECOLOGY

R.J. Putman and
S.D. Wratten

CROOM HELM
London and Canberra

© 1984 R.J. Putman and S.D. Wratten
Croom Helm Ltd, Provident House, Burrell Row,
Beckenham, Kent BR3 1AT

Croom Helm Australia Pty Ltd.,
28 Kembla Street, Fyshwick,
ACT 2609, Australia

British Library Cataloguing in Publication Data

Putman, Rory J.
    Principles of ecology.
    1. Ecology
    I. Title   II. Wratten, Stephen D.
    574.5   QH541
    ISBN 0-7099-2016-4
    ISBN 0-7099-2050-4 Pbk

Printed and bound in Great Britain
by Billing & Sons Limited, Worcester.

Preface

As Ecology teachers ourselves we have become increasingly aware of the lack of a single comprehensive textbook of Ecology which we can recommend unreservedly to our students. While general, review texts are readily available in other fields, recent publications in Ecology have tended for the most part to be small, specialised works on single aspects of the subject. Such general texts as *are* available are often rather too detailed and, in addition, tend to be somewhat biased towards one aspect of the discipline or another and are thus not truly balanced syntheses of current knowledge. Ecology is, in addition, a rapidly developing subject: new information is being gathered all the time on a variety of key questions; new approaches and techniques open up whole new areas of research and establish new principles. Already things have changed radically since the early '70s and we feel there is a need for an up to date student text that will include some of this newer material. We have tried, therefore, to create a text that will review all the major principles and tenets within the whole field of Ecology, presenting the generally accepted theories and fundamentals and reviewing carefully the evidence on which such principles have been founded. While recent developments in ecological thought are emphasised, we hope that these will not dominate the material to the extent where the older-established principles are ignored or overlooked. Rather we have tried to present both traditional and new ideas in complement, in a balanced and integrated way. Such synthesis permits in addition the development of new ideas and extensions from the marriage of new and old; while this book is essentially conceived of as a review textbook, we have included some original, and in places frankly speculative material.

Our approach throughout is critical and evaluative; neither old nor new theories are taken on trust but are examined rigorously and often critically: the student should be aware of what material is sound and which untested or controversial. Those principles which survive such analysis we have put together in a new and current synthesis of 'the state of the art'. The emphasis throughout is of course on a functional approach and it is the underlying principles of ecology that are to be examined. We have deliberately avoided a descriptive, 'habitat' or taxonomic approach to the material, feeling that Ecology must be shown to be a quantitative, exact science, with certain underlying principles and laws as fundamental as the laws of pure physics.

Preface

The text is aimed at 2nd or 3rd year University undergraduates — but should also be suitable for any interested advanced student of Ecology. Throughout, consideration of each topic is complete in itself and does not rely on any background information. (Where such background knowledge is to be assumed, the relevant background information will be reviewed briefly in the text before it is extended upon.) Treatment of each topic should therefore be complete and self-contained. Throughout the text, reference will be made to the original papers upon which the material is based; the reader should be encouraged to return to the original material for detailed study. This volume, while it attempts to be comprehensive, cannot be particularly detailed; it concentrates on principles and is intended to provide a framework to which the student may relate his own further reading.

In preparation of such a text we have obviously had a great deal of help from a great many people. We would like to take this opportunity to thank the many friends and colleagues with whom we have thrashed out our ideas, particularly mentioning Peter Edwards, Michael Fenner, Peter Hopkins, Andrew Johns, Stuart Pimm, Ian Spellerberg, Paul Vickerman and Allan Watt; each read many of the chapters in draft form and the revised manuscript benefited tremendously from their generous and constructive criticism. We thank them most warmly for helping to create the book; any mistakes that remain are our own.

Much of SDW's work on the text was carried out during a visit in 1982 to the Department of Zoology of the University of Canterbury, New Zealand. Thanks are due to Professor R.C. Pilgrim and colleagues there for their hospitality and facilities. Finally we would thank Dawn Trenchard and Alison Hamlin for undertaking our typing, and Raymond Cornick, Paul Chester, Nickie O'Rourke and Claire Wratten for help with the drawings.

Introduction

Ecology is the study of organisms and their environment — and the interrelationship between the two. The term has in fact taken on a variety of other meanings in addition. Most commonly (and for want of any alternative) it has come to be used, in lower case, to denote the way of life of a particular organism: a convenient term for the relationship itself between organism and environment. More debasedly, the word 'Ecology' is used politically, and usually uninformedly, as an easy analogue for the environment and its natural function or as an abbreviation for 'Human ecology': a sense in which we shall endeavour to avoid using it in this volume. Ecology, as we shall examine it — or the study of the 'ecology' of different organisms, may be considered the *science* of natural history — the scientific study of how animals and plants live, and why they live the way they do: a study aimed at understanding the basic underlying principles of operation of natural systems.

This interaction between organism and the environment is the basic relationship for another biological phenomenon too. Any organism is under continuous evolutionary pressure to be optimally adapted to its environment. The same interaction with the environment which is the province of Ecology constitutes the selection pressure powering evolutionary adaptation. The study of Ecology is thus in one sense the study of selection pressures, and the results of selection past: in adaptation. It is important therefore to retain throughout this evolutionary perspective in Ecology. Not only is what we study the 'stuff' of evolution; the things we do study are themselves not constant but changing as evolution itself continues.

1 The Organism and its Environment

1.1 The Organism and its Abiotic Environment: Limits to Tolerance

The very word 'environment' conjures up an impression of a structural, physical 'stage-set' upon which background biological processes are acted out. In practice, of course, any organism's environment includes other organisms, as well as this physical 'set'; living organisms of its own or different species are as much a part of any creature's environment as are more structural abiotic components as we shall later consider. However, the physicochemical, *abiotic*, background is perhaps the most obvious, and fundamental facet of the environment. Abiotic relationships are the primary factors determining whether or not any organism can exist in a certain environment, and, while biotic interactions may later modify how an organism lives within that structural environment, may alter the fine detail of its ecology, if the organism is not initially adapted to its physico-chemical environment, it cannot exist there in the first place.

Temperature, light intensity, concentration of oxygen, carbon dioxide, wind, exposure, chemical nature of bedrock or waterflow, speed of waterflow in streams, structure of stream bed, soil type, availability of nitrogen and other chemicals — these and a host of other characteristics of the abiotic environment exert a profound influence on organisms living — or trying to live — in that system. Because physiological processes proceed at different rates under different conditions any one organism only has a limited range of conditions in which it may survive; within those limits is a further restricted range where the organism may operate at maximum efficiency. The very fact of having such *tolerance limits* restricts markedly the conditions under which an animal or plant can operate efficiently and provides the mechanism whereby influences from the abiotic environment may affect its ecology in such a fundamental sense.

Curves of performance may be drawn for any organism for any particular physiological process representing its efficiency of operation over a range of some physicochemical parameter. Such curves, known as 'tolerance curves' (Shelford, 1913), are typically bell-shaped, with their peaks representing optimal conditions for a particular physiological process, and their tails representing limits of tolerance (Figure 1.1). From consideration of such tolerance curves for individual physiological processes, we may derive tolerance limits for the organism as a whole in regard to any particular environmental variable. It is usual to define such limits as a series of 'nesting' inner limits. Thus, within an upper and lower *lethal limit*, at which death occurs, we may recognise a pair of inner limits, the *critical maximum and minimum*, outside which the organism, though not dead, is ecologically inviable (let us say, for an animal, a range beyond which it is no longer able to move). Within this we may define yet another, narrower, *'preferred range'* – and within this again, an *'optimum range'* (Figure 1.2).

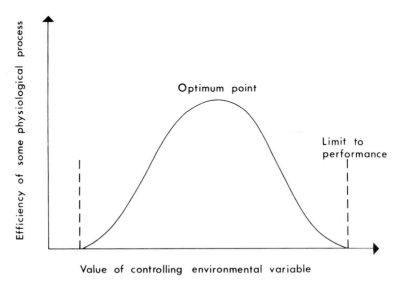

Figure 1.1: Limits to Performance of Physiological Processes.

Similar curves can be drawn as envelopes of the tolerance limits of a number of individuals to define the limits to tolerance of a whole species: the original conception of Shelford (1913) who then used such species tolerance curves to explain observed biogeographic distributions of animals and plants. In this sense the curve represents the numbers of individuals within the population which may be found occupying environments at a given value of the chosen physicochemical factor. Shelford's 'population', or 'species tolerance' curves are defined in terms of a central zone (the 'bell' of the normal distribution) where the majority of individuals occur: the optimum, or preferred range for the species as a whole, with the two 'tails' of the curves, occupied by relatively few individuals operating under less than optimum conditions, considered the stress zones.

Some individuals and species have narrow, high-peaked tolerance curves for any particular environmental variable; such are described as stenotopic organisms. (The prefix *steno* (narrow) may also be applied to specific responses, thus: steno-thermic, steno-haline: narrow tolerance to temperature, salinity.) In other organisms (or in the same organism for a different environmental variable) curves may be broader and flatter. (Broad tolerance curves are given the prefix *eury* (thus eurythermic, euryhaline etc.)).

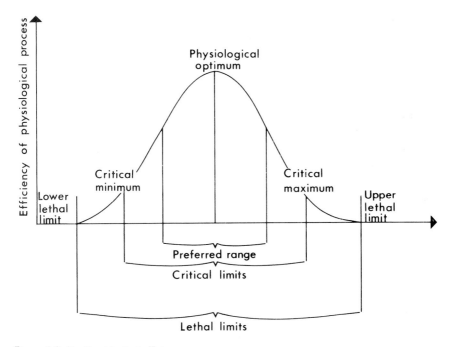

Figure 1.2: Nesting Limits to Tolerance

Such tolerance curves are not, as is sometimes presumed, immutable. Their position within the environmental gradient, and their breadth may be shifted to a certain extent by genetic or evolutionary change, or by physiological or behavioural changes during the organism's lifetime. Thus an organism may adjust over time to a slight shift in environmental conditions, may even evolve to exploit extremes of conditions: as have for example the organisms adapted to living in the hot springs of volcanic geysers. But such adaptation must be at some cost. In general the broader the tolerance, the lower the overall efficiency at any one point. There appears to be something in the nature of a trade-off: a jack of all trades can be master of none, while organisms that 'choose' to specialise and become extremely efficient over a narrow range of conditions, like our hot-spring denizens, lose the ability to operate at *all* under other conditions (Figure 1.3).

1.2 Interactions between Environmental Variables

There is, in fact, considerable interaction between the effects of different environmental variables, so that it is somewhat misleading to consider responses to any one abiotic factor in strict isolation. Tolerance to temperature for instance in many terrestrial organisms is intricately bound up with tolerance to relative

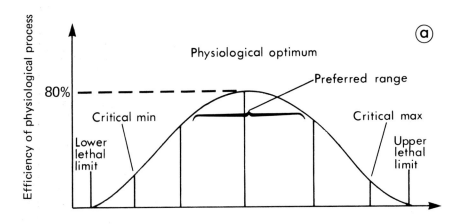

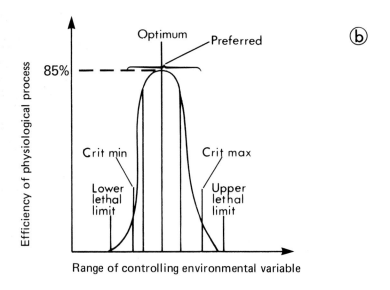

Range of controlling environmental variable

Figure 1.3: Tolerance Curves of a Generalist (a) and a Physiological Specialist (b). The cost of specialisation — whether for increased efficiency at optimum, or in order to be able to operate under extremes of conditions — is reduced lability.

humidity because the physiological processes affecting ttemperature regulation are themselves controlled by water availability. In more general terms, one may frequently observe interdependence of tolerances to pairs or groups of environmental variables which affect the same physiological process. As a result, organisms which are stressed along any one environmental gradient are less able to tolerate a wide range of conditions along other environmental gradients as well. Thus organisms near the limits of their range in one environmental variable, or those which have 'chosen' to become physiological specialists in order to exploit an extreme environment, become less able to tolerate variation in conditions in other abiotic parameters as well as that one actually under stress. (This reduced tolerance to *all* abiotic variables by organisms at extremes of tolerance to one variable, is clearly a factor of extreme importance in conservational exercises. For such organisms, *variability* in environmental conditions may become as important a limitation as the actual values of the abiotic variables themselves.) Pianka (1978) offers a clear illustration of this interaction between tolerances to two different variables. He considers the fitness of a hypothetical organism in various microhabitats as a function of relative humidity (or vapour pressure deficit), as shown in Figure 1.4a. Fitness varies similarly along a temperature gradient (Figure 1.4b). Figure 1.4c combines humidity and temperature conditions to show variation in fitness with respect to both variables simultaneously. The range of thermal conditions tolerated is narrower at very low and very high humidities than it is at intermediate and more optimal humidities. Similarly, an organism's tolerance range for relative humidity is narrower at extreme temperatures than it is at more optimal ones. The organism's thermal optimum depends upon humidity conditions (and vice versa). Fitness reaches its maximum at intermediate temperatures and humidities.

Nor is such interrelation of variables purely hypothetical. In the real world, Haefner (1970) has shown interactive effects of *three* factors on determining the 'optimum range' for the sand shrimp *Crangon septemspinosa*. Defining tolerance limits in terms of percentage mortality, Haefner derives interactive zones for tolerance to salinity and temperature by eggbearing female sand shrimps to produce a figure for a real example very much of the same form as Figure 4c. Once again fitness reaches its maximum at intermediate levels of both variables (Figure 1.5a). Haefner's elegant experiments actually extend to consider the effect of a third environmental variable – dissolved oxygen, comparing the temperature/salinity tolerances curves of sand shrimps in water at low concentrations of dissolved oxygen (Figure 1.5a) with those in aerated water (Figure 1.5b). Once again the tolerance zones differ under the influence of yet another interactive variable. Such examples emphasise that the concept of a single fixed optimum is in some ways an artifact of considering only one environmental dimension at a time (Pianka, 1978).

Such physiological limits to tolerance of a whole range of abiotic parameters clearly have a profound influence on the distributions of animals and plants, and the efficiency of their operation once they are there. Indeed it is probably fair to say that much of the observed distribution of living organisms is determined

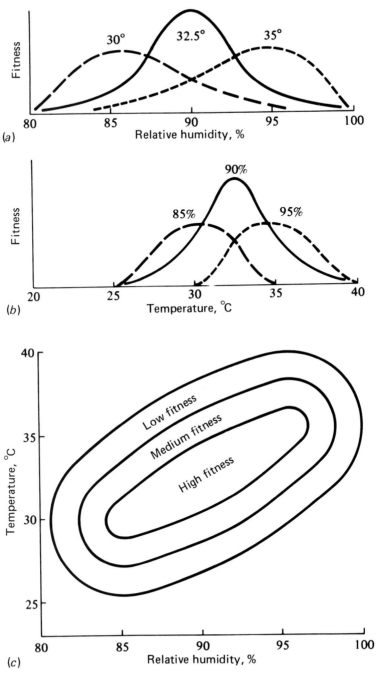

Figure 1.4: Hypothetical Response Curves, showing how Two Variables can Interact to Determine an Organism's Fitness. Fitness is reduced at extremes of either temperature or humidity, and the range of humidities tolerated is less at extreme temperatures than at intermediate ones — see text for details. Source: From Pianka, 1978.

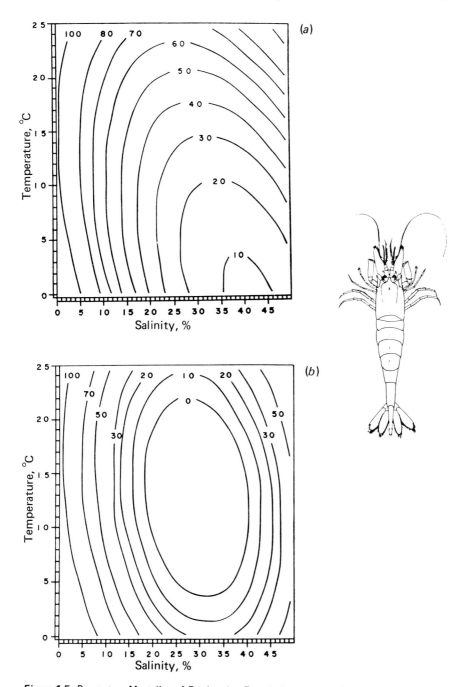

Figure 1.5: Percentage Mortality of Egg-bearing Female Sandshrimps (*Crangon septemspinosa*) at Different Temperatures and Salinities. (a) Shows results at low concentrations of dissolved oxygen; (b) shows results in aerated water. Source: After Haefner, 1970.

by such abiotic considerations. More strictly it is fair to say that abiotic factors
− through limits to tolerance − may determine where an organism may *not*
occur; but in practice will not dictate where such an organism *will* occur. For
within those systems where abiotic conditions might permit any organism to
exist, its actual occurrence may then be limited by other factors − accidents
of biogeographic history for example, or biotic interactions. Relationships with
other organisms may exclude an organism from many systems in which physiologically it could survive − through competition or predation perhaps. Thus we
may distinguish between the range of conditions under which a species can exist
best in isolation (*physiological* range; *physiological* optimum) and its actual
observed range in nature where it grows in association with other species (*ecological* range; *ecological* optimum (Figure 1.6)).

There is, of course, considerable interaction between the two types of process, biotic and abiotic. An organism will be less competitively able, populations
will be less well able to buffer the depredations of predators or parasites, if they
are operating in any case at or close to the limits of physiological tolerance.
There have been in addition a number of experiments which demonstrate, by
converse, how the tolerance range or optimum for some *abiotic* factor may be
modified by competition (e.g. Harper, 1964; Ellenberg, 1958) (Figure 1.7). In
short as we have noted, an organism's physiological relationships with its abiotic
environment cannot in themselves explain its distribution; fairer to say that
considerations of abiotic influence enable us to say where an organism may *not*
live.

1.3 Macro-environment and Micro-environment

The abiotic factors which we may consider to affect living organisms are broadly
speaking characteristic of a relatively large regional area. Different geographical
areas may be classified in terms of a set of physicochemical characteristics: soil
type, geological formation, climate etc. etc. And from such classification certain
general ecological conclusions can be drawn about what organisms may or may
not occur. Thus we may recognise certain general climatic regions which will in
turn be characterised by certain ecological associations or *biomes*. We talk about
the tropical, temperature or tundra *biomes*, linking areas of similar abiotic
environment and thus ecological structure (Putman, 1983a). However, such
conclusions can only be of the most general nature: for an organism is influenced
only by its own immediate environment, and gross, general characteristics of a
geographic region may have been considerably modified by the time they
impinge on any one organism. Little local features of topography, shade, exposure, slope may translate a 'general' climate for the region into a host of
markedly different micro-climates in so far as they affect living organisms; local
changes in soil type, drainage or other variables, modify other elements of the
environment at the smaller scale. Yet it is the characteristics of these micro-
environments that affect the organism; these are the abiotic factors which

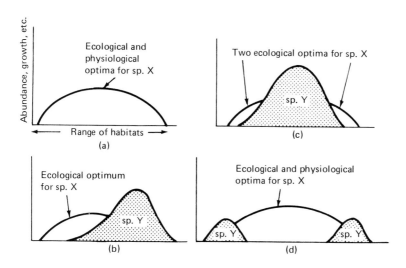

Figure 1.6: Examples of How the Ecological Optimum may be Different from the Physiological Optimum due to Competition. In (a) species X grows alone; laboratory experiments determine that its physiological optimum lies in the centre of the curve as shown. In nature, it faces competition from other species, which displace it from habitats it could grow in, alone; consequently its ecological optimum is displaced from its physiological optimum (b). In (c) species X competes poorly, and is displaced from the middle of its range. In (d) species X competes poorly at its limits to tolerance, with the result that the ecological optimum still coincides with the physiological optimum, but the range is restricted. Source: From Walter, 1973; after Barbour, Burk and Pitts, 1980.

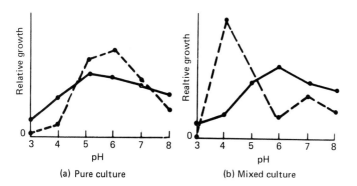

Figure 1.7: The Relative Growth of Wild Radish (*Raphanus raphanistrum*) (solid line) and Spurry (*Spergula arvensis*) (dotted line) as a Function of Substrate pH. (a) Plants grown in isolation, (b) plants grown in competition with each other. Source: From Ellenburg, 1958, after Barbour, Burk and Pitts, 1980.

influence its distribution and performance — not the gross 'area-descriptors' which may have been so much modified by the time they 'reach' the organism itself. And it is extremely important in considering the interaction between organisms and their abiotic environment that we recognise that we are dealing with such micro-environments. (There is for example, no point in measuring temperatures 20 m up an oak tree to relate to the activity and distribution of a beetle that lives its life in the leaf litter at the soil surface.)

Many of the factors translating macro- into micro-environmental variables are themselves abiotic. Slight changes in geology, soil type, slope, exposure to the sun, all affect local physicochemical conditions. Little hollows sheltered from air movements create different conditions of temperature and humidity. Mountains and mole-hills have their windward and lee sides, sunny and shaded sides. But the biotic community that develops in these environments may further modify the micro-environment. Plants offer shade beneath them; the structure and tangle of stems and leaves changes air-currents, insulates the ground below. Their respiratory activity, coupled with the restricted air movement creates local changes in humidity and in gas concentration. The litter they drop with their leaf fall acts like a blanket on the soil before it is decomposed, insulating the soil surface. Animals, too, are involved in this translation of macro- to micro-environment. Those that dig burrows create a new environment for others to exploit. Patterns of feeding by herbivores affect vegetational structures, and the micro-environments they create. Patterns of dung deposition affect local soil conditions — and in themselves create a new microhabitat for saprophytes etc. etc.

The importance of these micro-environmental variables, and the degree to which these may differ from macro-environmental characteristics may perhaps be illustrated in consideration of that set of factors influencing micro-*climate*. In a study of the micro-environment offered by a fallen tree trunk Schimitschek (1931) analysed the use of such a habitat by the bark beetle, *Ips typographus*. Even within such an entity as a fallen trunk, Schimitschek recognised five distinct microhabitats with markedly different environmental characteristics (Figure 1.8). In analysis of the use of these different segments of the trunk by bark beetles for reproduction, Schimitschek noted that segment 4, alone, proved suitable for normal development, since in segment 1 — that most exposed to the sun — no eggs could be laid; in segment 2, eggs were probably laid, but through exposure to the sun and lack of humidity they dried up; in segment 3 larvae developed, but since the areas were still well heated, died before reaching maturity; and in segment 5, below the log, conditions became too damp, and larval mortality was between 75 and 92 per cent. Another delightful study of micro-climatic effects and the translation of macro-climate to micro-climate — due in this case to the insulating properties of plants — is that of Coe (1969). Temperature measurements were taken in three zones within giant tussocks of grass (*Festuca* spp.) on Mount Kenya. Temperature fluctuations in the outer air, were buffered within the plant to a fluctuation between 0.3°C-13.6°C in the outer leaves, 1.8°C-11.7°C in the inner leaves; the basal part of the tussock showed a temperature change of only 2.1°C around a mean of 7.0°C. Such vast alteration of climatic effects by the time they reach the small scale at which they influence the organisms of the environment, stresses again the importance in ecological terms of studying abiotic factors at the level of the micro-environment.

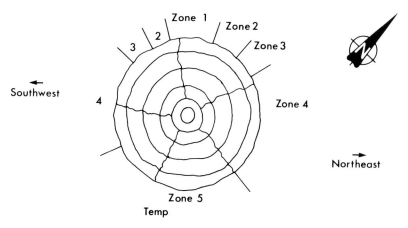

Figure 1.8: Microclimatic Variation in a Decaying Tree Trunk. Source: After Schimitschek, 1931.

1.4 Adjustment of Tolerance Limits

As we have noted, the tolerance limits of any one organism are not immutable. Through evolutionary time, tolerance limits and range of optima may be shifted, displaced or expanded. Even over a shorter time scale various minor adjustments may be made.

1.4a Acclimation

Organisms may adjust very slightly, their range of operation with respect to some environmental parameter, or suite of parameters through *acclimation*. Prolonged exposure to conditions slightly to one side of the optimum (but usually still within the 'preferred range') results in displacement of the tolerance curve to produce a new optimum at the ambient value and new upper and lower critical limits (Figure 1.9). This process of acclimation involves changes in enzyme systems (for it is the restricted range of conditions under which enzymes operate efficiently which imposes the tolerance limits in the first place), but the details of such changes are uncertain. Such acclimation may be completed within a relatively short time — as low as 24 hours for most small animals. (We should perhaps distinguish here between two terms which often appear in the literature as *acclimated* and *acclimatised*. For our purposes we shall define *acclimation* as a physiological compensatory mechanism that may be induced under experimental conditions, and *acclimatisation* as a physiological compensatory change that is

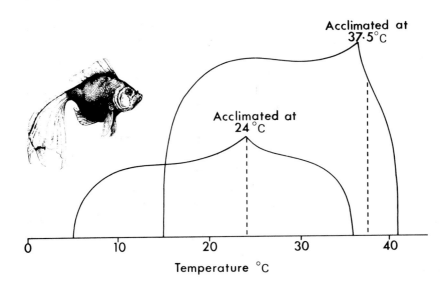

Figure 1.9: Thermal Tolerances in Goldfish Acclimated to Different Temperatures. Upper and lower lethal temperatures, defining the total thermal range of the fish, are shown for individuals acclimated at 24°C and 37.5°C. Data from: Fry, Brett and Clawson, 1942.

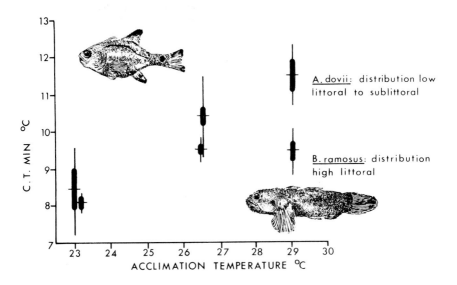

Figure 1.10: Influence of Thermal History (Acclimation Temperature) on the Critical Minimum Temperature of *Apogon dovii* and *Bathygobius ramosus*. Horizontal line is mean, blocks are 95% confidence limits of the mean, vertical lines are ranges of observed values. Source: Graham 1972.

induced (and usually over a longer time scale) under natural environmental conditions.)

Examples of such acclimation are legion. Adjustment of critical minimum temperature of water skinks (*Sphenomorphus*) from South Eastern Australia has been demonstrated by Spellerberg, 1972; the work of Graham (1972) demonstrates acclimation of the lower lethal temperature in tropical fish species (Figure 1.10) and more importantly shows a differential capacity for such acclimation between species, related to littoral distribution. Differential capacity for acclimation is also evident from the work of Billings *et al.* (1971) on alpine sorrel. Alpine sorrel (*Oxyria digyna*) seeds were collected from a range of habitats, germinated and grown in a uniform greenhouse for four months, then subdivided into three growth chamber environments: warm (32/21°C day/night), medium (21/10°C), and cold (12/4°C). After five to six months in the chambers, replicates of each collection were measured for net photosynthesis at a range of temperatures, from 10 to 43°C, and the optimum temperature for photosynthesis was noted. Table 1.1 shows clear acclimation of photosynthetic optimum but also marks a clear distinction between the capacity for such acclimation of different ecotypes.

Table 1.1: The Effect of Acclimation Temperature on the Optimum Temperature for Photosynthesis of Alpine and Arctic Ecotype Extremes of Alpine Sorrel (*Oxyria digyna*). Source: Adapted from Billings *et al.* (1971).

| Population site | Ecotype | Optimum temperature for photosynthesis at each acclimation regime (°C) | | | |
		Warm (32/21°C)	Medium (21/10°C)	Cold (12/4°C)	Ra
Sonora Pass, California	alpine	28	21.5	17	11
Pitmegea River, Alaska	arctic	21	20.5	20	1

Further adjustment is possible in the development of special 'back-up' physiological systems brought into operation only in extreme conditions. Such mechanisms are usually invoked merely in expansion of lethal limits: the organism is able to survive extreme conditions, but rarely remains active. Thus certain ectotherms are able to supercool: the body temperature may drop well below that temperature at which body tissues freeze without the tissues freezing. The tissue temperature of the codling moth, *Cydia pomonella*, for example can go down to −25°C; the hymenopteran *Bracon* can supercool to −40°C and regularly has to withstand tissue temperatures of −15°C in its normal environment. Among plants, lichens show an equivalent ability to survive desiccation without damage. Such mechanisms clearly enable the organism to survive well beyond its normal lethal limits. (Supercooling ability is in itself subject to acclimation. Sømme and Conradi-Larsen (1977) have made some detailed studies of cold-hardiness in collembolans and found that in some species the extent of supercooling can be altered by acclimation to low temperatures. Of more importance from an ecological point of view, Sømme and Conradi-Larsen were able to show that the ability to supercool varied on a seasonal basis: that is during the winter animals had an increased ability to supercool.)

1.4b Dormancy

The concept of 'opting out' — of becoming inactive, is in itself a useful physiological mechanism for withstanding, temporarily, suboptimal environmental conditions. Conditions which would be well outside preference limits, if not lethal limits, were the organism active, may be accommodated by inactivity. The animal or plant enters a phase of dormancy, or suspended animation and in this state may withstand a far wider range of conditions. Extreme examples may be cited: of *Amoebae* encysting when a temporary pond dries up — or even more complicated organisms such as *Chirocephalus* whose eggs may be dormant for many years. Seeds, too, may enter such dormant phases — withstanding extremes of conditions and remaining viable for up to 1000 years (Table 1.2).

But even under less extreme circumstances, seasonal dormancy may be a valid way of colonising an environment otherside beyond one's tolerance ranges. Many insects enter a state of *diapause* in unfavourable weather — where metabolic rates are reduced to one-tenth of that of animals not in diapause: and such insects show an increased ability to withstand cold. Endothermic animals too, which regulate their own body temperature by physiological means relatively independently of prevailing conditions, may 'opt out' when environmental temperatures exceed the range within which they can effectively regulate, by entering a state of *torpor*. In such a state they become functionally ectothermic, allowing body temperatures to change with environmental temperature. (For many ectotherms, cold temperatures directly curtail activity and induce dormancy for *diapause*; true *torpor* is the endothermic equivalent of this. The more complex phenomena of *'hibernation'* and *'aestivation'* are triggered instead by intermediate stimuli (changes in photoperiod for example) or are linked to an endogenous cycle so that the animal can prepare in advance for the dormant period by building up food reserves.) Plants, too, can withstand extremes of conditions by temporary 'withdrawal'. Many tropical and subtropical trees shed their leaves during seasonal periods of drought; while temperate and arctic broad-leaved trees shed their leaves in the autumn to avoid desiccation. Moisture frozen in the soil is unavailable to plants; if these trees kept their leaves through the winter, transpiration of water from the leaves would rapidly dehydrate them.

The advantages of such 'opting out' mechanisms are easily demonstrated. A very detailed study of torpor responses in the California pocket mouse *Perognathus californicus* allows for a precise examination of the energy saving made possible by this state. If we take an extreme case, that is if one of these mice were to enter torpor at 15°C and then immediately arouse, the process would take 2.9 hours. The cost of maintaining normal body temperature for this length of time has been calculated as 11.9 ml O_2/g. The oxygen consumption during entry and arousal from torpor is 6.5 ml O_2/g (Bartholomew, 1968, 1982). Thus even the shortest period of torpor for this species can be viewed as an energy-saving device.

Table 1.2: Dormancy in Seeds.

Seeds may enter a resting state and remain dormant in the soil for very considerable periods until such time as conditions become favourable for germination. The current (substantiated) record is held by the Egyptian lotus, *Nelubium speciosum*; of a number of seeds excavated in Manchuria and radio-dated at over 1000 years old, over 80% were still viable. This is obviously an exceptional example, but potential dormancy of up to 30 years is quite common. Table 1.2, below, lists, from a number of sources, dormancy periods for a selection of common plant species. The table presents the overall limit to dormancy (beyond which the seeds are not viable) together with 'half-lives' (the period beyond which less than 50% viability remains).

Species	Limits to germination (time in years)	50% germination (time in years)	Authority
Avena fatua	1	1	a
Helianthus annuus	1	1	a
Polygonum scandens	1	1	a
Agropyron repens	1–3	1	a
Panicum virgatum	3	1	a
Plantago major	{ 10	3	a
	15	—	b
Stellaria media	{ 10	6	a
	30	—	b
Pastinaca sativa	16	1	a
Phleum pratense	21	10	a
Plantago lanceolata	21	3	a
Chrysanthemum leucanthemum	30	10	a
Polygonum persicaria	30	10	a
Verbascum glauca	30	—	b
Portulaca oleracea	{ 30	1	a
	40	—	b
Anagallis arvensis	32	—	c
Papaver rhoeas	32	—	c
Amaranthus retroflexus	40	—	b
Boehmeria nivea	39	21	a
Capsella bursa-pastoris	39		a
Chenopodium album	39	21	a
Convolvulus sepium	39	39	a
Potentilla norvegica	39	21	a
Brassica nigra	50		b,a
Polygonum hydropiper	50	—	b
Oenothera biennis	80	21	b,a
Rumex crispus	80	10–12	b,a
Verbascum blattaria	80	—	b

Sources: (a) Toole and Brown, 1946; (b) Darlington and Steinbauer, 1961; (c) Brenchley, 1918.

Many physiological changes accompany dormancy (Prosser, 1973). The onset of hibernation in mammals is anticipated by the accumulation of a specific type of fat, with a low melting point, that will not harden and cause stiffness at low temperature. Heartbeat is reduced (hibernating ground squirrels have heart rates of 7-10 beats per minute by comparison with normal active rates of between 200 and 400 beats per minute (Svihla *et al.*, 1951); with a reduced rate of blood flow, blood chemistry must change to prevent clotting. In ectothermic cold diapause, water is chemically bound or reduced to prevent freezing and metabolism drops to near zero; in summer diapause, drought-resistant insects either allow their bodies to dry out and tolerate desiccation or secrete an impermeable outer covering to prevent drying. Plant seeds and the spores of bacteria and fungi have similar dormancy mechanisms (e.g. Koller, 1969; Wareing, 1966).

1.4c Circadian and Other Cyclic Compensatory Changes

As we start to examine the compensatory adjustments which are used over a longer period of time we find that the changes tend to be rhythmic. Organisms may show seasonally different physiological optima, as acclimation fits them to seasonally changing external conditions. Even the *capacity* for adjustment may show seasonal change: so that an organism may show a greater ability to acclimate at certain times over others, or greater ability to make other compensatory adjustments. Sømme and Conradi-Larsen (1977) for example demonstrated seasonal changes in the supercooling ability of Collembola (above, pp. 27-8).

Many of these cyclic changes in compensatory ability reflect cycles in the environment of relatively long periodicity. Thus many are related to seasonal change in temperate latitudes on an annual basis, or to wet and dry seasons in more tropical climes. Many littoral organisms, however, show changes in tolerance based on tidal and lunar cycles, and we may observe rhythms in, for example, temperature tolerances or temperature thresholds for different kinds of activity, even over the course of a day. (Even such short-term changes have been shown to involve true acclimation exactly as observed in longer-term adjustment.)

These rhythmic changes in tolerance, or changes in optima selected are for the most part exogenous. They do not necessarily imply some internal rhythm but can result quite simply from continuous adaptation by the organism to cyclic changes in environmental variables. There is evidence however that *some* of these cyclic changes in tolerance, or acclimation ability — both over a long time scale, and over a diel period — may be at least partially due to endogenous rhythms within the animal itself. Thus, lacertid lizards (*Lacerta sicula*) show a pronounced diurnal cycle in critical minimum temperature associated with normal daily cycles of locomotor activity; critical minimum temperature (defined in terms of loss of righting reflex when turned upside down) shows a change under natural conditions from 4.5°C to 7.5°C over a 12 hour period. Spellerberg and Hoffman (1972) have shown that the lizards retain this cycle in temperature tolerance even in a constant environment (although the degree of tolerance change is less pronounced: 6.5°C to 7.5°C over the 12 hour period). Spellerberg and Hoffman conclude that there does exist some endogenous cycle of critical minimum temperature, but that, under natural conditions, this cycle is governed at least as much by short-term acclimation during the cool night hours.

1.5 Homeostasis: Avoidance of the Problem

A more fundamental mechanism for overcoming the restrictions placed upon an organism by physiological tolerance limits may be developed in assuming a greater 'responsibility' for controlling one's own internal environment — divorcing it to a greater or lesser extent from external conditions in the evolutionary development of sophisticated mechanisms of homeostasis. By maintaining their own internal state in some way independent of external conditions, such homeostatic organisms are capable of tolerating a far wider range of values of environmental variables. There are, still, overall limits, beyond which homeostatic mechanisms are ineffective in controlling body state — or beyond which they become too energetically expensive to maintain. But the development of some degree of homeostasis does permit at least partial release from the *immediate* control of environmental conditions.

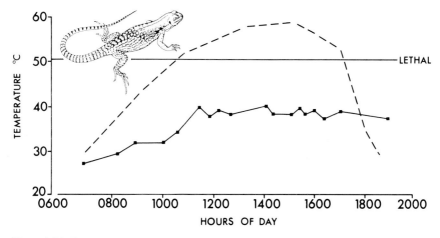

Figure 1.11: Thermal Regulation in the Desert Iguana (*Dipsosaurus dorsalis*). Use of a variety of behavioural mechanisms enables the animal to maintain a relatively constant body temperature (solid line) despite wide variations in ambient temperature (dotted line represents temperature of ground surface in full sun). Source: After McGinnis and Dickson, 1967.

Such homeostasis may have a physiological and/or behavioural basis. Thus, many animals have a degree of homoiothermy — control over body temperature. Control of body temperature may be achieved through physiological control of heat production and heat loss by those animals which are capable of independent heat production as metabolic heat (endotherms); those animals dependent on the environment as heat source and sink (ectotherms) must of course

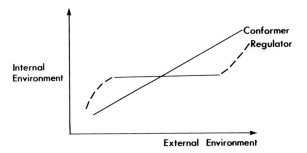

Figure 1.12: Conformers and Regulators: Effects of Changing External Conditions on Internal Environment. See text for details.

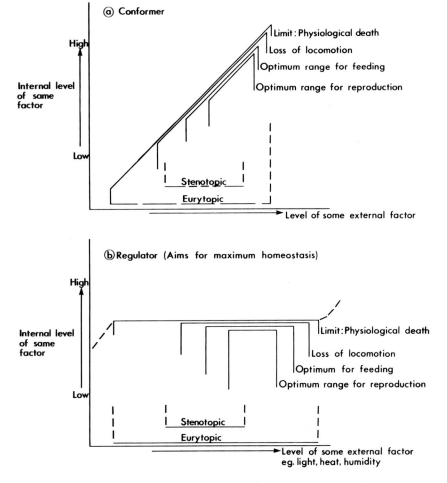

Figure 1.13: Different Significance of Tolerance Limits in Conformers and Regulators.

regulate their temperature largely by behavioural means, but are nonetheless able to achieve a remarkably sophisticated control of body temperature (Figure 1.11); homoiothermy is by no means, as is widely mis-believed, restricted to endo-therms. The same sort of control may be exercised by different organisms over internal salt-concentration (through a variety of mechanisms of osmoregulation), and other internal states.

Such homeostasis is a primary mechanism in expansion of environmental tolerance and is widely used by a variety of organisms. But, in practice, even if it does enable organisms to expand their tolerance range, it does not enable them to escape entirely from the imposition of environmental limits – for the organism remains limited by the restricted range at which homeostasis can be maintained. In effect, a homeostatic organism expands its tolerances to become eurytopic.

In detail however, the actual cause for the limitation is fundamentally dif-ferent between homeostatic and non-homeostatic organisms. Perhaps the most useful way of viewing what becomes a fundamental distinction in the way organisms respond to their abiotic environment is by defining organisms as *conformers* or as *regulators*. (In this regard we may summarise the effect of changes in external conditions on the animals' internal state as in Figure 1.12.) Such a distinction is an important one, for, although as we have noted, there still exist limits of tolerance for homeostatic organisms, as much as for 'conformers', the *reason* for such limitation is different. The tolerance limits of conformers are defined simply in terms of the temperature ranges over which specific enzyme systems are operative; for regulators, tolerance limits are defined as those limits outside which it is not possible to maintain internal homeostasis. Thus the 'nesting tolerance curves' of Figure 1.2, for conformers and regulators, are better viewed as in Figure 1.13.

But although the form of the relationship may differ, the fact remains that even the most adept regulators are *still* effectively limited by physiological tolerances. And the rules of Section 1.2 still apply. Tolerance limits may be adjusted through acclimation, or long-term evolutionary change; tolerances to different abiotic factors may be interrelated – and the broader the tolerance curve, the lower the overall efficiency of performance will be, even at optimum (page 17). As we have noted above, homeostasis merely offers a way of develop-ing eury-tolerance.

1.6 Behavioural Mechanisms for Homeostasis

Although a number of complex morphological and physiological processes may be developed to achieve homeostasis, by far the simplest, and most general way is through behavioural adaptation: the organism seeks to avoid unfavourable conditions.

Animals and plants may adopt a variety of behavioural mechanisms to achieve internal constancy in a range of external conditions. Plants may redistribute chloroplasts within their leaves to keep the chlorophyll within the optimum

Figure 1.14: Behavioural Homoiothermy. Heat uptake and heat reduction postures of the Mallee Dragon (*Amphibolurus fordi*). (a) Lizard on rock surface, warming up: orientation sideways to sun's rays, body pressed close to rock surface. (b) Lizard on rock surface now minimising extra heat uptake: orientation parallel to sun's rays; body raised from ground on stilt legs. Photos from Cogger, 1974.

Figure 1.15: Under extremes of posturing the animal reduces ground contact still further. The lizard now stands on tiptoe with rear left, front right legs raised from rock surface.

range of intensities for efficient photosynthesis in conditions of poor light intensity or conversely very bright light; they may open and close leaf stomata to control transpiration. Ectothermic animals similarly employ a range of curious behaviours to maintain a constant body temperature. In the cool of a morning a desert lizard orients itself broadside to the sun, presses its body close to the warming rock to take up heat so that it may, as quickly as possible, raise its temperature to an operational optimum (Figure 1.14a). As the heat of the day progresses the lizard changes its stance: stands head on to the sun (to present to it the minimum surface area), stands on tiptoe to allow air circulation around its body to relieve it of heat (Figure 1.14b). In some species contact with the ground is further reduced as the animal lifts two feet from the surface in alternating pairs (Figure 1.15). Such positioning enables the animal to maintain a degree of homeostatic control of body temperature within a limited range of environmental temperatures.

Outside the range over which such posturing is effective a different type of behavioural mechanism comes into play, and the animal 'shuttles' between areas of comparative heat and comparative cool, neither optimal: operating in the cooler temperatures with the residual heat of the over-hot zone, and vice versa. Such behaviour introduces a third, and more general class of behavioural adaptation for expansion of tolerance limits: where the animal occupies a particular environment only at that time of day at which it is suitable, or occupies different geographical areas within its range by turn, as each becomes temporarily favourable. The mosquitoes *Anopheles billator* and *A. homonculus* of the Trinidad rain forests demonstrate both these mechanisms. Both species favour a specific degree of air humidity, and achieve this by concentrating at different heights within the forest canopy at different times of day. *A. homonculus* always remains nearer ground level however, and does not utilise the vertical gradient in humidity so extensively; instead it restricts its activity to moister times of day. In the same way, the Mallee dragon, *Amphibolurus fordi*, of Figure 1.14 forages on the soil rock surfaces of the mallee community only at those times of day when the substrate temperature is between 43°C and 50°C (Cogger, 1974). These three behavioural mechanisms of posturing, location seeking or 'shuttling' permit a remarkable degree of homeostatic control of the internal environment, and can markedly extend the period over which an organism may remain active.

Still greater variation in environmental conditions can be accommodated by yet other behaviour: in this case by creation of one's own micro-environment. Hyraxes on Mount Kenya can withstand temperatures well below −10°C by withdrawing into a burrow in the soil. For at a depth of only 10 cm, buffering of external temperature is such that temperatures do not range beyond +1 to +4°C (Coe, 1969). Still greater independence is achieved by the various mound-building termites. Temperatures can be maintained at 30°C ± 0.1°C and relative humidity at 98% by *Macrotermes natalensis*, despite the fact that outside the mound, temperatures may range from 22 to 25°C (Lüscher, 1961). The structure of the nest itself permits such regulation (Figure 1.16). With an outside wall of 0.5 m thick, the nest is in any case kept fairly well insulated from external conditions. Further, because of the heat generated within the nest from the metabolism of the termites themselves and that of their fungus gardens, the colony is effectively endothermic. Temperature regulation may be achieved by controlling air flow through a number of air shafts in cooling vanes on the outside of the nest (Figure 1.16); while humidity level within the nest is maintained by special water-bearer termites which carry up water into the nest mound — sometimes from depths of 50 m or more. A similar control of temperature is achieved within the nest mounds of the Australian 'incubator bird' *Leipoa ocellata*, which uses a combination of solar heat, and the heat of vegetable decomposition to incubate its eggs, rather than its own body heat (Figure 1.17) and can maintain the nest mound at a temperature of $33^{\circ}C \pm 0.5$ (Frith, 1959), while of course the greatest exponent in the creation of his own suitable micro-environment, to enable him to survive under a remarkable range of external conditions, is Man himself.

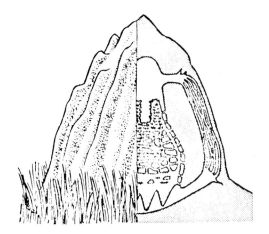

Figure 1.16: The Air-conditioning of a Termite Nest. The cooling fins, visible externally, contain ventilation shafts through which air passes downwards into underground chambers. From here the whole nest is aired. Source: After Lüscher, 1961.

1.7 Adaptive Suites

Adaptation to displace or expand tolerance ranges — whether major or only small-scale adjustments are involved — may thus be physiological or behavioural. However, adaptation to abiotic environmental conditions is not usually restricted to one single mechanism or another. More commonly, a whole range of interactive adaptations will be involved. Further, as we noted in Section 1.2, many environmental variables are interrelated, even synergistic. Thus adaptations to a particular set of environmental conditions too must necessarily show interrelationships amongst themselves — as is clear in the above examples. Such sets of co-adaptive characters have been termed 'adaptive suites' (Bartholomew, 1982) and are perhaps best examined amongst organisms adapted to the greatest extremes of environment.

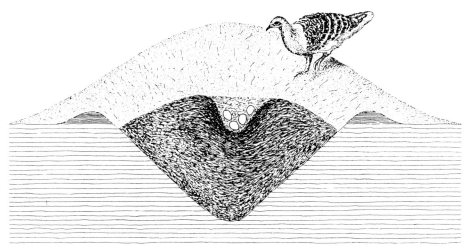

Figure 1.17: View and Cross-section of Incubator Mound of *Leipoa ocellata*. This species does not incubate its eggs with its own body heat, but uses a combination of solar heat, and the heat of vegetable decomposition. In preparation for breeding, the male collects a mass of wet vegetation and buries it in a pit some 3 m deep. As it starts to compost, he digs it over, airing it and working it until it reaches the right temperature. The female then lays her eggs. From this time on, the temperature of the mound must be maintained and must not be allowed to rise or fall more than 1.0 deg. from 34.5°C. As the summer proceeds, solar radiation makes a major contribution to the heat of the pile during the day; heat from the decaying vegetation is required only at night. The male megapode therefore digs into the mound in the morning to create ventilation tunnels, allowing composting heat to escape. In the evening the holes are stopped up. As time wears on, decomposition slows; the birds must rely entirely on solar radiation to maintain the temperature in the mound. However, the sun's heat during the day, added even to the residual heat of decay, is too great; at night, decomposition is not sufficient to maintain the temperature. The male megapode changes his approach, and now covers the mound with an insulating layer of sand to prevent overheating during the day and reduce heat loss at night. Incubation must continue for many weeks, well into autumn, when, not only is the heat of decomposition exhausted, but even the sun is losing energy. To maximise heat uptake during the day, the megapode scrapes away the sand covering the eggs so that the eggs are only a few centimetres below the surface, and receive all the available heat. In preparation for the cold night, he spreads the sand from the mound in a thin layer over the ground, so that it, too, may heat up. In the evening, he collects together all this sand (some 20 m^3!) and buries it in the nest mound, or uses it as an insulatory cover, to maintain nest temperature at night. This amazingly laborious and complex process succeeds in maintaining a nest temperature of 34.5°C throughout incubation.

1.7a Desert Plants

Because most studies of extreme environments have been restricted to the hot, arid end of the scale, we may pick our examples from adaptations to hot deserts.

Many arctic and temperate plants show variously developed adaptations to possible drought: reduced water availability from heat or frost. We have already made mention of the advantages in such circumstances of a deciduous habit, but even amongst evergreens, much can be achieved through toughening of the leaf cuticle, reducing stomata, or curling the leaf in upon itself so that the stomatal openings of the under surface open into an air pocket relatively well enclosed by the curled leaf and in which relatively high humidity can be maintained.

Under really extreme conditions, however, such adaptations are insufficient. Under such conditions the most persistent of the xerophytes are the succulents: plants which can store water (obtained when supplies are available) in swollen leaves, stems, or roots. During drought periods no water absorption occurs; the plant survives on stored water, and reduces water loss in transpiration by opening the stomata only in the cool of night. Gas exchange can thus occur at night when the stomata are open; succulents are distinguished by a number of metabolic peculiarities which have been developed to overcome this problem. During that period when stomata are open carbon dioxide is absorbed, and incorporated into organic acids which are stored in the tissues. These acids are decarboxylated during daylight to release the carbon dioxide required to permit at least a low level of photosynthesis.

1.7b Desert Animals

The adaptations of animals to desert life are essentially related to problems of both heat regulation and water conservation; of these problems water conservation is again undoubtedly the most critical.

Animals in deserts are not only hampered by low initial availability but also experience considerable difficulties in retention of that water. The three main sources of body water loss — during respiration, during nitrogenous excretion and from transpiration through the body surface — are exacerbated in the hot, dry conditions: particularly since most vertebrates rely upon evaporation for thermal regulation. Basic mechanisms for water conservation may be adopted in reduction of each potential source of loss. Thus, desert vertebrates can reduce evaporation through panting and sweating as succulents may reduce transpiration. Most desert rodents do not sweat at all and camels only sweat, so to speak, in an emergency. Respiratory losses, too, can be reduced. While the rate of evaporation of water from the human in dry air is 0.84 mg/ml O_2 absorbed, amongst desert rodents this water loss is reduced to only 0.5 mg/ml. Such reduction of evaporation is achieved by reduction of the air temperature to $24°C$ as it leaves the nose, so that it requires less water for saturation (Schmidt-Nielsen and Schmidt-Nielsen, 1952). Losses during nitrogenous excretion can be reduced by resorption in the kidney of as much water as is possible. Because of the high solubility of urea and its relatively low toxicity, extremely concentrated urine can be produced. (Indeed birds can resorb water to the extent that uric acid is precipitated and excreted as crystals through peristalsis of the ureters.) The structure of the mammalian kidney, however, is such that the kidney tubules or pelvis would be blocked by solid waste, so it is essential to keep urea in solution; nonetheless water loss can be reduced significantly in this way.

Low water intake can be compensated for by eating succulent foods, or if oily material is available, by enhanced ability to assimilate fats and break them down for their metabolic water.

By using essentially these three methods, kangaroo rats (*Dipodomys* species) have got water economy to such a pitch that even feeding on a completely dry, seed diet, they can obtain enough water for their needs and do not drink at all. Antelope such as the Oryx (*Oryx beisa*), Grant's gazelle (*Gazella granti*) and Gerenuk (*Litocranius walleri*) similarly satisfy their water needs purely from the vegetation ingested, and have no need to drink directly.

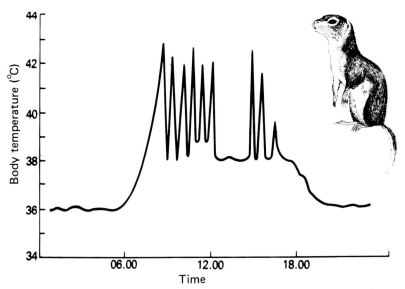

Figure 1.18: Fluctuation in Body Temperature of the Antelope Ground Squirrel (*Citellus tereticandus*) as it Periodically Retreats to its Cool Burrow after Foraging Trips on the Hot Surface. Source: from Hudson, 1962, after Louw and Seely, 1982.

Reduction of water loss from evaporation, however, intensifies the already extensive problems of heat regulation, which must therefore be overcome in other ways. Behavioural adaptations of burrowing, or occupation of the environment only under congenial conditions are extremely important here, but are complemented by a variety of physiological mechanisms. The most general of these is to become, within limits, a 'conformer' — that is permit the body temperature to range over a far greater range than would be normal for a homeo-thermic organism. The Antelope Ground Squirrel (*Citellus tereticandus*) for example has enzyme systems which can operate over a far greater range than most other mammals, and thus permit considerable lability in body temperature. The squirrel solves its heat-loss problems by allowing the body temperature to reach extremely high levels (ca 42°C), actually higher than that of the ambient temperature, so that heat may still be lost to the surrounding environment. When the body temperature reaches these peak levels it retreats to an underground burrow to cool off (Figure 1.18). Similar adaptation is shown among the desert antelope mentioned earlier. Taylor (1972) has established that antelope such as oryx, eland and Grants' gazelle also show extreme thermal lability, recording rectal temperatures of 45°C in oryx and 46.5°C in Grant's gazelles. Use of the body mass as a passive heat store enables the animal to continue to function effectively at high temperatures; with a greater body mass than *Citellus* the antelope can absorb a great deal more heat in this way, and remain active for prolonged periods without *Citellus*' regular retreat to shade. (The heat load accumulated during the day, can easily be dissipated during the cooler desert night.) By allowing its body temperature to continue to rise during the day the animal reaps an added advantage. It *decreases* the temperature differential between itself and the environment, and therefore decreases further heat uptake. (Body temperatures above 43°C rapidly produce brain damage in most mammals, but Taylor observed that Grant's gazelles maintained rectal temperatures of 46.5°C for as long as 6 h with no apparent ill effects. He has suggested that these antelope can keep brain temperature below body temperature by using a counter-current heat exchange to cool blood before it reaches the brain. In ungulates the blood supply to the brain passes via the external carotid arteries (Figure 1.19). At the base of the brain these arteries break into a *rete mirabile* that lies in a venous sinus. The blood in the sinus is venous blood, returning from the walls of the nasal passage where it has been cooled by the evaporation of water. This cool venous blood thus cools the warmer arterial blood before it reaches the brain. A mechanism of this sort has been demonstrated in sheep and goats and is probably widespread.)

The dromedary (*Camelus dromedarius*) is perhaps everyone's classic desert animal: an animal whose coordinated suite of adaptations to this extreme environment have so often been extolled that one begins to suspect exaggeration. Nonetheless, in truth, camels possess a really remarkable suite of co-adaptations permitting them to exploit the desert, and provide delightful illustration of how these various strategies so far viewed in isolation may be coordinated within a single animal. Water input is maintained from feeding on succulent materials or fog-dewed browse in the early mornings. Water output is reduced to a minimum through concentration of urine, while water balance may be maintained by the production of metabolic water from fat stored in the hump and body cavity. Like desert antelope, dromedaries use the body mass as a heat store during the day allowing general body temperature to rise. In the morning the body temperature is approximately 34°C, by mid afternoon it may have risen to nearly 41°C (Cloudesley-Thompson and Chadwick, 1964). Raising some 450 kg through seven degrees takes up a great deal of heat; while as already noted, reduction of the differential between body temperature and ambient decreases additional heat uptake. Reduction of insulation by subcutaneous fat by relocation of fat into a central hump permits faster heat loss when cooling does become possible (and by converse allows the animal to store more fat than it could otherwise accommodate — increasing its reserves for production of metabolic water!) The camel's thermal lability is, however, somewhat more restricted than that of the oryx or Grant's gazelle; the animal cannot permit the body temperature to rise above 40.7°C. Once this upper limit is reached, the camel begins to sweat. Yet sweating increases water loss and recreates problems of water conservation. Such is the camel's remarkable suite of adaptations, it has an answer for this, too — by permitting itself to be desiccated. In most mammals desiccation results in loss of water from blood. This then becomes more viscous, putting considerable strain on the heart. At a stage where 20% body weight has been lost, the blood cannot circulate fast enough to carry away metabolic heat from tissues, and temperature rises rapidly to cause *heat death*. The Camel by contrast ensures that water is lost primarily from interstitial and intercellular fluids of the tissues and plasma volume is maintained. (In a total water loss of 50 litres only 1 litre is lost from the plasma.) In addition a modification to the structure of the red blood cells protects them in turn from damage through plasmolysis, even when the blood water level *is* reduced. The same structural adaptation preserves the cells from bursting when blood water levels rapidly increase — permitting the camel the extra adaptation of being able to rehydrate at a single drinking bout — and adding the final touch in real life to perhaps a more remarkable suite of coordinated adaptations than one could devise in imagination!

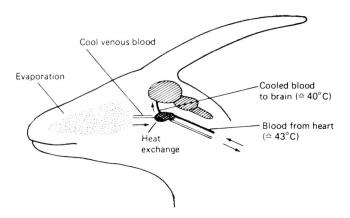

Figure 1.19: A Counter-current Heat-exchange Mechanism that May Cool Blood Going to a Gazelle's Brain. The blood supply to the brain in ungulates travels via the external carotid arteries. At the base of the brain, each of these arteries breaks into a capillary network (a rete mirabile) that lies within the venous sinus. This sinus is filled by venous blood that has been cooled by evaporation of water from the walls of the nasal passages. The close contact between cool venous blood and warm arterial blood permits an exchange of heat that cools the arterial blood going to the brain. In this way a gazelle may keep its brain temperature several degrees cooler than its core body temperature for long periods. Source: redrawn from Taylor, 1972.

1.8 Organism and Abiota: a Two-way Interaction

Whatever their mechanisms for adjusting, expanding or gross displacement of their tolerance ranges, animals and plants can never totally escape from the problem. Tolerance limits may be altered — but never removed; whatever adjustment may be made, organisms are still restricted in performance and distribution by their tolerances to abiotic conditions. As we noted above, such restriction may determine limits to distribution: will dictate, if not where an organism does or does not occur, at least where it *cannot* occur.

But the relationship between organism and environment is *not* one-sided. Animals and plants may also alter the fabric of their environment. We have already seen how biotic elements of an ecological system may become a primary consideration in determining the physicochemical characteristics of a micro-environment. Plants and animals may contribute to the physical structure of the environment — creating physical diversity and providing, purely passively, new habitats for each other. Plants, in particular, may modify physical conditions, creating beneath their canopy areas of altered insulation, humidity, altered air-flow etc. etc.

In addition to these, rather passive effects, organisms also influence their environment in a more active sense through their use of it. Gas concentrations under the canopy of a forest, or beneath the lower canopy of herbaceous plants may differ markedly from those in the upper atmosphere (Woodwell and Dykeman, 1966; Monteith, 1968). Nutrient and oxygen levels in marine and fresh waters can fluctuate over wide ranges as a result of biological activity — with important implications for the nature of the communities present. Various plant species can affect the chemical composition of soils: nitrogen-fixers like the Leguminosae may markedly alter nutrient status; species like gorse can acidify near-neutral soils by extracting cations and 'locking them up' in slowly decomposing litter (Grubb and Suter, 1970). Large herbivorous animals which have distinct feeding and distinct latrine areas, may be responsible for gross shifts of soil nutrients within a community and resultant patchy distribution (Putman *et al.*, 1982a). All these factors affect the physicochemical nature of the environment and exert an indirect influence on the biotic communities supported.

In brief, the relationship of organisms with their abiotic environment is a two-way process, a true interaction. Initial, dominant, abiotic factors restrict distributions and performance of animals and plants. But they, once established, themselves influence the environment causing it to change. The implications of such continuous interaction in, ultimately, causing a change in the biotic community itself, is taken up further in Section 4.6.

2 The Ecological Community

2.1 Introduction

No one organism lives in simple isolation, interacting, according to selfish physiological requirements, with its physical and chemical — abiotic — environment. Such a straightforward relationship is shattered by the fact that the individual shares its environment with other organisms of the same or different species; the simple interplay of organism and environment becomes confounded by a whole host of *biotic* factors. In brief, each organism is part of a complete *community* of creatures, each interacting with each other as well as with their abiotic environment, and each affecting each other's use of the resources that they share.

2.2 Communities and Ecosystems

A community may be very simply defined as an assemblage of animal and plant species occurring together in a particular area. The term embraces the living parts of a contained ecological system; the whole system — of both biotic and abiotic elements — comprises an ecosystem. As happens so readily when terms are loosely used, the words community and ecosystem are regularly confused. Yet an ecosystem is a self-contained ecological entity of both organisms and their complete biotic and abiotic environment. It is quite simply the smallest functional ecological unit: an independent self-contained and self-sufficient block. By contrast a community embraces only a set of interacting organisms, and concerns itself only with biotic relationships.

The distinction is in part one of scale: individuals aggregate into populations, populations assemble into communities, communities, with their physical environment combine to form ecosystems. Beyond this, ecosystems are linked into whole regional classes or 'biomes', the tundra biome, the temperate forest biome etc. — and the scale expands. At each stage, however, we see an increase in complexity. An ecosystem is more than just a collection of communities or of communities and their abiotic environment; a community is more than a simple assemblage of species. The whole is something far beyond the sum of its component parts, but is a complete new entity with characteristics and properties of its own.

Thus the organisms of a community do not simply co-occur, continuing their independent lives in total disregard of each others' presence. They interact. A community is, as we defined it, an assemblage of animal and plant species occurring together in a particular area. But more importantly it is an assemblage of species which are all linked to and interact with one another in a variety of biotic relationships to bind the community into a functional whole. And that is the essence of the community, that it is a *functional* unit, which can be described only in terms of its operation, never frozen as a flat species mosaic. Indeed it is very hard to define a community except by reference to these rela- tionships − it is in practice the interactions within the community which create and shape it and not the community which shapes relationships between its component organisms.

2.3 Biotic Relationships

The biotic relationships which link the organisms of a community are many and varied. An animal or a plant may rely on another animal or plant for its habitat requirements: thus many species of birds nest in holes in trees, *Calliactis* sea anemones attach themselves to the shells of the hermit crab (*Pagurus*), and South American bromeliads support a unique fauna of invertebrates and even small lizards in the water trapped at the base of the dense leaf-rosettes (Figure 2.1). The relationship may become more intimate: as many plants rely on insects for their pollination, and mammals or birds for the dispersal of their fruits and seeds. Some flowers have even developed to rely on mammals for pollination too (Figure 2.2). Plants may interact with other plants in *competi- tion* for light or space. Animals, too, may compete amongst themselves for food, shelter, or other resources; indeed competition is frequently cited as a major force in the shaping and maintaining of communities and populations, as we shall see later. And of course the organisms within the community may rely upon each for their food.

Feeding relationships are, in fact, by far the most common route of interac- tion between the different organisms of the community. Such relationships may be *commensal, mutualistic, parasitic* or *holozoic*. In true commensalism, two organisms associate in feeding 'at the same table'; one derives benefit from the association, the other neither gains nor loses. Thus the *Calliactis* anemones attached to the shells of hermit crabs neither help nor hinder the crab; but they themselves feed on the debris cast up by the foraging crabs. In mutualism the relationship has progressed to one of mutual benefit: both organisms gain from their association. In extremes, neither could in fact survive for long without the other. Many of the bacteria of the ruminant stomach cannot survive outside the comfortable medium of the rumen; without these symbiotic bacteria, the ruminant would itself be unable to digest the cellulose of its food. Lichens, too: a hardy and resistant life-form, rely in fact on a mutualistic relationship between an alga and a fungus (Section 11.4). In parasitic relationships one organism feeds upon the other to the benefit of itself but at the expense of its host: feeding directly on the hosts' tissues or body fluids, or at least side-tracking some of the foodstuffs ingested and intended for the host's own purposes. Clearly, however, it is not really to the parasite's advantage to damage its host in any serious way: should the host die, the parasite must itself die or seek a new host. Thus a well-adapted parasite enjoys a secure relationship with a host upon which it inflicts the minimum of burdens; indeed it is often hard to distinguish so refined a parasitic relationship from commensal or even mutualistic associations. Finally, animals feed directly one upon the other or on plants in holozoic (predator-prey, herbivore-plant) fashion consuming directly portions of the food organism.

Figure 2.1: The large hermit crab *Eupagurus bernhardus* in a whelk shell to which is attached the sea anemone *Calliactis parasitica*. Source: Reproduced by permission of D.P. Wilson/Eric and David Hosking.

These various feeding interactions may be seen to embrace a range of relationships in which in all cases the consumer gains, but in which the 'consumed' may also gain, may lose, or may sustain neither advantage nor disadvantage. Such relationships may be coded as ++ (mutually beneficial); +0 (beneficial to one participant, not affecting the other); +− (beneficial to one, damaging to the other) — and in fact *all* the various biotic relationships which may occur between the organisms of a community may be viewed in the same way. Analysis of feeding relationships, by such a scheme, offers a template for a system by which it may help to view all biotic interactions within the community. Under such a

a

b

Figure 2.2: (a) Captive *Aethomys namaquensis* Foraging for Nectar on a Flowering Head of *Protea amplexicaulis*. The anterior half of the snout has accumulated a dense pollen load which appears whitish in the photograph. (b) Diagrammatic representation of flower-pollinator 'fit' when the pollinator laps nectar.
Source: From Wiens & Rourke, 1978.

scheme, the various interactions discussed so far can be simplified, and, more importantly, grouped into relationships of similar *effect*, if different origin, as in Table 2.1. Such rationalisation is a particularly useful one, for, as well as summarising all the various possible interactions and classifying together relationships of like type, it also carries, implicit within it, an indication of the effect that any particular relationship may be expected to have on the 'performance' of the participants — increasing (+) or decreasing (−) population growth. This assumes considerable importance in modelling the dynamics of single species populations and the effects upon those populations of interactions with other species (Chapters 6, 8 and 9).

Table 2.1: A Survey of the Various Types of Biotic Relationships which may Link the Organisms of a Community. Relationships are grouped as mutually beneficial (++); one-sided (+−); mutually damaging (−−) and so on (see text).

Effect of relationship on participants	Examples of relationship
+ 0	Use by one organism of another as habitat
	Commensal feeding relationships
+ +	Pollen-feeding and thus pollination by animals, or fruit-feeding by animals resulting in seed dispersal
	Mutualistic feeding relationships (e.g. symbiotic algae in green hydra)
+ −	Heterotrophic, and parasitic feeding relationships
	Secretion of allelopathic chemicals by plants, preventing others growing around them
− −	Competition
− 0	Incidental damage: non-deliberate damage to individual organisms or their environment (e.g. trampling by animals causes damage to vegetation species)

Table 2.1 also offers a convenient summary of the types of biotic interactions that may be observed within a community. The organisms of the community as well as interacting each on their own with their abiotic surroundings, are also interrelated with each other by a complex series of these biotic links. And the complexity of these interrelationships cannot be over-exaggerated. Considering only one type of relationship (heterotrophic feeding) in an extremely small and simple system (a freshwater pond) Figure 2.3 gives some conception of the complex web of interaction.

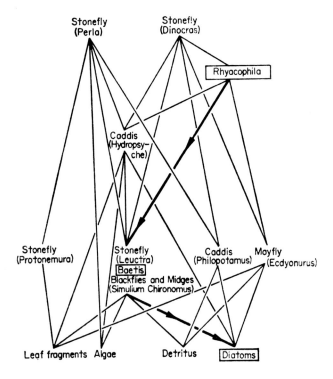

Figure 2.3: Part of the Foodweb of a Fresh Water Stream in Wales. Source: Based on Jones, 1949 after Phillipson, 1966.

2.4 The Organism in the Community

Any one organism's position within the community is defined in terms of these various interactions with biotic and abiotic environment: its position, or niche (a concept to which we shall return) thus means more than just where it may occur, the habitat or community in which it may be found, but also implies a consideration of the role it actually plays within that community. This *niche* was in fact originally defined merely in terms of the type of community in which an organism might be found, the particular habitat and micro-environment it may occupy, the sorts of organisms with which it may be associated — establishing merely what has been described as the organism's 'address' within the community: a rather passive classification. However, as Odum (1959) points out, if used in this way the term 'niche' becomes almost synonymous with 'habitat', for the place where an organism lives is defined not only by a series of abiotic parameters, but should also include biotic elements: 'the habitat of an organism, or a group of organisms includes other organisms as well as the abiotic environment; the concept of niche includes much more than this' (Odum, 1959). Thus it is generally accepted (after Elton, 1927) that niche means more than just the 'address' of a particular organism, but also embraces its 'profession' — a consideration of the various interrelationships it has with other organisms around it, the role it plays in the operation of the community as a whole (its feeding relationships, relationships as prey, competitive interactions etc. etc.).

2.5 The Community Level of Organisation

While a community may of course be considered as the sum of all the interactive populations it comprises — a jigsaw puzzle of the niche relationships of all its component organisms or species, it is not purely a passive 'envelope' in this way. Communities function in their own right as dynamic structures and have certain common characteristics of structure, design and function. Yet, in any attempt to evaluate the characteristics, structure, mode of operation of communities themselves the terrific complexity of each individual 'niche jigsaw' hides the wood amongst the trees. Every attempt to study ecological systems at this community level of organisation has therefore tried to rationalise the picture in one way or another, so that similarities (or differences) of design and function of different communities may be appreciated. But we should not forget that our conclusions cannot extend beyond the limits of the level of analysis we employ, the method of rationalisation of community complexity that we adopt. 'Choice of the appropriate "macro" descriptors or aggregate variables may be vital to progress; community ecology both depends upon and yet is simultaneously constrained by the identification of such conceptual building blocks' (Pianka, 1980).

2.6 Tropho-dynamic Analyses

Of all the relationships between the organisms within a community the most important are probably heterotrophic, feeding associations. As a result it has become customary to define the structure and operation of the community largely in regard to these feeding relationships: dividing up the member species in terms of the *trophic role*.

It is important not to forget that feeding relationships are not the only associations within the community, that many other interactions of mutualism, competition, and so on, occur around them. Thus a rationalisation in terms of trophic relationships will leave quite a few organisms unaccounted for, quite a few links omitted where creatures rely on other members of the community for shelter, dispersal or whatever. Feeding relationships, as we have seen, are not the be all and end all, and it is important not to neglect the other possible biotic interactions. Nonetheless, every community consists of three basic types of organisms:

Producers. Autotrophic organisms (mainly green plants, but including some chemosynthetic bacteria) are so-called because they are the only group within the community that can actually synthesise organic compounds and thus *produce* food for the community. (*Autotrophic* organisms are of course those which can synthesise organic compounds from inorganic precursors, using external energy sources, as distinct from *heterotrophs* which cannot synthesise organic matter but can only re-organise it, and thus must feed directly on organic material.)

Consumers. The next type of organism within the community are heterotrophs that feed upon the producers or upon each other. Within the community, they may be separated out into *primary consumers* — those that feed directly on the producer level (= herbivores), *secondary consumers* — carnivores feeding on the primary consumers, tertiary consumers, and so on. (Since the primary consumers may be regarded as 'producing' food for the secondary consumers, they are sometimes alternatively referred to as secondary producers: and the separate terminologies equate as:

Producers   = Primary producers (autotrophs)
Primary consumers   = Secondary producers
Secondary consumers = Tertiary producers, and so on.)

A final, distinct group of organisms within the community are the

Decomposers. These are a group of organisms, largely saprophytic, which, as their name suggests, feed upon dead and decaying materials from the other two groups.

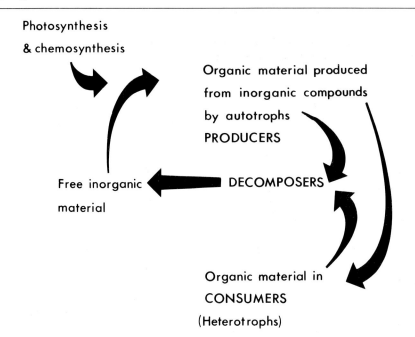

Figure 2.4: Schematic Diagram of the Cycle of Material Through the Ecological Community.

Organic matter, synthesised by the producers, is eaten by a series of consumer levels. With the aid of the decomposers, all the organic materials incorporated into the bodies of the consumers (and, indeed, unconsumed producers) are eventually broken down again and returned to the producers for reuse (Figure 2.4). Thus all available matter cycles through the system and there is no overall loss or gain.

Of course, the system is not quite as simple as all that: as we have noted, this takes into account only one of the many types of interaction which exist between the different members of a community. Further, not all organisms fit tightly within one category. Many overlap, being partial herbivores and partial carnivores, for example. But this sort of rationalisation does enable us to draw certain general conclusions about the characteristics and properties of communities as a whole.

2.7 Community Structure

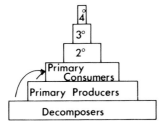

Figure 2.5: Stylised Pyramid of Numbers/Biomass in a Terrestrial Community. The area of each bar in the diagram is related to the number or biomass of organisms in that particular trophic level.

Such overview of the community level of organisation allows us to draw a number of conclusions about the structuring of such 'assemblies'. For instance, on the producer-consumer side of the system, because not all the material taken in by one trophic group is passed on to the next (some will be lost in respiration or in faeces and urine, and in any case many organisms in any one trophic level will escape consumption), there is in this subsystem a net loss of material as we pass from the producers through the ascending scale of consumers (Figure 2.5). As a result we find a pyramidal structure of trophic levels: the total numbers of organisms in each trophic class decrease as we ascend the trophic scale. One pigeon eats a great deal of grain during its life-time, one peregrine eats a great many pigeons: each organism relies on more than one organism in the previous level to support it. Not only does all this result in a 'pyramid of numbers' of organisms (Figure 2.5) as noted by Elton in 1927, but it also offers an explanation for Elton's further observation that there seemed to be a limit to the number of possible links in any 'food-chain' (= number of trophic levels). Since much material is lost between one trophic level and the next, and any one organism needs a multitude from the trophic level below to support it, there must be a limit to the number of trophic levels in a community. The available food decreases so rapidly that ultimately insufficient remains in the system to support a further level above (although alternative explanations for this observation may be offered, as for example page 77 and page 352). Indeed, in most terrestrial communities, the number of trophic levels rarely exceeds 4 or 5: producers and three or four consumer levels.

More importantly communities differ in the number of trophic levels they *do* contain and this in itself can be related to differences in quality or type of community function.

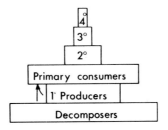

Figure 2.6: Trophic Pyramid for a Detritus-based Aquatic Community.

We can also find interesting differences between communities in terms of the relative importance of each trophic class within the community. Thus the pyramid presented in Figure 2.5 is typical for terrestrial green-plant-based communities. Aquatic, plant-based communities usually show a marked reduction in the consumer element, while certain other aquatic systems, which are not in themselves self-contained, but rely on the continuous input of material from outside in the form of detritus (many rivers, the lower, hypolimnion layers of deep lakes) show a totally different balance of trophic elements, with the consumer levels supported largely by a dominant detritus-based decomposer level (Figure 2.6). These differences in trophic emphasis reflect different methods of operation within separate communities, and can be used to understand their function.

In fact, sheer numbers of organisms in any one trophic category is not the most useful yardstick in community analyses of this sort: since what we are really interested in is *impact* and, to be slightly facetious, as a herbivore, one elephant has a lot more impact than one vole! Odum *et al.* (1962) from their studies of grassland ecosystems, provide more formal evidence that numerical density is a poor basis on which to found analyses of ecological function, showing that 16,000 grasshoppers were not 1000 times as significant in terms of impact as 16 mice, but only 2½ times as important. In consequence the relative balance of the different trophic levels within a community is usually assessed in terms of *biomass*, that is, the total weight of living material (numbers of organisms multiplied by the average weight of each). Such calculations still give rise to pyramids of trophic levels, although the decrease in biomass from one trophic level to the next is usually considerably more rapid than mere decrease in numbers.

Even considerations of trophic biomass need to be treated with some caution, however, for they represent only what is happening at one instant in time. They reflect what is called the '*standing crop*' — the biomass actually present at the time of sampling — and do not take into account changes within that biomass in time, or more importantly, differential changes in biomasses of different trophic levels over time. For the biomass of any one trophic level is always in a state of flux: after all, it is composed of populations of organisms which are continually growing, being born, and dying. Thus, although the standing crop biomass of a given population of small mammals at any one time may be in one example 554 g ha^{-1} (Bobek, 1969, small rodents in Polish forests) it may in fact reflect a much greater impact if that biomass is continually turning over. The actual *production* of voles and mice, in a given area or by the vegetation of an area in a year, is not just the standing crop of small mammals at that time, but the total number that have been produced — including those that have died as well as those that are now alive. Since Bobek has shown that the small mammal populations of his study turnover completely about 2.5 times a year (Bobek, 1969) this is a production in fact of 2.5 times the standing crop. The biomass of rodents supported by the vegetation is thus in reality 2.5 times the measured standing crop.

| ZOOPLANKTON & BOTTOM FAUNA | 21 |
| PHYTOPLANKTON | 4 |

Figure 2.7: Inverted Pyramid of Biomass in Marine Plankton. Biomass values (g^{-2}) from Harvey, 1950 (after Odum, 1959).

In order fully to appreciate the role any particular organism/trophic level is playing within the actual operation of the community therefore we need to take some account of this flux; we need to weight standing crop biomasses by some measure of the speed of *turnover* of that standing crop. Otherwise we are faced with unanswerable paradoxes such as the now famous ecological 'Paradox of the Plankton' (Hutchinson, 1961). An analysis of trophic biomass within marine plankton reveals an apparently *inverted* pyramid of biomass, with consumers far out-weighing producers (Figure 2.7). This apparent reversal is, however, readily resolved when it is appreciated that the turnover time within the phytoplankton is many times more rapid than that of the zooplankton — a diatom may survive in the plankton for only a few days, while the larvae of various marine crustaceans may persist in the zooplankton for months. In practice, the actual annual *production* of the phytoplankton is far in excess of the annual production of the consumer levels, even though instantaneous standing crop biomass may be much lower.

2.8 Analyses of Food Web Design

However informative such a tropho-dynamic rationalisation of the community may have proved in understanding the community as a level of organisation, it tells us little about the detailed structuring of such communities. It is, we must accept, an oversimplification. As we noted on p. 49 'Community ecology both depends upon and yet is simultaneously constrained by the identification . . . of the appropriate "macro"-descriptors for rationalisation' (Pianka, 1980).

Treatment of the community as a series of trophic classes obscures consideration of its structure and organisation *within* each trophic level: the relative numbers of species and relative importance of each to the trophic class. In addition although we know that not all organisms of one level necessarily interact with all those of another, but that relationships between trophic levels are made up of many separate interactions between individual pairs of organisms in the different levels, this complexity is crushed into a single overall relation between one trophic block and the next. Yet again, as we have already noted, many organisms are, in any case, not restricted to a single trophic level.

In such simplification, then, much resolution has been lost; much information about the structure of the community is unavailable. It is possible to gain some insight into the details of this more subtle structuring, while maintaining some level of simplicity in a study of food web design in different communities. Such study permits examination of community structure at a slightly more biologically meaningful level of complexity than does pure trophic level analysis: still seeking information about community organisation but without returning to the confusing complexity of considerations of individual species. (It should be noted that consideration is still restricted only to a single type of interaction between organisms: ignoring non-feeding relationships.)

Study of food web design (as May, 1973b; Pimm and Lawton, 1978; Pimm 1979a; Paine, 1980) does however permit us greater understanding of how communities are put together, greater insight into the actual relationships which do occur between trophic levels: the nature of those relationships and the relative importance of the different relationships in maintaining the community structure (for some particular types of relationship prove to have far greater import in determining and maintaining the community structure than do others). Such considerations assume particular importance in discussions of the stability of ecological communities (Chapter 13). The approach is a new one and much remains to be explored, but Paine (1980) suggests 'The central significance of webs is derived from the fact that the links between species are often easily identified and the resultant trophic scaffolding provides a tempting descriptor of community structure. If this structure is in any fashion related to the persistence of natural communities or their stability, however defined, then we are dealing with issues of vital ecological importance.'

2.9 Subcompartments in Community Structure

It is often possible to identify within the overall framework of the community discrete subunits. These may be distinguished from the rest of the community spatially (by occupying a different micro-habitat, a different structural layer of the vegetation) or temporally (Chapter 4). Alternatively they may be recognised as relatively independent ecological units.

One of the most common devices for rationalising the complex species structure of the community is in recognition of *guilds* of species within the community (Root, 1967). These guilds comprise assemblages of species (usually, but not necessarily, taxonomically related) which exploit a common resource in a similar way. Thus we may identify guilds of insectivorous birds (MacArthur, 1972), guilds of desert lizards (Pianka, 1978, 1980) etc. Such guilds are composed of species which are ecologically extremely similar, which indeed differ from each other essentially only by specialising in a slightly different portion of some major shared resource, differing only along one dimension of the niche (Chapter 5), and separated out for example merely by size of insect prey taken. These guilds are usually continuous (the resource spectrum exploited by one species abuts closely that of the next) and may in fact be objectively defined as a group of species separated from all other such clusters by a difference greater than the difference between neighbouring guild members.

The existence of such guilds raises a number of intriguing questions. Are they important functional units, or are their boundaries totally arbitrary — defined merely because of gaps in the distribution of resources? What are the effects of such guild structure on community organisation, on stability? What are the effects on competition within the community? Members of such guilds would interact strongly with each other but relatively weakly with other members of their community. The questions are intriguing — the answers unknown: the analysis of guild structure is a recent development, but has begun to attract increasing interest (Feinsinger, 1976; Holmes *et al.*, 1979; Pianka *et al.*, 1979; Pianka, 1980; Hairston, 1980).

Guilds may be recognised as subcompartments of the community's structure *within* one trophic level, but we may also be able to identify *relatively* independent subunits within the complete trophic structure of the community: bits of food webs that seem tightly interconnected within themselves but only weakly connected with other parts of the community as a whole. The implications of such compartmentalisation for stability have been examined by May (1972, 1973) who argues that for a given species number and web connection, model food webs have a higher probability of being stable if the interactions between them are arranged into 'blocks' or compartments.

Pimm and Lawton (1980) examine alternative models, which they claim incorporate biologically more realistic assumptions. Unlike those of May, these models do *not* necessarily predict that food webs are more stable if divided into blocks. Further, in analysis of published food webs, Pimm and Lawton conclude that in natural systems such food web subcompartments seem primarily to be a function of the distribution of a community between different habitats or microhabitats, with each subunit relating to one particular habitat; they claim there is no evidence for such subcompartments within habitats. They conclude that there are at present neither adequate theoretical nor convincing empirical grounds for believing that food webs are divided into compartments.

2.10 Common Denominators of Community Design

Analyses of these, and other elements of community structure have provided hints of characteristics of design common to all communities. Examination of trophic organisation, of food web design and various other facets of the structure of communities, does in fact leave one with a powerful impression that there may in fact be a number of general principles of community organisation: general rules of conformation obeyed by all communities, or others providing outside limits on design, or constraints on variability. Thus we may determine that, for whatever reason, there is a limit to the number of trophic levels that may occur in any one community. (Although we may have a number of alternative explanations, the fact remains that, whatever the community, it may have no more than five trophic levels.) Further, there seems to be a remarkable constancy of trophic structure in communities of similar type: the 'trophic pyramid' is of remarkably similar form in all communities of like 'function'.

Table 2.2: Evidence for stability of trophic structure

Island	H		S		D		W		A		C		P		?		Total	
E1	9	(7)	1	(0)	3	(2)	0	(0)	3	(0)	2	(1)	2	(1)	0	(0)	20	(11)
E2	11	(15)	2	(2)	2	(1)	2	(2)	7	(4)	9	(4)	3	(0)	0	(1)	36	(29)
E3	7	(10)	1	(2)	3	(2)	2	(0)	5	(6)	3	(4)	2	(2)	0	(0)	23	(26)
ST2	7	(6)	1	(1)	2	(1)	1	(0)	6	(5)	5	(4)	2	(1)	1	(1)	23	(27)
E7	9	(10)	1	(0)	2	(1)	1	(2)	5	(3)	4	(8)	1	(2)	0	(1)	23	(27)
E9	12	(7)	1	(0)	1	(1)	2	(2)	6	(5)	13	(10)	2	(3)	0	(1)	37	(29)
Total	55	(55)	7	(5)	13	(8)	8	(6)	32	(23)	36	(31)	12	(9)	1	(3)	164	(140)

The table is after Heatwole and Levins (1972). The islands are labelled in Simberloff and Wilson's (1969) original notation, and on each the fauna is classified into the trophic groups: herbivore (H); scavenger (S); detritus feeder (D); wood borer (W); ant (A); predator (C); parasite (P); class undetermined (?). For each trophic class, the first figures are the number of species before defaunation, and the figures in parentheses are the corresponding numbers after recolonisation. The total number of different species encountered in the study was 231 (the simple sum 164 + 140 counts some species more than once).

A particularly clear − and classic − example of this apparent constancy of community structure may be drawn from the studies, by Simberloff and Wilson (1969), of the faunal communities of mangrove islets in Florida Keys (further analysed by Heatwole and Levins, 1972). A number of mangrove islets were totally defaunated − and the structure of the arthropod community which re-established upon the islands determined, and compared with that *before* defaunation. Heatwole and Levins classified the arthropods into eight functional types and compared the structure of the communities before defaunation and after recolonisation. Results (Table 2.2) again suggest a striking constancy of structure. While actual species composition of the new communities altered quite significantly, the basic trophic structure of the new communities showed a remarkable similarity to that before defaunation. The 20 or so species develop-ing on each island were by no means the same species as those which were there before defaunation; yet, despite this, the relative numbers of organisms in each trophic class remained very much the same.

Further investigation reveals even more striking constancy. Not only does there appear to be a constancy of overall trophic design between equivalent communities (as if there were only one efficient structure for operating in a particular way). In addition there appears to be a remarkable constancy even of the detailed structure *within* each trophic level. There is a striking similarity of niche-form and niche-structure, of the way 'the jobs are partitioned out' in 'parallel' communities. In a comparison of coral reef fish communities of the Western Atlantic and Central Pacific, Gladfelter, Ogden and Gladfelter (1980) reveal striking similarities in trophic organisation in equivalent systems, and show that the constancy of community structure of these discrete systems extends beyond this to a constancy of niche-structure, even of actual species composition and species abundances. They demonstrate convincingly that particular reef structures support *constant* communities.

Even where these communities are in different biogeographic regions and thus cannot share the same species, there remains this striking similarity of niche-form and niche-structure in equivalent systems. Cody (1974, 1975) for example has shown that the 'potential vacancies for insectivorous birds' are divided up in almost exactly the same way in Californian chaparral, Chilean matorall and South African macchia (Figure 2.8). These various grassland systems may be considered ecological analogues within three different con-tinents. Each is colonised by taxonomically distinct groups of insectivorous birds; yet the way the resources are divided up in each between specific niches shows this striking constancy of pattern, suggesting once more a unique efficient design. Similar results have been shown for assemblages of lizards in montane areas of Chile and California by Fuentes (1976).

This is not of course to say that this unique structure is in some sense pre-determined to the extent that the design rules are deterministic, and that the community develops to 'seek a preconceived structure'. Such a constancy of design would also result, and more probably does, by virtue of the fact that it produces a stable community — which thus persists: in short that only the stable communities survive, and there are certain common features that promote such stability (see Chapter 13).

There is in practice considerable controversy about this possible 'fixity of design' and views as to how the data should be interpreted are widely divided. Cody's conclusions have been disputed by Ricklefs and Travis (1980), while Simberloff (1978b) reviews the evidence of trophic constancy presented by Heatwole and Levins (1972) (Table 2.2) and claims that these results, too, are statistically inconclusive. Such arguments themselves are, however, rather negative and inconclusive; they cannot *prove* that the apparent constancy is spurious. More powerful doubts are cast by a number of recent experimental studies equivalent to those examined above, but which show no such constancy.

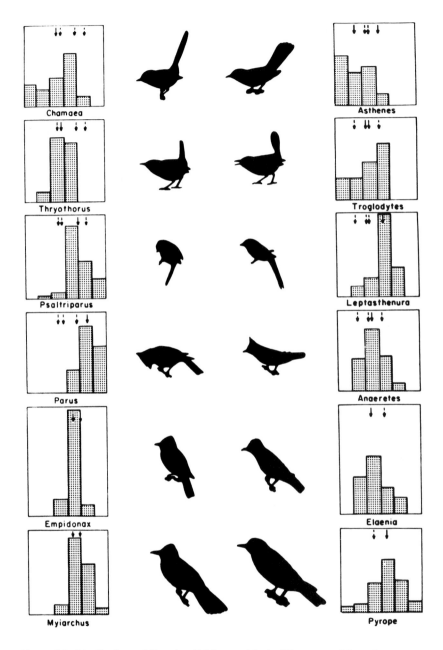

Figure 2.8: Distributions of Foraging Heights, and Body Silhouettes of Four Canopy Insectivores (above) and Two Sallying Flycatchers (lower) in Californian Chaparral (left) and Chilean Matorral (right). Foraging height intervals are, from the left: ground; ground − 6 in; 6 in − 2 ft; 2 ft − 4 ft; 4 ft − 10 ft; 10 ft − 20 ft. Arrows above each distribution indicate its mean, and those of the frequency distributions of neighbouring species. Source: From Cody, 1974 after May, 1976.

Lawton (1982), for example, in an extremely elegant comparison of the communities of herbivorous insects feeding on the above-ground parts of bracken (*Pteridium aquilinum*) in the north of England and in New Mexico and Arizona, found marked differences in community structure between the two systems. While 27 species of insects feed regularly upon the above-ground parts of bracken fronds in Britain, only five species were found on bracken in New Mexico with a further two species in Arizona. With such a small 'pool' of 'potential' species in Arizona and New Mexico, individual local communities of bracken herbivores are also species-poor by comparison to equivalent communities in Britain. In necessary consequence, niche relationships between the bracken herbivores of the two continents show little similarity (Lawton, 1982). The results of such studies – taken together with the growing doubts about interpretation of the work of Cody (1974) or Heatwole and Levins (1972) certainly weaken the case for postulating a constancy of design at this level; we must for the present reserve our judgement.

2.11 Species-abundance Relationships

The structure of a community is effected not only by the actual relationships between species, but by the relative numbers of organisms in those different species. While, clearly, the relative abundance of individuals in particular species may have a marked influence on the nature and function of the community, the distribution of individuals between species within the community, even irrespective of what those species are, has an effect on the 'balance' of the community, and ultimately its stability.

In natural communities, it has been observed, almost as a curiosity, that there are many more rare species than there are abundant species. That is, although there is a relatively large number of species which are represented within the community at low or medium abundance, there are relatively fewer species with large numbers of individuals representing them. If this should prove a biological imposition – rather than a mere mathematical artefact from random assortment – analysis of species-abundance distributions might offer another approach towards revealing something fundamental about the basic principles and restrictions underlying community design. And if species-abundance relationships differ between different communities, those distributions might prove useful comparative 'descriptors' of particular communities, summarising their different design and characteristics.

Such observations have led to a number of detailed analyses of species abundance distributions of both real and idealised communities. Thus, some of the distributions proposed are derived as purely empirical fits to observed data (with no explanation of why the species-abundance pattern should show such form). Others are derived from theoretical considerations about how the abundances of species in a community *should* be related to each other.

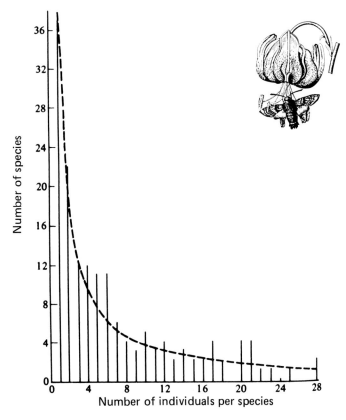

Figure 2.9: The Frequency Distribution of the Abundance of Various Species of Moths in a Light Trap at Rothamsted Experimental Station in England (1935). The expected frequency distribution from a logarithmic series is shown as a dotted line. Source: From Williams, 1964.

In the 'subsample' of a community provided by the assemblage of night-flying moths attracted to a light trap, for example, the frequency distribution is as shown in Figure 2.9. To these empirical data and other data of similar type, have variously been fitted logarithmic and lognormal curves (Fisher, Corbet and Williams, 1943; Williams, 1964; Preston, 1948), in an attempt to describe such patterns mathematically (Figures 2.9 and 2.10). Both mathematical distributions fit the observed data relatively well and can be extended to analyses of other communities or species assemblies, most of which have remarkably similar patterns in species abundance to those shown in Figure 2.9. But such a fit is not necessarily very informative. We do not know *why* communities should display such distribution.

A different and more theoretical approach was adopted to tackle the problem the other way around. If various assumptions about the rules of community organisation are made, and from these the species-abundance relationship that would result is extrapolated, the fit of such a relationship to empirical data may be used to justify or refute the assumptions about community design on which the original 'model' was based. In this way a 'reasoned fit' of some mathematical distribution to community data might be derived.

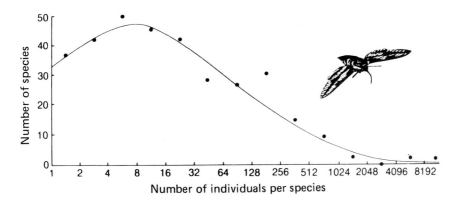

Figure 2.10: Relative Abundances of Moth Species Attracted to a Light Trap near Orono, Maine. Size classes (octaves) which increase by a factor of two from one class to the next are on the horizontal axis. The number of species in each class is plotted on the vertical axis. The distribution of abundance is best fitted by a log-normal relationship. Source: After Preston, 1948.

2.11a MacArthur's Broken-stick Model

MacArthur (1957) was the first to attempt to develop a model based on a set of hypotheses about how the organisms in a community might be structured, and to test these hypotheses by trying to match observed species-abundance distributions with those expected if the model were true. MacArthur's model is based on the idea that the resources of a community may be divided randomly into a number of pieces as a stick may be broken into a finite number of random sections. If each piece is thought of as the resources used by one species, the model postulates S species dividing the environment into S non-overlapping niches of randomly allocated size. The expected abundance of any species is then a percentage of the total number of individuals in the community. In fact, the expected percentage abundance of the j th species is

$$E\frac{Nj}{n} = \frac{1}{S} \sum_{i=1}^{j} \frac{1}{S-i+1} \qquad (2.1)$$

Such a model results in a distribution of species as in Figure 2.11 (which has come to be dubbed, familiarly, the 'broken-stick' distribution).

The most thorough treatment of the statistical properties of the relative abundances of a group of species which apportion randomly amongst themselves a fixed amount of the various environmental resources, is due to Webb (1974). As pointed out by MacArthur (1957, 1960), if the underlying picture is one of intrinsically even division of environmental resources, the statistical outcome will be the broken-stick distribution. However, the model has its limitations. It assumes implicitly that the environment is dominated by some major single resource to be shared and that this single governing resource is to be shared randomly. Further, the same mathematical distribution can be derived from totally different initial premises. Cohen, for example (1966) postulated a multi-dimensional resource, available to S species, divided into S subdivisions. The rarest species occupies one subdivision, the second rarest two subdivisions and so on. The commonest species occupies all the subdivisions. (In contrast to MacArthur's model there is considerable overlap in the resources utilised by the various species, even though the species do not occupy exactly the same niches.) This set of hypotheses leads to exactly the same broken-stick distribution of species abundance. Thus, as pointed out by Cohen (1968) and reviewed by Pielou (1969) the observation of a broken-stick distribution does not validate the very specific model initially proposed by MacArthur (1957, 1960). Where observed, however, it *does* indicate that some major resource dominates the community structure and is being roughly evenly apportioned amongst the community's constituent species (May, 1975b).

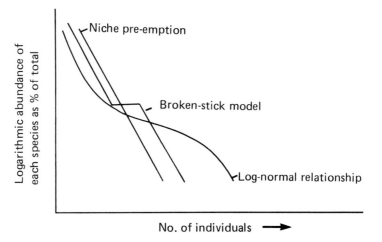

Figure 2.11: The Expected Shape of Species-abundance Relationships Conforming to a Log-normal Relationship, the Broken-stick and Niche Pre-emption Models.

2.11b A Niche Pre-emption Model

An alternative approach was suggested by Motomura (1932) reviewed by Poole (1974). Suppose the proportion of the total available resources used by a species is determined by the success of a species in pre-empting for its own use part of the available resources. The less successful species occupy the resources left. The most successful, i.e. dominant, species occupies some fraction k of the resources of niche space. A second species is able to occupy a similar fraction k of the remainder left by the dominant species, and the third most important species k per cent of the remaining resources left after species 1 and 2 have taken their share. If this set of postulates is true, the percentage of the total number of individuals of all the species, or some other measure of importance such as biomass, ranked from commonest to rarest, forms a geometric series

$$I_n = Nk(1 - k)^{n-1} \tag{2.2}$$

where I_n is the percentage importance of the nth species, N is the total number of individuals and k is equal to $1 - c$, where c is the ratio of the importance values of a species to that of its predecessor in the series of ranked species. If the species ranked from commonest to rarest are plotted against the logarithm of importance, the predicted percentage importance or abundance of the ranked species forms a straight line if the species-abundance relationship fits the niche pre-emption hypothesis (Figure 2.11).

Species-abundance relations similar to the predicted values of the niche pre-emption hypothesis are most often found in communities strongly dominated by a single species. Some plant communities, particularly those occurring in severe environments with a small number of species, are well fitted by the geometric series of the niche pre-emption model (Poole, 1974). Once again, however, the assumptions are oversimplified (of a constant pre-emption fraction for each species for example).

2.11c Theoretical Analyses of Species Abundance: a Conclusion

A whole host of other models have been developed, but in practice it is fair to say that the analysis of species-abundance relationships has not been as fruitful as might have been hoped. The fact that empirical data can generally be fitted by a lognormal or logarithmic distribution is striking, but a biological interpretation of why this should be the case is obscure. Nor have attempts to approach the problem the other way around — by making various assumptions about the rules of community organisation, extrapolating the species-abundance relationship that would result, and using the fit of such relationship to empirical data to justify or refute the initial assumptions about community design — been of great value. For two different distributions derived from conflicting premises can frequently both adequately fit an observed set of data. So that even if a hypothesised distribution does fit the observed species-abundance relationship — that fit neither proves nor disproves the postulates of the model.

Perhaps the best synthesis of the whole species-abundance problem is that offered by May (1975b). He notes that broken-stick, geometric or log series distributions of species abundance result, in theoretical analysis, from considerations of relatively simple systems with only a single factor considered to be involved in determining the organisation. All reflect some basic dynamic aspect of the community and will be characteristic in the real world, of relatively simple communities whose dynamics are indeed dominated by some single factors. The broken-stick is the statistical expression of an ideally uniform pattern of distribution; at the other extreme, the log series is the expression of an uneven distribution (such as would be produced by a 'niche pre-emption' or competitive process).

But as soon as several factors become significant a log-normal relationship is to be expected. Where the distribution of abundance is liable to be governed by many more-or-less independent factors it will take a log-normal form. (This tendency is in fact a function quite generally of all multiplicative processes. Merely as a function of a statistical law of large numbers, they tend towards a log-normal distribution. Thus the log-normal distribution need not necessarily have a biological basis, although this is not to say it may not have such a basis – see Sugihara, 1980 and Chapter 12.)

May concludes therefore that for real communities a logarithmic distribution of species abundance may be expected amongst relatively small sets of species where competition may result in an uneven distribution; while amongst small sets of species which randomly apportion amongst themselves available resources, MacArthur's broken-stick model may be applied. For large or heterogeneous assemblages of species (such as complete communities) a log-normal pattern of relative abundance will be expected – merely as a mathematical reflection of the Central Limit Theorem (May, 1975b).

Stenseth (1979) has suggested the further thesis that the shape of the species-abundance distribution is related to environmental circumstances – being log-normal for communities in stable environments, logarithmic even for whole communities, in unstable systems. Evidence for this may be adduced from Patrick's (1963) investigation of species-abundance relationships among diatoms in natural and polluted streams (Chapter 13, Figure 13.1). Such suggestion is perfectly in keeping with May's conclusions – for unstable systems are liable to be influenced by a few dominant biotic factors, while communities of stable environments are affected equally by a multiplicity of factors.

All these conclusions do suggest that the species-abundance distribution of a community does reflect some underlying characteristic of that community, but may merely be a passive indicator of the way in which it is ordered rather than a positive requirement of a particular community design. That indicator principle is useful in its own right: if we encounter a community or species assemblage which displays let us say, a broken-stick distribution of species abundances we may conclude that this assemblage is a relatively simple one, and one predominantly ordered in regard to a single dominant ecological resource; where we encounter logarithmic patterns of abundance, we may conclude that a small community engages in strong and uneven competition for a number of resources, while a log-normal distribution suggests a complex and heterogeneous assemblage ordered by a large number of ecological pressures in conjunction. But these conclusions are merely an empirical spin-off; the study of species-abundance distributions still does not permit us to say *why* a particular community or assemblage should display its characteristic abundance-distribution.

2.12 Species Associations

Other 'general principles' of community organisation have been suggested. There seem to be certain well-established relationships between species — that if one is found so will be another (see for example page 66). Such relationships may be positive or negative, and are not just restricted to associations between different trophic levels (i.e. that a species cannot exist in a community if its food species are absent) but may exist between organisms of the same trophic status. Some of these principles of organisation have been formalised by Diamond (1975) from a study of bird assemblies in the islands of the New Guinea area. Diamond asserts that:

'If one considers all the combinations that can be formed from a group of related species, only certain ones of these combinations exist in nature.'

'Permissible combinations resist invaders that would transform them into forbidden combinations.'

'A combination that is stable on a large or species-rich island may be unstable on a small or species-poor island.'

'On a small or species-poor island, a combination may resist invaders that would be incorporated on a larger or more species-rich island.'

'Some pairs of species never coexist, either by themselves or as a part of a larger combination.'

'Some pairs of species that form an unstable combination by themselves may form part of a stable larger combination.'

'Conversely, some combinations that are composed entirely of stable sub-combinations are themselves unstable.'

Examining data from 147 species of land birds distributed in various combinations over 50 islands in the Bismarck Archipelago near New Guinea, Diamond argues that much of the explanation for these 'assembly rules' has to do with competition for resources and with harvesting of resources by permitted combinations so as to minimize the unutilized resources available to support potential invaders. He claims that 'communities are assembled through selection of colonists, adjustment of their abundances, and compression of their niches, *so as to match the combined resource consumption curve of all the colonists* to the resource production curve of the island.' Such assembly rules have been modelled, and tested, using data on a guild of foliage-gleaning birds in spruce forests of islands of the coast of Maine, by Haefner (1981). Diamond's conclusions are however controversial. Connor and Simberloff (1979) claim that most of Diamond's observed distributions might be expected were the species distributed randomly on the islands, and that thus the data offer no support for any 'assembly' rules. Indeed they conclude (Connor and Simberloff, 1979): 'Every assembly rule is either a tautological consequence of the definitions employed, a trivial logical deduction from the stated circumstances, or a pattern which would largely be expected were species distributed randomly on the islands.'

2.13 Niche Relationships and Design Rules

Another possible approach to gaining some insight into the 'rules' of community organisation may come through analysis of the details of niche structure and organisation. Limits to the design of individual niches in size or shape, limits to similarities of adjacent niches in the same communities, limits to overlap and to the method of 'packing' of niches within the community may have profound implications for the way communities are structured. This will be considered further in Section 5.7.

The community however is a complex entity. For many years ecologists have considered it, essentially, as a composite of component populations and have concentrated their efforts on unravelling the dynamics of those populations themselves. 'Community ecology' says Pianka (1980) 'is in its infancy': it will be some time before all its 'rules' are discovered.

2.14 The Structure of Particular Communities

So far we have been discussing generalities of community structure, principles applicable to any community, differences and similarities in general terms. But what actually dictates the particular structure of a specific individual community? What dictates the number of trophic levels it shall have, the diversity of species, and complexity of web structure or indeed the very species composition? Clearly the constraints of the physical and chemical environment limit to a large extent which organisms can survive in any particular system. But within that a variety of biotic considerations combine to dictate the final system achieved. Any community is built upon its primary producers, and the species array within the plants influences much of the structure and species composition of the community which develops upon this level. Thus for any set of environmental conditions — climate, temperature, wind, moisture, soil type, subsoil, physical and chemical structure of top soil, exposure etc. etc. — there is a characteristic set of dominant plant species which may occupy an area. These few dominant species modify the environment (provide shelter, humidity pockets, shade, or whatever) and create suitable conditions for the establishment of a second series of dependent plant species. Associated with this particular species spectrum of plants will be the herbivorous animals, whose food species are represented. Associated with them will be the carnivores which prey on those particular herbivores and so on (Figure 2.12). Once the selection of dominant plants species is made the rest follows automatically, and it is surprising how constant, both in the plant world, and in the animal/plant context, species associations are, at least in the lower trophic stages. (Higher up the trophic scale, such relationships are not so rigid: higher consumers are able to exploit a wider variety of food species and thus tend to be much less community-specific, but can occur in a variety of different contexts.)

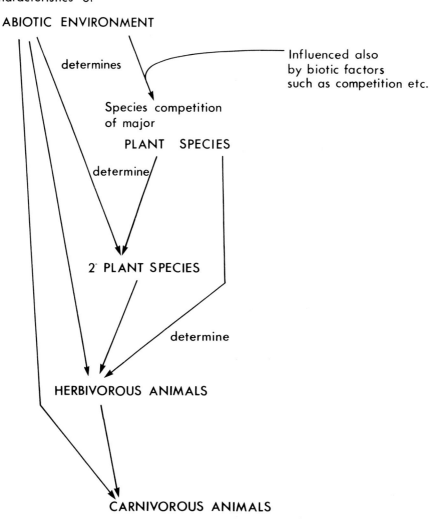

Characteristics of

ABIOTIC ENVIRONMENT

determines

Influenced also
by biotic factors
such as competition etc.

Species competition
of major

PLANT   SPECIES

determine

2° PLANT SPECIES

determine

HERBIVOROUS ANIMALS

CARNIVOROUS ANIMALS

Figure 2.12: Determination of Structure in a Typical, Terrestrial, Green-plant Based
Community.

At each stage, the selection of plant and animal species is of course still
restricted and influenced by abiotic factors (Figure 2.12). But the framework
outlined in the figure is deceptively oversimplified. For although it is true that
there are well-established species associations between both plant and animal
species, it is also true that these frequently have alternatives. Primary plant A
can either provide shelter for plant B or plant C. Equally, either species X or
species Y may occupy the invertebrate carnivore niche in the aquatic community
of a waterfilled beech tree-hole. In Britain, *Metriocnemus* and *Prionocyphon*

species are equally valid but mutually exclusive alternative occupants of the carnivorous midge-larva niche in such tree-hole communities (Kitching, 1977). What decides which of these alternatives will be represented in the community? In this analysis it is almost a matter of chance — scientifically explicable chance, but chance nonetheless. In the case of the tree-hold midge larvae, whichever species is first to arrive, establishes itself, and out competes the later-arriving second species.

This establishes another, more general principle and one true of the entire species structure and number within community — that the particular species present within a community at any one time are the net result of conflicting pressures of colonisations and extinctions within the system. At any one time one species may become extinct, and be replaced by another. These interactions occur continuously, and thus an attempt to define the community's species composition as a static entity is unrealistic: as we noted above, on page 44, the community is in practice always in a state of change.

2.15 Community Flux

And so we return to perhaps the most important feature of community structure: that they are never static. Species are continually lost and replaced; the ecological characteristics of those species and of each species already within the community are also continually changing as the result of evolutionary pressures. In brief the overall impression is of a shifting changing pattern of species definable only in terms of the interrelationships between them, and a community definable only in operation. But while from the standpoint of the populations of individual species, one may gain an impression of ceaseless change and flux, dominated by environmental vagaries or the accidents of history, at the same time, from the standpoint of trophic structure and community function the picture may be one of steadiness and pattern (Ricklefs, 1971).

3 Community Dynamics

3.1 Introduction

The previous chapter stressed the fact that a community is a functional entity: this complexity of structure and organisation lacks a dimension, until it is seen in operation — for it is a system which only exists in motion. Clearly the operation of any community is a function of the interaction of the various processes affecting its component organisms: competition, parasitism, predation and so on, and these individual processes will each be examined in detail in later chapters. But to stick first with our 'overview' let us consider the operation of the community as a whole: the net result of all these separate interactions combined.

This is of course, somewhat more easily said than done, and for many years ecologists lacked this overview. They had founded their science on meticulous observation and field recording; with such an approach, a picture of community function only could arise as the product of the interactions of its component species. It was not possible to study the dynamics of the whole, for suitable techniques were simply not available; even *within* the community it was often not possible to form more than a qualitative, subjective, estimate of the roles of the various organisms, and there were many problems that could not be tackled at all, or others which, through their complexity, could only be partially resolved. In short, both in terms of understanding relationships within the community, and in understanding the dynamics of the whole, ecologists were looking for a more rigorous unifying approach, a method of quantification with respect to some very basic common theme. And they took their cue from the theoretical physicists, by turning to a study of energy flux.

3.2 The Community as a System of Energy Transformations

It is indeed valid to consider any biological system — individual, population or community — as a system for the transfer, storage and dissipation of energy. It is, what is more, a system that obeys all the basic laws of thermodynamics: that energy can neither be created nor destroyed, and that each transfer, or transformation of energy cannot be 100 per cent efficient: energy will be lost and there will be an increase in entropy at each stage. Animals and plants during their lifetimes act as stores of potential energy. In feeding or in being fed upon, that energy is transferred to and from other organisms. Not all the energy is retained as potential energy within the body: the life processes of maintenance – basal metabolism, or whatever you like, require energy expenditure; still more energy is released in 'work'. Thus some of the original energy is dissipated from the system in respiration.

Within an ecological system, of course this 'energy' is not free energy but is the chemical energy incorporated into the structure of complex organic materials – the energy of molecular bonds. Animals, and certain heterotrophic plants (fungi and many bacteria) take in their energy ready-stored within the organic molecules of their food, but what is the ultimate source of energy for the ecological system? For this we must look back to the Producer level: to those autotrophic organisms that are able to synthesise organic compounds from inorganic precursors, using free energy in the process. Certain bacteria use energy derived from the low potential chemical energy of inorganic chemical bonds – in 'chemosynthetic' processes. The majority of autotrophs however are photo-synthetic, using the energy input from solar radiation to form the bonds of their synthetic compounds. Thus, in one way or another 'inorganic' energy is taken from the environment and used to synthesise 'organic' energy, by the autotrophs – the producers. They use the energy thus derived both for growth and respira-tion. That used for growth may be taken in by the primary consumers, and passed on up the trophic scale. However, not all the consumed material is usable by the herbivore; some passes direct to the decomposers in faeces and urine. What is retained by the consumers (or *assimilated*) may then also be used for growth and respiration. If they in turn are then eaten by other consumers, the small amount of energy they have stored in tissue production is again diminished (Figure 3.1). At each stage the amount of energy passing to the trophic level above decreases and at each stage, energy is dissipated through respiration (for that, too, is the ultimate fate of that material which passes to the decomposers). Ultimately all the energy fixed by the producers is dissipated in respiratory activity, so that eventually all the solar (or chemical) energy entering the system is released in the heat energy of respiration – conforming with the first law of thermodynamics noted earlier: no energy is lost and none gained. (Note that while nutrients and basic elements cycle through ecosystems the flow of energy is strictly linear. Odum (1959) has coined a simple catchphrase as a mnemonic 'Matter circulates, energy dissipates'.)

So energy does flow in a logical way through biological systems, and this provides the unifying link which enables us to study the mechanics of the system or its operation: for we may analyse the flow of energy through the system, quantifying the efficiency of transfer from organism to organism, the speed and direction of that transformation and the use to which energy is put within the organism or trophic level; and we can compare communities with respect to these various parameters. As always we have again the opportunity of looking at the community as a jigsaw of its component species, or in viewing it as a single functional whole. Here we may study the energy relationships of individual organisms, and by linking them into their appropriate context, build up a picture of the community, or we may look at the general flow of energy through the community in its own right.

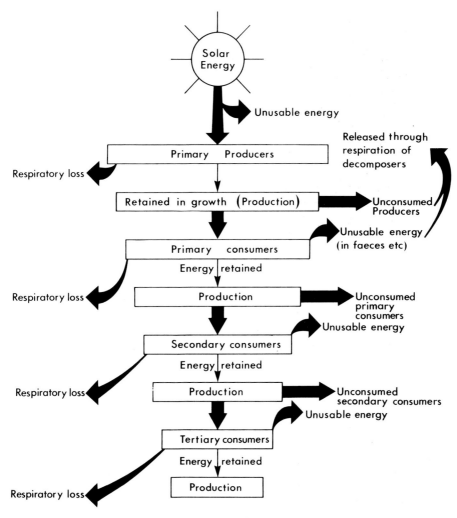

Figure 3.1: Losses of Energy in Passage Through the Different Trophic Levels of a Community.

3.3 Energy Relationships of Individuals

The energy relationships of an individual organism (or species) may be defined in terms of its Consumption, and its Assimilation (the amount of energy retained after excretory losses — in animals, faeces, urine, gases of fermentation). Thus

Assimilation = Consumption − Faeces and Urine

or, (adopting the conventional terminology of Petrusewicz, 1967):

$$A = C - Fu \qquad (3.1)$$

Similarly, the use to which assimilated energy is put may be represented. Assimilated energy may be used for Production (P) in both individual growth (Pg) and reproduction (Pr), and for respiratory costs (R) of basic maintenance (R_m) and work (R_w). Thus

$$A = P + R \qquad (3.2)$$
$$\text{or } A = Pg + Pr + Rm + Rw \qquad (3.3)$$

All these individual quantities may be precisely determined in terms of calories, or joules (kgm^2s^{-2}).

Analysis of these values and their relationships may tell us a great deal about the operation of the organism itself. For example, the relationship of Assimilation to Consumption (*Assimilation efficiency*: A/C per cent) reflects efficiency of food use: the assimilation efficiency of most herbivores is low — of the order of 30-40 per cent — while that of carnivores is in general much higher (63-80 per cent). (The 'assimilation efficiency' of plants is surprisingly low. Green plants trap only 1-5 per cent of the sun's energy in photosynthesis.) A derivative of this relationship is the agricultural *Production efficiency* (P/C per cent). Analysis of the way in which the assimilated energy is apportioned between Production and Respiration (P/R) is also very illuminating in informing us whether or not this is a very active organism which in consequence has little energy to spare for production, or whether it needs little energy for respiration and puts most of its available resources into growth. This last is a relationship that changes rapidly through the life of an individual, too: younger animals and plants tend to be very fast growing and (since respiratory cost is related to body size) have relatively low respiratory expenditure; older animals have higher respiratory costs and less energy to set aside for growth and reproduction. We can learn a lot in detail about the use of the 'Production' energy: whether it is used in growth, purely in size increase, growth in replacement of insect larval cuticular exuviae, pupal cases etc. which are discarded, growth in reproduction. All this information can be quantified to draw interesting conclusions about the organism's use of energy and energy balances. From such knowledge we can even make predictions about its ecology. If we know the energy costs of foraging for a certain type of food, and the energy return (assimilation) from feeding on one such food item, we can make predictions about the way in which an animal may forage, its prey choice and pattern of hunting, whether it will accept or reject any given items and so on (Chapter 10). We can even make suggestions about the predators ideal size, and whether or not the species may show sexual dimorphism (Schoener, 1969). Further, the information can be used in an applied sense, too. From calculations of daily energy consumption and assimilation, knowledge of respiratory costs of maintenance and activity, we can estimate the net production surplus any individual animal will retain after satisfying respiratory costs from a measured assimilation. If we also know the energy costs of reproduction for that species we can predict, from a mere survey of food availability, whether or not that individual or species will be able to breed in any given season — a prediction which may be of great value if the species concerned is an economic pest.

3.4 Energy Relationships in the Community

From knowledge of the energy relations of an individual, we can, as we have noted, build up a picture of the energy relations of the whole community. For the Consumption of an individual or species must be taken from somewhere else in the community, while Production (if the system is in steady state) must be used either to sustain standing crop biomass in replacing those individuals which have died, or to supply food for other organisms in the system. Dead material, faeces and urine are inputs for the 'Consumption' of the decomposers; respiratory releases are the only loss. So for each organism we can refine the simple

$$C - Fu = P + R$$

to fit it into the community framework (Figure 3.2). If we repeat the exercise the full pattern of energy flow through the community will emerge. Such analyses have been made (e.g. Teal, 1959; Odum, 1959; Figure 3.3) but are perhaps of limited value — except insofar as a detailed analysis is required of some particular system for some specific purpose — for they permit of little generalisation. A detailed analysis is offered of one specific system, and is true only for that system; further, because of the complexity of interactions within a community, it is necessarily only very simple systems, or parts of systems, which can be treated in this way (and it is notable that the classic studies of Teal and Odum were both carried out on small freshwater springs). In addition, even within such restricted systems the complexity of relationships is such that it is not possible to draw any conclusions which might be applicable to communities in general. To do this we must adopt the same rationalisation as we invoked in the previous chapter; regarding the community as a system of discrete trophic levels. This approach — and indeed the whole concept of the community as a series of trophic classes — was first developed by Lindeman in 1942 in a study of . . . a freshwater spring!

3.5 Energy Flow Within the Community: the Tropho-dynamic Approach

This reduction of the community structure to a series of trophic levels makes analysis much more manageable. We may then examine various aspects of the passage of energy through the system: amounts of energy transferred, the speed of transfer through the system, and the *efficiency* with which energy passes from one trophic level to the next; we may also consider the use to which the energy is put within each trophic level.

$$C - Fu = P + R$$

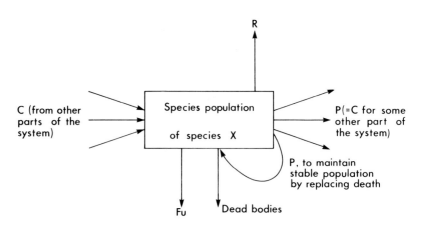

Figure 3.2: Energy Budget of an Individual Organism in Community-context.

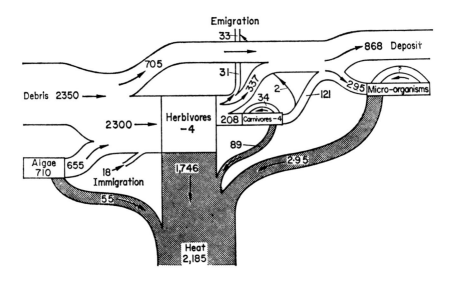

Figure 3.3: Energy Flow Diagram for Root Spring, Concord, Massachussetts in 1953-54. Figures in kcal m^{-2} yr^{-1}; values inside boxes indicate changes in standing crops. Source: From Teal, 1959.

3.5a Energy Relationships within a Trophic Level

Within any one trophic level, of course, the use and apportionment of energy can be examined by exactly the same analysis as applied to evaluation of the energy relationships of an individual organism or species population. Thus we may summarise the relationships (Section 3.4) in terms of the equations

$$C - Fu = A = P + R$$

In the same way, we can define trophic level assimilation and production efficiencies (A/C per cent, P/C per cent). Relationship of the production efficiencies of various trophic levels within the same, or different communities make interesting comparisons (e.g. Table 3.1). Other analyses of exactly the same type as those carried out for individual organisms or species population, offer figures for the amount of material held in living tissue within the trophic level — in standing crop biomass, rate of turnover and so on: all measures defining the structure and dynamics of the trophic class. But it is in the transfer of energy to and between the different trophic levels that the real keys to community function can be discovered.

Table 3.1: Net Production Efficiency (Net Primary Production/Gross Primary Production) of Several Plants and Plant Communities.

Plant community	Locality*	Net production efficiency (per cent)
Terrestrial		
Perennial grass and herb	Michigan	85
Corn	Ohio	77
Alfalfa		62
Oak and pine forest	New York	45
Tropical grasslands		55
Humid tropical forest		30
Aquatic		
Duckweed	Minnesota	85
Algae	Minnesota	79
Bottom plants	Wisconsin	76
Phytoplankton	Wisconsin	75
Sargasso Sea	Tropical Atlantic Ocean	47
Silver Springs	Florida	42

Source: From Odum, 1959.

3.5b Energetic Efficiency

One of the major considerations here, as we have already hinted, is of the *efficiency* with which energy is transferred through the ecological system. Not all the energy any animal takes in in its food is retained as potential energy in growth; much is lost in assimilation, and of that assimilated energy, more is expended in respiration. Nor is the entire production of a population of animals or a trophic level, passed on to the trophic level above; much is unavailable to the consumers and many individuals will die unconsumed. Of that which is consumed, again only a proportion will be assimilated. Thus if we look at efficiency of energy transfer between trophic levels as

$$\frac{\text{Energy consumed by trophic level n}}{\text{Energy consumed by trophic level below (n}-1)} \tag{3.4}$$

we will always find it considerably less than one. Indeed, in terms of percentage, this 'gross ecological efficiency' is usually somewhere between 7 and 14 per cent. It has been suggested that the efficiency is in fact a constant: that the range of figures derived between 7 and 14 per cent is due to necessary inaccuracy in measurement and that in natural ecosystems, the efficiency of energy transfer between trophic levels approximates to a constant 10 per cent: an axiom always attributed to Slobodkin (1961). But, while this may be a useful rule of thumb, it is in fact the very differences in efficiency which are most interesting. Efficiency of energy transfer may differ between different pairs of trophic levels within one community – offering an insight into the different mechanics and efficiency of operation of different trophic classes, to complement knowledge derived from analyses of energy flux within the trophic levels themselves. In the same way, efficiency of transfer may also differ between equivalent pairs of trophic levels in different communities, hinting at differences in methods of operation between the various communities (Table 3.2). Many of these differences in operation and differences in energy flow we shall explain in later chapters.

Table 3.2: Efficiency of Energy Transfer at Various Trophic Levels in Three Aquatic Ecosystems

| | Trophic Level Energy Intake Efficiency (per cent) | | |
Trophic level	Cedar Bog Lake, Minnesota	Lake Mendota, Wisconsin	Silver Springs, Florida
Photosynthetic plants (producers)	0.10	0.40	1.2
Herbivores (primary consumers)	13.3	8.7	16
Small carnivores (secondary consumers)	22.3	5.5	11
Large carnivores (tertiary consumers)	not present	13.0	6

Source: From Odum, 1959.

While gross ecological efficiency (or *Lindeman's efficiency*) is the ratio most commonly referred to in studies of community function, other relationships are also invoked. To avoid confusion in reading other literature these are summarised together with efficiencies relating to individuals or populations and thus by extension to trophic levels (Section 3.3) in Table 3.3.

Table 3.3: Various Types of Ecological Efficiency (based on Odum, 1959)

Ratio	Explanation
A Ratios *within* trophic levels	
$\dfrac{P}{A}$	Tissue growth efficiency
$\dfrac{P}{C}$	Production efficiency (agriculturally: food conversion efficiency)
$\dfrac{A}{C}$	Assimilation efficiency
B Ratios between trophic levels	
$\dfrac{C_t}{C_{t-1}}$	Gross ecological efficiency or Lindeman's efficiency (Energy intake by trophic level t/Energy intake of trophic level t−1)
$\dfrac{A_t}{A_{t-1}}$	Trophic level assimilation efficiency
$\dfrac{P_t}{P_{t-1}}$	Trophic level production efficiency
$\dfrac{A_t}{P_{t-1}}$	Utilisation efficiency

The depletion of energy availability up the trophic scale is frequently quoted, in explanation of Eltonian pyramids of numbers or biomass, as a more precise explanation than that offered on page 52. If only a small percentage of the energy entering one trophic level is effectively transferable to the level above it, then the biomass supported must be reduced. The same argument is then extended as explanation for the limit to the number of trophic levels themselves. As Pimm and Lawton (1977) have recently observed, however, this last claim is not easily reconciled with the fact that the number of trophic levels appears to be a constant — and essentially independent of enormous variations in the amount of energy flow and transfer efficiencies: 'food chains are not noticeably shorter in barren Arctic and Antarctic terrestrial ecosystems compared with a productive tropical savannah, or the fish guilds of a tropical coral reef'. Energy attenuation obvious ultimately provides an upper limit to the number of *possible* trophic levels in any given system. Pimm and Lawton suggest that the actual number of trophic levels observed is well below this limit, and that the actual limit to the number of trophic levels observed (and hence the apparent constancy) is determined in the dynamics of the various populations of the community. They use mathematical studies of the stability properties of variously constructed model food webs to argue that long food chains may typically result in population fluctuations so severe that it becomes impossible for higher order predators to persist. This suggestion is further developed by Saunders (1978), Vincent and Anderson (1979) and Pimm and Lawton (1978).

3.5c Speed and Amount of Energy Transfer

Other facets of the flow of energy through the community, besides pure efficiency, can provide useful pointers to important aspects of community function. Thus the *speed* with which energy passes through the system differs between different communities, whether measured as speed of turnover within one trophic level (Table 3.4) or through the whole community; and this may have important implications for community structure and stability. De Angelis *et al.* (1978) have shown that relatively unstable model food webs can be kept effectively stable provided that the transit time for a unit of energy through that web is relatively fast.

Table 3.4: Average Transit Time of Energy in Living Plant Biomass (Biomass/ Net Primary Production) for Representative Ecosystems

System	Net primary production $(g\ m^{-2}\ yr^{-1})$	Biomass $(g\ m^{-2})$	Transit time (yrs)
Tropical rain forest	2,000	45,000	22.5
Temperate deciduous forest	1,200	30,000	25.0
Boreal forest	800	20,000	25.0
Temperate grassland	500	1,500	3.0
Desert scrub	70	700	10.0
Swamp and marsh	2,500	15,000	6.0
Lake and stream	500	20	0.04 (15 days)
Algal beds and reefs	2,000	2,000	1.0
Open ocean	125	3	0.024 (9 days)

Source: From data in Whittaker and Likens, 1973 after Ricklefs, 1971.

Further the *amount* of energy passed through the system at any one time (defined by input, in terms of energy fixed by the producers and then passed on by them through the community) may also differ between communities. The total amount of material synthesised within the community by the producer level is referred to as the Gross Primary Production (GPP). That which is available to other elements of the community after respiratory expenditure by the producers (GPP − Respiratory cost) is defined as Net Primary Production (or NPP). That GPP − and thus the amount of energy passing through the whole community − can vary enormously between different systems is illustrated by the summary table (Table 3.5) taken from Odum (1959), while Odum himself has shown that the quantity and quality of energy flow through a community are indeed tremendously important features of that community, defining indeed its whole structure and mode of function. In 1975 he wrote

The species matrix adapts to the strength and variety of energy input and resource flows coupled with it. ... Furthermore, the quality of the energy in terms of utility and low entropy is as important as the quality. When one or a few sources of high utility energy ... are available ... low diversity (within the community) has advantages, a concentrated and specialised structure is more efficient in exploiting the bonanza that is a dispersed structure. Where energy is limiting or of low utility then a higher diversity appears to be optimum for (the operation of) a steady-state system (Odum, 1975).

This in itself clearly has important implications in terms of our considerations of rules of community structuring in Chapter 2.

Table 3.5: Gross Primary Productivity of Various Ecosystems as Determined by Gas Exchange Measurements of Intact Systems in Nature

Ecosystem	Rate of Production $(g\ m^{-2}\ day^{-1})$
Infertile open ocean, Sargasso Sea	0.5
Shallow, inshore waters, Long Island Sound; year average	3.2
Texas estuaries, Laguna Madre	4.4
Clear, deep (oligotrophic) Lake, Wisconsin	0.7
Shallow (eutrophic) Lake, Japan	2.1
Bog lake, Cedar Bog Lake, Minnesota (phytoplankton only)	0.3
Lake Erie, Winter	1.0
Lake Erie, Summer	9.0
Silver Springs, Florida	17.5
Coral reefs, average three Pacific reefs	18.2

Source: From Odum, 1959.

3.6 Limitations of Energy Analysis

To review: the study of the flow of energy through an ecological system enables us (1) to put the individual organisms of the community in context, and to define more exactly their actual function in the system, (2) to extend this if we wish to a detailed analysis of the entire flow of energy through specific systems and (3) to restrict the consideration to one of energy flow between trophic levels and apportionment within them.

However, in pursuing our analysis of community function purely in terms of the transfer and transformation of energy we must bear in mind the possible limitations of this approach. For consideration of the community merely as a system for the transformation of energy is of course a gross oversimplification. It restricts consideration to trophic relationships alone (all others are neglected) and further presupposes that energy is the single most important commodity passing through the system. In reality, while of course organisms need energy in order to function, they require other raw materials as well – nitrogen to build protein, potassium, phosphorus, calcium – a whole host of other nutrients. And in practice, at least in the producer and primary consumer levels of the community (which are, after all, the two major elements within the structure) it is very unusual to find energy in itself to be limiting. Far more likely to be in limited supply are these other nutrients, particularly nitrogen and potassium. Energy, as we have seen, has a linear flow through any ecological system; nutrients cycle and since the available 'pool' is limited in extent, bottlenecks within the system – in say, release of nitrogen from dung or carrion – are likely to have considerably greater effect in restricting community function than energy itself. While analysis of energy flow is conceptually simple and straight-forward, and so to speak 'embodies' the total operation and functions of an organism, population or trophic level, in a single, quantifiable measure of impact, it should always be regarded as a first approximation only. It is not 'sufficient unto itself'. Working on the energy dynamics of a species of grassland bug (*Notostira* sp.), Gibson (1977) found that, at the time when females began to prepare and develop eggs and at just that time when one might anticipate they should feed where maximum energy intake would be assured, they moved from a grass of high calorific value to one of considerably lower energy content: a grass in fact, upon which they could only barely satisfy their minimum energy requirements. In terms of energy alone: a paradox, resolved only when it is appreciated that the second grass is extraordinarily high in nitrogen. In another, similar study, Greenstone (1979) has shown that wolf-spiders (*Pardosa ramulosa*) feeding on three main prey species (the shorefly *Ephydra riparia*, the mosquito *Aedes dorsalis* and waterboatman *Trichocorixa reticulata*) seem to select prey with regard to aminoacid composition rather than energy, selecting a diet that optimises their own intake of essential dietary aminoacids. The importance of nutrients other than energy has also been recognised in recent studies by, for example, Westoby (1974), Pulliam (1975b), White (1978) and Stenseth and Hansson (1979). Nonetheless, studies of energy flux within communities give a good first approximation, a good first impression of dynamics and methods of function – information which could not be obtained by any other analysis. As a complementary approach to many ecological problems, consideration of energy flow will crop up in a number of later chapters, together with information derived from a more classical, observationally based approach, to supplement and clarify many of the points we will discuss.

3.7 The Flow of Nutrients Within Communities

As we have already noted, the flow of energy through any ecological system is linear, and relies on continuous new input at the producer level. By contrast most other materials cycle through the ecological system: there exists a 'finite' and specified amount of each nutrient within the system which cannot easily be replenished; in this case continued function of the system relies on the continuous cycling of each nutrient within it.

Maintenance of life depends on a whole host of different nutrients: water, carbon, oxygen, nitrogen, potassium, phosphorus, calcium, iron, magnesium, sulphur — and a variety of other minerals required only in minute quantities (so-called trace elements). Many of these nutrients are superabundant, but supplies of others are potentially limiting. It is the dynamics of these nutrients — required in large amounts in proportion to their overall availability — with which we are primarily concerned here: these are the nutrients likely to restrict community function; these are the nutrients for which efficiency of transfer and circulation is most critical.

3.7a 'Nutrient Pools' : Free Nutrients and Biota

We may consider, in essence, a limited quantity of each such material available within the system: a limited nutrient 'pool'. Some proportion of this may be immediately available as 'free' nutrients in the soil, ready to be taken in by the primary producers of the system as required, while the remainder of the eco-system's supply of each nutrient will actually be bound up temporarily within the living tissues of its component organisms: so to speak, currently in use. The relative sizes of the 'free nutrient' pool and that proportion bound up in the biota of the ecosystem are important considerations. If there is a relatively large stock of available nutrients then the continued operation of the system is not immediately limited by the speed of return of 'used' nutrients for re-use. It can, so to speak, live off its fat, and continue to operate efficiently by utilising the large stock of free-nutrients which will ultimately be replenished by the return of those nutrients presently 'in currency' when they are finished with. Nutrients do not move through living systems in smooth even flow, but in pulses, jerks and floods. The larger the stockpile of free nutrients, the greater the capacity of the system to accommodate the vagaries of return of those nutrients bound up within the tissues of living organisms. If, by contrast, very little of the available pool of a particular nutrient is 'free' within the system, that system becomes effectively limited in its operation by the speed of return and recycling of materials from the living component. In general, in relatively simple and undeveloped ecosystems, only a small proportion of the available nutrients are incorporated in the tissues of living organisms at any one time. As the community develops however, and becomes more complex, the biomass of living organisms supported increases dramatically; more and more of the total 'nutrient pool' becomes trapped in the biota (Chapter 4). In these more complex communities, the efficiency and speed of nutrient turnover indeed becomes limiting; in such communities the importance of the decomposers in releasing such nutrients from the dead bodies and excretory wastes of animals and plants cannot be over-emphasised (Section 3.8).

3.7b Cycles of Nutrients Between Biota and Abiota

However much of the available 'pool' of a nutrient exists as free nutrients, or is bound into animal and plant tissue, there must ultimately be an exchange between the two or the supply of free nutrients will ultimately be exhausted; the nutrient must cycle through the system. Such cycles are complex, for most nutrient elements can take a variety of forms in different chemical compounds. Most readers will have encountered such cycles at an elementary level: a simplified carbon cycle and nitrogen cycle are shown by way of example in Figures 3.4 and 3.5 and an able review of all the basic biogeochemical cycles is given in Scientific American (ed.) (1970).

The idea that a limited quantity of any given nutrient is continuously cycled through an ecological system is, however, somewhat oversimplified – for it implies a completely closed cycle with no potential losses or gains. In practice this is not the case. While on a global scale there can indeed be no loss or gain of material, at the level of the individual community or ecosystem this is not necessarily so. While on a global scale, the availability of, for example, the element nitrogen is, ultimately, finite, the proportion of that nitrogen which is in organic form and thus available to living organisms (depending as it does on nitrogen fixation by bacteria) may be altered, while the amount of this organic nitrogen in any one community may *also* fluctuate if material is imported from, or exported to another system. In fact, within a community or ecosystem we can recognise two distinct classes of nutrient cycle: those which *are* effectively closed, and those which emphatically are not.

While all nutrients pass through the biotic part of the ecosystem in very much the same way: fixed by the producers in organic compounds, eaten by consumers, passed to decomposers for degradation, we may distinguish two quite different forms of cycle in the abiotic phase – one characteristic of those elements which may occur in gaseous phase (such as carbon, oxygen, nitrogen), the other typical of elements (calcium, phosphorus, potassium etc.) which lack a major gaseous phase. The distinction between such 'gaseous' cycles and non-gaseous or *'sedimentary'* cycles is an important one, since the two different types of nutrient cycle have quite separate properties. Gaseous nutrients form what Odum (1971) has termed perfect cycles: that is, the cycles are effectively closed within any one system. Because the major reservoir for the element is in the atmosphere – an atmosphere common to all communities – the entire global supply of that nutrient is effectively available to each and every community. The 'open' cycle of nitrogen just considered is in fact illusory; because nitrogen engages in gaseous cycle, local perturbation due to 'export' is rapidly compensated for and equilibrium quickly re-established. The cycle is effectively closed. By contrast, elements involved in sedimentary cycles, because their major reservoir – in terrestrial systems – is the soil, tend to leach away from individual systems (whether specifically exported or not), draining into streams and rivers, and ultimately being deposited in the sea. Their cycle within any one system is not entirely closed: losses can occur and those losses cannot be

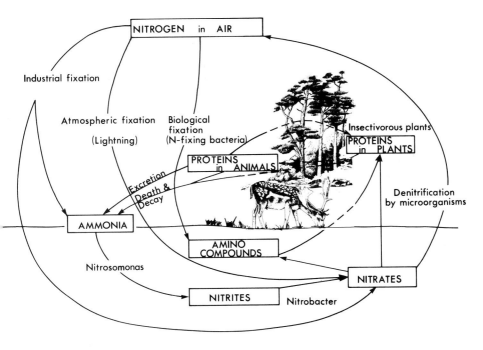

Figure 3.4: A Stylised Representation of the Nitrogen Cycle.

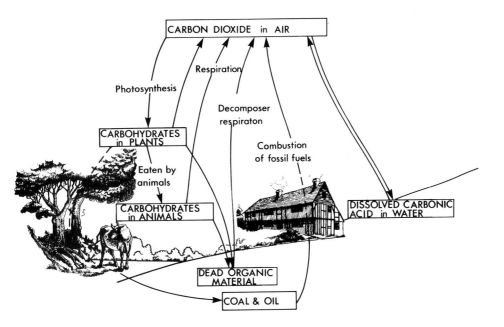

Figure 3.5: Diagrammatic Representation of the Carbon Cycle.

replaced. Further, since such nutrients tend to accumulate at the bottom of the oceans, they are not merely lost to one community or ecosystem but passed to another; they are effectively lost to the biosphere. Nutrients involved in sedimentary cycles are by definition continuously depleted. (This paradoxically enough has a rather ironic advantage: most pollutant materials that enter biogeochemical cycles of any sort enter sedimentary cycles, and ultimately are deposited at the bottom of the sea. The advantage is, however, questionable: an alternative interpretation of the same phenomenon forces one to conclude that any pollutant practice on land ultimately adds to the pollution of the ocean.)

3.8 The Importance of the Decomposers

All nutrients spend some proportion of their time incorporated into animal or plant tissues — by very definition, for nutrients are defined as those biochemicals required by living organisms. But continued operation of the ecosystem requires that as soon as these materials are 'finished with' through excretion or through death, they are rapidly released and passed back to the producers for re-use. As more and more of the nutrient pool becomes tied up in the biota and an ever-increasing biomass is supported by the same nutrient supply, such rapid release of nutrients from animal and plant waste becomes more and more critical.

As a trophic group within the community, the organisms of decomposition are frequently overlooked; ecologists in general tend to concern themselves more with producer and consumer levels. Yet in practice, the decomposers comprise by far the greatest biomass within the system after the producers, and in terms of their contribution to community and ecosystem function are far more significant than any other heterotrophs: all organic material synthesised within the community must ultimately pass to the decomposers before the nutrients contained may be released for re-use. The decomposer network is critical to continued nutrient cycling within the ecosystem and thus its continued operation. Without the decomposers, the community would rapidly run out of raw materials — and run to exhaustion.

(In fact while no community could function without its decomposers, in theory, a community could function perfectly satisfactorily with producers and decomposers alone. All the variety of consumers within the system are not essential to its function, and merely extend somewhat the cycle of operation. In most terrestrial communities in fact, consumers 'handle' only some 5-8 per cent of net primary production in any case; in excess of 90 per cent of primary production passes directly to the decomposers.)

Waste organic matter of any kind — leaf litter, animal faeces and carrion — supports a vast number of detritivores and scavengers. With such a vast resource available to them (ultimately equivalent to the community's Net Production) it is not surprising that many organisms have developed to exploit this dead material; a quick glance through any taxonomic textbook will confirm a remarkable impression that, in almost any animal group, the vast majority of species feed upon detritus of some form or another. In fact, despite their abundance, these detritivores contribute little directly to the decay of the material. Indirectly, through fragmentation of the material, increased aeration and surface area, they assist microbial decomposition, but it is the micro-organisms, the bacteria and fungi of decay, that are responsible for the full degradation of the biotic wastes.

This book is not, however, the place to examine the processes of decomposition in any detail; we wish merely to emphasise the importance of this crucial process in the dynamics of the community as a whole. A full review of the mechanisms of decomposition of plant materials will be found in Swift *et al.* (1979) and of carrion and dung, in Putman (1983b).

4 Temporal Change in Community Structure and Function

4.1 Introduction

Not only is a community always shifting and changing in the sense that it is an *operational* unit; as we noted at the end of Chapter 2, the community is in any case always in a state of flux — changing with time as species are continually lost and replaced, as their dominance or actual role in the dynamics of the community shifts and alters. Such changes over the time are not just due to major (and frequently man-made) perturbations of the system — through felling trees in primary forest, burning of heath or savannah grasslands, or spoilage of land and water with pollutants — but are part of the pattern of natural communities.

4.2 Short-term Cycles in Community Structure

Some of these changes are short-term — in the nature of cycles within the system. Thus a community may in fact be composed of a number of detailed sub-communities which replace each other in a regular pattern over time. Elton (1927) recognised different time-components of day, night, dawn and dusk in the community of animals occupying an English oak-wood (Figure 4.1). Other time-changes may be associated with the tidal flow in littoral communities, where certain organisms such as limpets, barnacles, sea anemones, peacock worms, mussels and a whole host of others, are only active when covered by water when the tide is in, becoming quiescent when the tide ebbs once more and the creatures of the exposed shore, hermit crabs (*Pagurus*), sea-slaters, bristleworms and sandhoppers reoccupy the area. Over a longer time scale, the components of a particular community may change — both in species composition and in relative abundance of the different species — with changing weather conditions, or the changing seasons of the year, as different temperature and light regimes favour different organisms. The small, pond-dwelling cyclops *Cyclops strenuus*, characteristic of small freshwater ponds in Britain during the winter, disappears from the community in the summer, to be replaced by different species, *C. fuscus* and *C. albidus* (Elton, 1927). Plants, too, may be present or absent from their communities at different times of year, as germination or emergence of the different species occurs at different times of year.

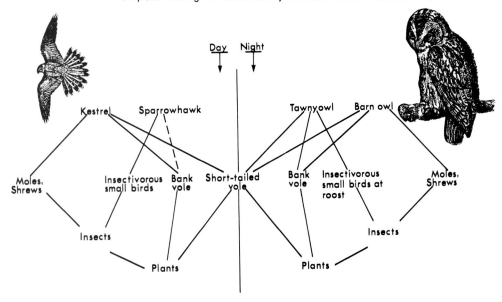

Figure 4.1: The Different Subcommunities of Day and Night Within an English Oak Wood.
Source: After Elton, 1927.

Within the plant community, the change may be more subtle. All species, or at least the majority, may be present together in the community at the same time, but because of some separation of growing season, or flowering time, the pattern of domination of the community (and equally importantly, the *physical structure* of the community) may show a cyclical change. And this emphasises another point: that while all these regular changes in community structure and composition may occur, with distinct subcommunities recognisable in different phases of the cycle, the different 'time shifts' need not be mutually exclusive, or absolutely distinct. The change is often a more subtle one of shifting dominance, or presence or absence of a few species against a backcloth of more constant features. Thus plants flower and senesce at different imes, but always within the context of the same total community of species. *Cyclops strenuus* may be replaced in English ponds in summer by *C. fuscus* or *C. albidus*, but other species of cyclops (*C. serratulatus, C. viridis*) are there year-round, as are many of the other organisms of the pond community. Even in the day-night shift of the oak wood, or the tidal shift of littoral zones, there is considerable overlap. The cycles are a shift in emphasis, a change in importance of different elements of the community web, rather than a change in the web itself.

4.3 Shifts in Community Structure: Colonisation and Extinction

The same shifting emphasis within communities occurs outside these cycles, too, with no regular return and repeat, but as permanent alteration. As species within the community change under selection pressures acting upon them, as some die out and are replaced and as new species become available in the surrounding area for new colonisation of the community, its species balance, species composition, shifts and slides. All the time local species populations are becoming extinct. The community may be recolonised perhaps by other representatives of the same species, so that in practice it appears to remain the same; alternatively the place of the lost species may be taken by other existing members of the community, so that the species balance of the system is altered; or they may be replaced by a new species altogether. The community remains in balance, but only as a result of an equilibrium between continuous colonisation and extinction, continuous evolutionary change even of more persistent species. And while it remains *in balance*, its structure changes constantly.

We owe our appreciation of this constant flux in communities to the extension, to communities in general, of principles first established for the ecological balance of islands. In an analysis of the relationship between the number of species on an island and island size, MacArthur and Wilson (1963, 1967) concluded that the number of species present on an island resulted from such balance between continuous colonisation and extinction. They suggested that rates of colonisation and extinction were simple functions of the number of species already on the island (Figure 4.2) so that the rate of immigration of new species to an island decreases as the number of established species increases (as more and more of the potential colonists are found on the island, fewer of the new arrivals constitute new species); rate of extinction is shown to increase with species number. The significance of this for a relationship between species number and island size is that extinction rates can be shown to be a function of island size as well as of species number (being lower on larger islands). Thus the equilibrium between colonisation and extinction is different for islands of different size. This supposition is borne out in practice (Figure 4.3). (See Chapter 12 for fuller treatment.)

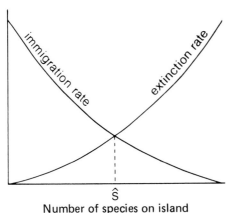

Figure 4.2: Equilibrium Model of the Number of Species on Islands. The equilibrium number of species ($\hat{S}$) is determined by the intersection of the immigration and extinction curves. Colonisation rate falls as number of species already on the island increases; extinction rate rises. Source: MacArthur and Wilson, 1963, 1967.

What is important in the present context, however, is that MacArthur and Wilson established that island communities are subject to continuous colonisation and extinction, that the rate of change of species is in itself a function of the size of the community, and that the size of the community expressed is determined by the breakpoint between these rates of immigration and extinction. Although such principles were established for island communities, they are equally applicable to mainland systems. For though it may not represent a physical island surrounded by seas, even a mainland system is in effect an 'ecological island', in that it represents a particular and discrete area of its type, surrounded by a 'sea' of different systems. Thus we may see how even mainland communities are subject to constant colonisations and extinctions and always in a state of change.

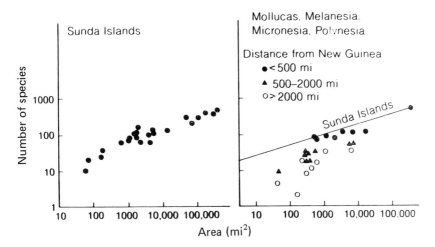

Figure 4.3: Species-area Curves for Land and Freshwater Birds on the Sunda Islands, Together with the Philippines and New Guinea (left), and Various Islands of the Moluccas, Melanesia, Micronesia, and Polynesia (right). The latter islands show the effect of distance from the major source of colonisation (New Guinea) on the size of the avifauna. Source: After MacArthur and Wilson, 1963.

4.4 Succession

Communities shift and alter in composition even when in equilibrium. And upon this constant flux is imposed yet one further element, directional change. For communities do not just suffer minor alterations to an essentially static structure, superficial changes to what is essentially the same concept of design; there is also always a progressive change, a grand and gradual shift in the whole basis of the community, its structure *and its function*, in the phenomenon of *Ecological Succession*.

Succession, then, is a directional change in community structure and function, an additional progressive change whereby all the minor changes of structure and operation accumulate over time so that the community itself develops a new emphasis, a different basis, a different function. It is a continuing process promoting continuous development, until the community reaches what is referred to as the *climax* of succession, when it stabilises in a persistent structure that appears to change (at least in this directional sense) no more (Clements, 1916).

As the community continues this development, it is possible to recognise a series of phases through which it passes which can be defined almost as distinct communities in their own right with their own particular structure and function. Thus, although it is accepted that the process is a gradual one of *continuous* change, it is customary to talk of succession in terms of development through these successive stages in its development, or 'seral stages'. (And, indeed it is equally common practice, when talking in terms of a relatively short time scale, to consider the various seral phases as communities in their own right. Thus it will be apparent that in the last two chapters on community structure and function, much of what we said in specific terms about the structure or characteristics of particular communities, should more properly have referred to them as 'seral phases'.)

Thus succession may be viewed as the development of a community from its inception (the 'pioneer' stage), through a series of recognisable intermediate seral stages, each almost equivalent to a community in its own right, to a climax. It is regarded as a process resulting in the greater stability of the community, and the climax community is considered the most stable community that can exist in that particular environment. (Such definition is however to some degree circular – derived in part at least from the very fact that the community ceases to change.)

As examples we may consider what happens to the community of organisms that initially colonises bare rock, or the series of changes that may be observed when a lake silts up to form first a marsh, then eventually a dry land community. A bare exposed piece of rock becomes colonised by resistant organisms such as mosses and lichens. As time progresses the rock flakes and crumbles, dead bits of moss or lichen decompose and particles of soil collect and are trapped by the leaves of these pioneer species. But with soil available the area is now suitable for a few scrubby grasses to take root. The accumulation of soil and humus proceeds apace. Taller plants shelter the ground from direct exposure to the sun, making it cooler, and reducing water loss through evaporation. The progression of species continues and the series of environmental changes until eventually what was once bare rock becomes a broad-leafed woodland. Or, as another illustration: water plants growing at the side of a lake also trap silt and soil particles between their roots and leaves. The edges of the lake grow inwards, and more and more becomes 'dry land'. As the lake margins progressively silt up, so the water plants move further and further into the middle of the lake leaving behind them wet, marshy, but solid ground. It is a continuous process and eventually the whole lake will be filled in (Figure 4.4). As the plants of the margin move ever inwards, they are replaced by plants which are characteristic of marshland. We find marshy-soil trees such as hazel and alder becoming established. Their root systems are strong and complex and bind more and more soil together. In addition the water they take up begins to drain the swamp. The level of soil rises and becomes drier and drier. So the hazel and alder are replaced by trees characteristic of drier environments, ash and birch, which are eventually themselves replaced by full broad-leaf woodland trees such as oak or beech. These changes occur in waves behind the encroaching water plants until ultimately the whole lake is filled and replaced by broad-leaved woodland.

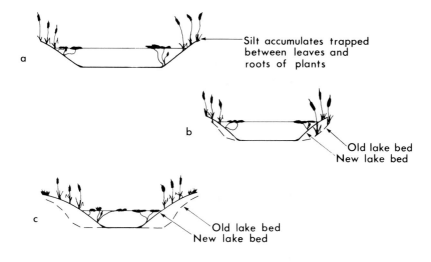

Figure 4.4: Successional Change During the Gradual Silting of a Freshwater Lake.

These examples are admittedly oversimplified and idealised, and there are many many others in the literature which are examined in more detail (e.g. Johnston and Odum, 1956; Odum, 1959, 1969; Whittaker, 1970). But they allow us to draw a number of general conclusions about succession.

4.5 Characteristics of Succession

Firstly we are dealing with a directional change, as already noted, and one that is very orderly and predictable, so much so, that if the progression is disrupted, and through interference set back a number of stages, the succeeding *secondary succession* goes back up very much the same pathway again, taking the same steps, encompassing very much the same intermediate stages, and certainly converges back to the same end point. This brings out a second point, that succession is a convergent process. Despite the very different starting points of bare rock and open water in our examples, the end point of the progression was in both cases, broad-leaved woodland. This is another important feature of succession, that communities from a variety of different starting points proceed towards a very limited number of end-communities or climax states. (These climax communities are characteristic of the gross ecological region. Their form and restricted variety are dictated by the gross climatological and geological characteristics of the region. Thus in temperate areas, we may find deciduous woodland or coniferous forest, while in the tropics we find rain forest, desert or savannah grassland.)

It should be emphasised that these climax states are only *potential* end points of succession. Quite frequently succession halts at what appears to be a pre-climax state. Such communities of *arrested succession* are characteristically those subject to periodic but major disturbance (which in effect 'pushes the succession back' to some earlier stage). The parkland savannah scene so typical of seasonally dry regions of Africa is in fact an example of just such an arrested climax: the fact that trees seed and grow there indicates clearly that succession could proceed ultimately to generate forest. But the erratic fires of the savannah, and the periodic ravages of elephant on the mature trees, serve to maintain the community in its preclimax condition.

All these observations, although somewhat qualitative, enable us to form a fairly clear idea of the process of succession — although they give us no indication of the mechanics of the process. In addition we may note a number of other, more quantifiable features of succession. We find in succession, as we proceed from early developmental stages towards climax:

(1) a progressive increase in the total organic matter of the community;

(2) a tendency for nutrients to be increasingly bound up within organisms rather than free or extrabiotic;

(3) there is increased nutrient conservation, and slow nutrient loss;

(4) the decomposer element of the community becomes larger and more important;

(5) the community itself becomes more diverse in terms of species number and balance;

(6) it becomes more *spatially* diverse and heterogeneous;

(7) the whole community becomes more complex — biotic interrelationships diversify (simple foodchains are replaced by complex webs);

(8) flow of material around the community becomes increasingly slower;

These changes are summarised (after Odum, 1969) in Table 4.1.

Table 4.1: A Tabular Model of Ecological Successtion: Trends to be Expected in the Development of Ecosystems

Ecosystem attributes	Developmental stages	Mature stages
Community structure		
Total organic matter	Small	Large
Inorganic nutrients	Extrabiotic	Intrabiotic
Species diversity — variety component	Low	High
Species diversity — equitability component	Low	High
Biochemical diversity	Low	High
Stratification and spatial heterogeneity (pattern diversity)	Poorly organised	Well organised
Nutrient cycling		
Mineral cycles	Open	Closed
Nutrient exchange rate, between organisms and environment	Rapid	Slow
Role of detritus in nutrient regeneration	Unimportant	Important
Community energetics		
Gross production/community respiration (P/R ratio)	Greater or less than 1	Approaches 1
Gross production/standing crop biomass (P/B ratio)	High	Low
Biomass supported/unit energy flor (B/E ratio)	Low	High
Net community production (yield)	High	Low
Food chains	Linear, predominantly grazing	Weblike, predominantly detritus
Overall homeostasis		
Internal symbiosis	Undeveloped	Developed
Nutrient conservation	Poor	Good
Stability (resistance to external perturbations)	Poor	Good
Entropy	High	Low
Information	Low	High

Source: After Odum, 1969.

Clearly, a large number of these features are interrelated. Increased nutrient conservation is clearly related to the increase in size and importance of the decomposer level within the community. The fact that nutrients tend to be bound up within organisms, and that the amount of 'free' nutrients declines, is a function of the progressive increase in organic matter of the community, and its greater complexity; these last two factors also explain why flow of material around the community becomes far slower, etc. etc. But, still more significantly, *all* of them reflect a change in the general economy of the system: a change, so to speak, in 'policy'.

All these separate changes are in fact a direct result of the primary increase in organic matter in the community. The community emphasis shifts from the rapidly growing, rapidly changing and expanding system characteristic of 'pioneer' communities, and early seral stages, to a slower, more stable, and fully exploited system. During the early stages of succession, the total gross productivity of the community (GPP) is far in excess of that required for the respiratory needs of the community at all trophic levels. Since community production exceeds community respiration, there is a surplus of energy/organic matter which may be devoted to growth; the community accumulates biomass. Gross Primary Production (GPP) minus total community respiration is sometimes referred to as Net Community Production (NCP). In these terms, NCP of early seral stages is positive and biomass accumulates. As time goes on, and biomass increases within the community, so community respiration increases until it balances exactly the production (NCP = 0) and the whole process must stop. This one feature alone explains all the other observed characteristics of succession that we have mentioned in Table 4.1 and in addition offers a very much simpler definition of the whole successional process as 'the accumulation of biomass by a community' (Cooke, Beyers and Odum, 1969; Odum, 1969).

Table 4.2: Changes in the Characteristics of Individual Organisms Associated with Different Successional Stages

Characteristic of organism		Developmental stages	Mature stages
Niche style		Broad	Narrow
Size of organism		Small	Large
Life cycles		Short, simple	Long, complex
Growth form	selection pressure for	Rapid growth	Feedback control
Production	selection for	Quantity	Quality

Although we have so far restricted our discussions to the changing characteristics of the community as a whole during succession, these changes are also echoed by the organisms which occur within the community. Thus (Table 4.2) there is a marked change in the size of organisms, a tremendous increase in niche-specialisation (as competition from other organisms in the increasingly complex community becomes more intense) and a radical change in the pattern of life cycles. Organisms of early successional communities tend to be relatively short-lived, but have relatively rapid generation times, and massive reproductive potential; organisms characteristic of later seres tend to be longer-lived and have a much reduced reproductive output. We shall return to these changes in the strategies of individual organisms in Chapter 10. At present we are concerned with parameters of the community as a whole.

4.6 The Mechanics of Succession

While we may now offer both subjective and objective descriptions and analyses of the changes that occur in succession, we have as yet little understanding of the mechanics of process. We have established, it is true, that any community is subject to a series of changes in structure and composition, as new species invade the community, and existing species become extinct, or change in evolutionary time, but what imposes the directionality of succession?

Any one organism has a limited range of environmental conditions in which it can exist (Chapter 1) and within those, certain optimum conditions which suit it best. Yet, as a organism lives in an environment it necessarily changes that environment: and any change may mean that the new environment favours that organism less well than it did originally. Thus, either the environmental conditions become so altered that the original coloniser can no longer exist within the community, and becomes extinct, with its position later taken over by some new immigrant species, or the new coloniser — already present within the community and better suited to the changed environment — outcompetes the original species from an environment to which it is no longer well-adapted. This is the basic mechanism underlying *all* species changes within a community (Section 4.2), but what causes the directional changes of succession is this continued alteration of the environment. The community of organisms occupying a particular environment at any one time have a significant impact upon the environment, which thus becomes less suitable for them and more suitable for other species (which could not successfully have colonised the original environment). This new 'community' alters the environment yet again and is itself replaced, species by species.

Amongst the community of scavengers and decomposers which occupy the decaying carcasses of animals, a clear successional sequence is established. Thus, in an analysis of the decomposition of sheep carrion in Australia, Fuller (1934) showed that fresh carcasses were readily colonised by the blowfly species *Lucilia cuprina*, *L. sericata*, *Calliphora stygia* and *C. augur* which rapidly established huge populations of larvae within the carrion mass. But maggots have a profound effect on carrion tissues. The enzymes which they secrete externally to digest the tissues cause a progressive liquefaction of the material; the mechanical disturbance through the tunnelling of the larvae within the carrion alters the matrix of the carcass and in assisting dissemination of bacteria and fungi, and in aeration of the whole carrion mass, stimulates microbial decay of the material. Finally, continuous excretion of nitrogenous wastes alters the pH of the medium. Within a few days the carcass has become a completely different environment with completely different characteristics. Now new species may colonise and the carrion mass becomes dominated by maggots of different blowflies: *Chrysomyia rufifacies* and *Sarcophaga* spp. The process is repeated and a third set of blowflies colonise the carcass in a later phase. Similar changes occur within the microflora and fauna, and all the other organisms associated with carrion systems.

This is of course a rather peculiar example of succession within a detritus community, without the primary-producer-consumer trophic structure with which we are more familiar. It is also a successional sequence which occurs over a much shorter time scale than is usual, because of the impermanence of the carrion system. But exactly the same principles are involved in the succession of plant-based communities.

Here, the changes in the community are normally mediated in large part through a change in species composition of the dominant plants, accompanied secondarily by a change in the animal species associated with them. Indeed succession is frequently regarded primarily as a botanical phenomenon — a change in the plant part of the community alone; a conception which is as erroneous as it is misleading. Animals and plants are inextricably interrelated in any community; any process which affects, or is affected by one, will influence or be influenced by the other. In this case however it is true that the most obvious successional changes within the community are caused by a change in the species composition of the primary plants. Any green-plant based community may be defined with regard to the species pattern of its producers; for although the spectrum of potential consumers and decomposers is dictated in part by physical and chemical facets of the environment, the actual species that occur in the primary consumer level will be controlled by the presence or absence of their particular food plant. Similarly, of all the potential secondary, tertiary and quaternary consumers, those that will actually be found within the community will be those whose favoured prey species are available there (Section 2.12). Thus any change in the species pattern of the producers will have repercussions in species composition throughout the rest of the associated community (although as we have noted, higher consumers, being characteristically more polyphagous than those of lower trophic levels, may be able to exist in a number of different communities, or seral stages, and thus may change more slowly than other species).

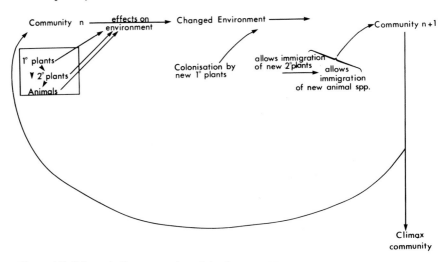

Figure 4.5: Schematic Representation of the Process of Succession.

Occupation of a particular environment by a community modifies, as before, the physical and chemical environment. Such modification will of course make the physical and chemical conditions more favourable to different types of organisms and will result initially in a change in the species pattern of the producers (Figure 2.12). Associated with that change will be a change in the community as a whole. The process of modification continues, the spectrum of producers changes continuously, and the animal part of the community 'tracks' this change (Figure 4.5). But note, although the effect of succession upon the community is mediated through the plants, with changes in the animal composition of the community largely in response to a change in the species composition of the producers on which they depend, the modification of the environment which leads to a change in plant species composition in the first place is *effected* by the whole community, of both animals and plants. Further, although in large part the successional change is mediated in this way through the green plants, *it is not restricted to such change*, and other changes may accompany this larger scale movement, due to successional events within the animal part of the community alone.

We have noted that succession may proceed by continued modification of the environment by an existing community, and its subsequent replacement in this altered environment by a new community more suited to the new conditions. But this isn't the whole story: pioneer species are outcompeted not solely because the environment no longer suits them, but also because they are by nature poor competitors. We noted in Table 4.2 that various characteristics of the organisms of a community change during succession: that the features of organisms of early seral stages differ from those of communities later in the succession. One of the most important differences claimed is that organisms characteristic of early seral stages are supposedly less competitively able than those found in communities later in the succession (Chapter 10). They are adapted to colonise harsh and unstable environments and by that adaptation do not have to face severe competition; as soon as a relatively stable community develops in which competition between organisms for available resources becomes more intense, such organisms cannot persist. Such subtlety refines the picture of succession of Figure 4.5; organisms of each successional stage give way to those of later communities not so much because the environment no longer suits them, but because it now does suit other organisms which are competitively superior.

Such conclusions of course rely on the presumption that pioneer species are indeed less competitive than species of more established communities. While it is often assumed that colonising species have poor competitive ability, this assumption is essentially deductive, based on observations that the situations in which these species occur in the field are, by definition, situations of low competitive stress; there is very little experimental evidence that pioneer species are indeed poor competitors. Working within the vegetational part of the community alone, Fenner (1978a and b) however has now demonstrated at least in grasslands, that colonisers have markedly poorer competitive ability than closed turf species. Fenner takes these observations further in suggesting that susceptibility to shade may be a key factor in the lack of competitive ability amongst these pioneer plants; that shade-intolerance (due in turn to an inability to germinate in light which has been filtered through leaves) is probably the factor preventing pioneer species persisting after the canopy closes (Fenner, 1978b, 1980). This rather elegant series of experiments illustrates nicely the importance of relative competitive ability in combination with environmental modification, in contributing to successional change, *and* its directionality.

One single problem still remains. If the processes of succession are so continuous, with continual replacement of a species here or a species there within the community, how is it that successional communities appear to pass through distinct and discretely recognisable seral phases? The pattern of succession which emerges in any set of circumstances, is, it would appear, a function of the type of community, and the characteristics of the organism involved. The gradual changes in environmental conditions which accompany, indeed are responsible for, succession could in theory result in a number of different patterns of change in the organisms of the community: either continuous, and imperceptible, or with discrete quantal shifts in community structure, depending upon the ecological lability of individual species and the strength of their reliance upon others — the different levels of dependence of one part of the community upon another. Thus, Whittaker (1970) presents a number of different patterns for community change along a continuously changing environmental gradient (Figure 4.6). (In original, this is argued for replacement of communities along an environmental gradient in space (or *ecocline*). However the same argument may be applied to an environment changing continuously in time as during succession).

1. Dominant species may be somewhat evenly spaced out and may replace each other at critical points along the environmental gradient. Subordinate species may show close correlation to these dominant species assemblages occupying a particular segment of the environmental gradient. Within these segments the composition of the assemblage changes little with change in environmental conditions. At critical points a small change in environmental conditions leads to replacement of one species assemblage by another.

2. Dominant species may be somewhat spaced out but may gradually replace each other along the environmental gradient. Subordinate species may show close correlation to the pattern of success and distribution of these dominants, thus forming recognisable species assemblages occupying a relatively distinct portion of the environmental gradient. Thus, within these segments changes in environmental conditions result in gradual changes in species composition and activity. A more rapid shift in composition from that characteristic of one assemblage to that of another occurs at critical regions of the environmental gradient.

3. Dominant species may be somewhat evenly spaced out and may suddenly replace each other at critical points along the environmental gradient. Groups of subordinate species may show similar patterns not, however, correlated

with those of the dominants. Changes in environmental conditions may thus produce appreciable changes in composition and activity of subordinates with little change in dominants, or vice versa.

4. Both dominant and subordinate species may show more or less bell-shaped curves of abundance and activity as in (2), the peaks of which tend toward a non-coincidental pattern for groups of similar resource requirements. Changes in environmental conditions consequently produce a gradual and relatively predictable degree of change in community characteristics.

5. Both dominant and subordinate species may show more or less bell-shaped curves randomly located along the gradient. Changes in environmental conditions thus produce a degree of change in community structure varying randomly around some mean value.

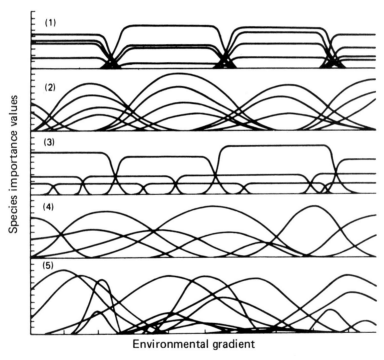

Environmental gradient

Figure 4.6: Five Hypotheses of the Pattern of Change in Abundance of Species Along an Environmental Continuum (Environmental Gradient with Conditions Changing at a Gradual Uniform Rate from One Extreme to the Opposite). Panel 1, dominant species evenly spaced and sharply replacing each other with subordinate species distributions strongly correlated to those of dominants. Panel 2, dominant species evenly spaced but gradually replacing each other with subordinate species strongly correlated with dominants. Panel 3, dominant species evenly spaced and sharply replacing; patterns similar for subordinate species but not correlated with dominants. Panel 4, both dominant and subordinate species show bell-shaped abundance curves that tend to be non-coincident for species with similar requirements. Panel 5, both dominant and subordinate species show bell-shaped curves varying randomly in height, extent, and location. See text for further discussion. Source: Modified from Whittaker, 1970 after Collier, Cox, et al., 1973.

We can see from this how succession in certain communities or certain parts of communities may proceed as a continuous development with no clearly defined intermediate seres, where the communities have little internal reliance between species (e.g. 4) or how in other circumstances, with close interdependence between primary and subordinate species, succession may develop almost through a series of precise steps, leading to the presence of sharply defined seral stages (1, or probably more typically 2). The close interdependence of primary consumers upon primary producers we have emphasised above explains the fact that plant and herbivore parts of a community usually do form distinct seres in succession (2), while the relative independence of higher consumers promotes a pattern of successional change more in the style of trace 4.

4.7 What Stops the Successional Process?

The process of succession within the community continues until a climax community is developed: a community which, it is supposed, remains unchanged with time. Succession, from a wide variety of starting points, converges upon one of the few climax states characteristic of the particular biome. But why is succession convergent and why does it stop? Succession, as we have defined it, is an orderly series of changes in the environment brought about by the community which colonises it, and reflected by subsequent changes in community composition. It is perhaps only logical to suppose that there will come a point when the community comes up against facets of the environment which are immutable. There will be a point when nothing is left to change but what cannot be changed by biotic intervention: the ultimate physical and chemical characteristics of the environment such as geological structure, soil and substrate structure and chemistry, overall climatic features such as rainfall, gross temperature regimes, ice, and so on. These immutable features will eventually block the chain of succession. And since the limitations of this gross environment will be optimal only for a limited range of organisms a given biome can support only a limited diversity of end-communities: hence convergence and an end to succession.

This is perhaps a rather passive approach, and would favour the use of a word such as 'end-community', rather than climax, for the end is, so to speak, thrust upon it from outside, rather than as the climax of events within. There is however an alternative way of conceiving the end point of succession. The definition of succession we have derived in terms of energetic balance within the community was that succession was the accumulation of biomass within a community. In early seral stages, when $CP > CR$, matter may accumulate; as matter accumulates CR increases until it equals CP. When $CP = CR$ and there is no more opportunity to accumulate biomass, succession has reached its climax. Surely this is a far crisper explanation? If succession proceeds as the accumulation of biomass, when no more biomass can be accumulated, succession must stop.

Such rationalisation also offers a solution for situations of *arrested climax*

(p. 92) where succession appears to stop short at some pre-climax stage, adopting a 'pseudo-climax'. If such communities are to have halted in successional development, they must somehow have reached in some pre-climax phase, an energetic balance (CP = CR). And it may be noted that a characteristic of all such situations is a regular and major export of material from the community. In African park-savannah this export of material is by fire or elephant damage; in other examples, marshes or carrs may be maintained in their pre-climax stages by erosion and washout of accumulated material.

4.8 Climax Communities

In Section 4.4 we defined climax communities not only as the end point of succession but as a persistent structure that appears to change no more, the most stable community that can exist in that particular environment. Just how static and stable are climax communities?

Climax communities like all other communities or seres change in time — in all the ways we have discussed in the earlier sections of this chapter. In addition they show a component of the directional change of succession. Climax communities do not necessarily represent a halt to successional change. Modifications of the environment by the community, resulting in changes in the community itself, continue in many cases, but resolve themselves into cyclic change. A classic example of this is that of cycles of vegetation within the climax community of heathland (Watt, 1947). In upland heathland systems a regular cycle of subcommunities may be observed within what is essentially a climax system of heather. Such communities are dominated by the single species, *Calluna vulgaris* (ling heather) which forms an almost monospecific cover in prime heath. But as *Calluna* matures, at an age of 12-15 years, the shrubby plants become 'leggy' and open-based, exposing the soil at the base of the plant. This clear soil permits colonisation by lichens (almost as if it were the start of a completely new successional sequence, but that the mature *Calluna* plant remains as well). Ultimately the heather plant dies completely. In the space where it grew the lichens die out and the soil is rapidly colonised by other plant species (e.g. *Arctostaphilus*). Amongst these are young *Calluna* plants which as they develop outcompete the earlier colonists to return the 'patch' to pure *Calluna* heathland (Figure 4.7a). Similar cycles can be observed within mature beech woodland when individual old trees fall, leaving a space within the canopy, and other examples of such climax cycles have been reported by Forcier (1975), and Yeaton (1978) (Figure 4.7b).

In terms of stability, a full discussion of the stability of climax communities by comparison with pioneer stages must be reserved for Chapter 13. Here at least we might note that productivity of climax communities is often low by comparison to that of earlier stages in the succession; due to the complexity of web-design, cycling of material through the system is extremely slow (Table 4.1). Both these are cited as beneficial features — but are they? There are many

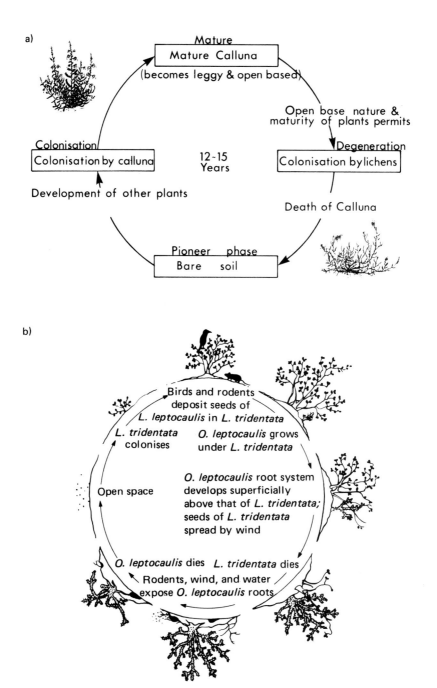

Figure 4.7: Cyclic Changes in Climax Vegetation (a) Within a Heath Community (Source: After Watt, 1947) (b) In a Desert Scrub Community in Texas (*Opuntia leptocaulis* and *Larrea tridentata*). Source: From Yeaton, 1978 after Barbour *et al.*, 1980.

examples of far more productive, indeed far more diverse communities characteristic of earlier, pre-climax seral stages. Is climax in fact tantamount to over-
maturity?

4.9 Succession as a Necessary Mathematical Consequence

Succession may be regarded as the natural result of an imbalance within the
energy relations of the community resulting in the accumulation of biomass
by the community. Such rationalisation robs the process of some of its nebu-
lousness and complexity. It accounts for the various phenomena recorded in
Table 4.1, of increasing biotic complexity, increasing importance of biotic
relations and interactions in the functioning of the community. It reveals the
driving force for succession itself and offers a clear explanation for why succession must ultimately stop. In the same way the underlying reasons for the
phenomena of directionality and convergence in succession can also be greatly
simplified. In these pages we have reviewed the various traditional explanations
put forward to account for these characteristics of succession, accounting for
them by biological argument. More recently various authors (e.g. Horn, 1975,
1976) have suggested that such explanations are a heavy-handed and laboured
attempt to account in a biological way for something which may well not have a
biological cause at all. These authors point out that the peculiar properties of
directional change, and repeatable convergence on the same climax community
from any of a variety of starting points is not unique to the phenomenon of
ecological succession, but is shared by a class of statistical processes known as
'regular Markov chains'. A Markov chain is a stochastic process in which transi-
tions among various 'states' occur with characteristic probabilities that *depend
only on the current state and not on any previous state* (Kemeny and Snell,
1960). A Markov chain is 'regular' if any state can be reached from any other
state in a finite number of steps and if it is not cyclic. A chain is cyclic if every
state necessarily returns to itself after the same fixed number of steps greater
than one or an integral multiple of that number.

The fundamental property of a regular Markov chain is that eventually it
settles into a pattern in which the various states occur more or less randomly
with characteristic frequencies that are independent of the initial state. It is
argued that this final 'stationary distribution' of states is the analogue of the
climax community and that climaxes *must* occur by the statistical certainty
that the Markov process always settles into a stable pattern. It is further suggested that different climax communities imply different probabilities of transition among the states, rather than different initial communities, thus convergence too is a necessary statistical artefact (Horn, 1975). Various authors have
modelled succession within different communities with Markovian processes
and find a remarkably close fit (e.g. Waggoners and Stephens, 1970; Leak,
1970; Botkin, Janak and Wallis, 1972; Horn, 1975). The most sophisticated
analysis is perhaps that of Horn, who concludes 'Several properties of succession

are direct statistical consequences of a species by species replacement process, and have no uniquely biological basis.' The process of succession must stabilise by statistical necessity in a 'climax state'; the fact that different pioneer communities may converge to the same climax could also arise merely as a statistical necessity. If a community is temporarily disturbed, something like the original community returns. This too is a function of Markovian processes. Finally Markovian developments, like succession, are characterised by rapid changes followed by undetectably slow changes. (Hence stability, in the naive sense of 'absence of change' increases tautologically as succession proceeds.) None of these characteristics is necessarily of biological origin (Horn, 1975). Of course such a conclusion does not mean that there is no biological reality about succession; it does however suggest that many of its characteristics are not necessarily biologically determined. Thus with imbalance between community Primary Production and Community Respiration, with its resultant accumulation of biomass providing the driving force, the community will undergo a directional successional change as a statistical necessity. The biological explanations we have presented for the mechanics of the process may prove no more than observations of how the statistical necessity is accommodated by the biological system, and not in themselves the explanation for the process. And the idea of succession as a stochastic Markovian process perhaps explains why climax seres are *not* always the ideal communities they are supposed to be: why climax states often *do* appear overmature.

5 The Concept of the Niche

5.1 Introduction and Definition of Niche

From considerations of the operation of the community as a whole we move to examine the position of the individual within the community: its *niche*. The term, as we have already noted, means more than just the 'address' of a particular organism (when, after all, the word 'niche' would merely become synonymous with 'habitat') but also its 'profession' (Elton, 1927): the position it occupies within the community, the various interrelationships it has with other organisms around it, the role it plays in the operation of the community as a whole. Although the concept is intuitively a useful one – and indeed an easy one – more rigorous definition of what is meant by a 'niche' is surprisingly difficult. The term as we wish to define it embraces not only the organism's community-type, habitat and physical requirements, but includes some element of its relationship with all other components of the community, of its own role in the dynamics of the community – almost every facet of its population dynamics, periods of increase and decrease, factors controlling its numbers and distribution, almost every element of its interaction with other parts of the community, feeding relationships, relationships as prey, competitive interactions and so on (Odum, 1959). In fact, niche as a concept embraces almost the whole ecological functioning of a particular organism within its community. Perhaps the only definition which can satisfactorily embody this amorphous concept is that of Hutchinson (1957) who presents the niche as a 'multidimensional hypervolume, *defined by the sum of all the interactions of an organism and its (abiotic and biotic) environment.*' Despite its bulkiness and air of non-commital, this definition in fact proves a very useful one. However, any definition encompassing almost every aspect of an organism's ecology is rather unwieldly for general use and it is convenient to subdivide this multi-dimensional space and talk in terms of its separate dimensions in turn, e.g. food niche, nesting requirements, etc. Only when this is done is it practicable to examine the characteristics and features of the niche in more general terms.

When reduced to single dimensions the niche may be defined in terms of the set of resources (within that particular dimension) which are occupied by any organism. Such niche is, of course, ultimately constrained by the physiological limitations of the organism concerned. The entire set of resources which could, physiologically, be used by any one organism is normally referred to as its *'fundamental niche'*. In practice, however, the organism rarely occupies this entire fundamental niche, the limits of which are effectively set by abiotic factors alone. Various biotic factors — competition, predation, relationship with its own food resources — combine to restrict it to a far narrower area of operation; its actual position — defined as the *'realised niche'* — is usually only a small part of the potential, fundamental or 'preinteractive' niche (terms after Hutchinson, 1957; Vandermeer, 1972). (These descriptors, related to an organism's niche, have close parallels in the concepts of physiological range and ecological range applied in the context of *biogeographical distribution* in Chapter 1.)

5.2 Parameters of the Niche

In any dimension the realised niche may be defined as we have said in terms of the set of resources exploited by an organism (or by a population of organisms) and the relative extent of use made of each of the different portions of that resource set. We can portray this diagrammatically. If a set of resources is visualised in terms of some continuum (perhaps prey size in mm) along one axis of a graph (Figure 5.1) and extent of exploitation of various portions of that resource continuum is represented by the other axis, the niche may be seen upon this plot as a defined area, enclosing the range of the resource continuum used, and showing relative proportion of resource use at each point.

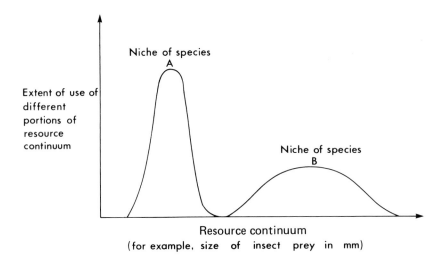

Figure 5.1: Diagrammatic Representation of an Organism's Niche on a Single Resource Dimension (Prey Size). See text for explanation.

Ecologically speaking the most important features of a niche are its position on the resource continuum, its spread, and its overall shape and form. The niche position and form relate to the role of the organism within its surrounding community, describing the relationship of that organism to all others which may utilise the same resource continuum, while one of the most important descriptors of the niche itself is its 'spread'. This *niche width* is a measure of the breadth of exploitation of a given resource by an organism; it is a complex function of how much of a generalist or specialist the organism may be in that particular element of its ecology. In the diagrammatic representation of the niche in Figure 5.1 these same niche parameters of *position*, *width* and *form* may be defined quite precisely in terms of the distribution mean $(\overline{x})$, the standard deviation (s) and the actual distribution of y.

(Note: These niche descriptors — position, width, niche shape — must of course be separately defined for the niche of a species, or population, and that of any individual organism within such population, and we must distinguish clearly between the separate concepts of individual niche and population niche. Niche position and niche form of the *individual* are determined by its own individual abiotic tolerances, and its own competitive or other interactions with other organisms around it, of its own or different species. The niche position and niche form of the *population* or *species* is derived as the envelope of the separate niches of its component organisms. And because of variability between different individuals within such a population, the niche described for the entire population may have different limits and a different form from those described for any one individual. It is important to stress this distribution between individual and population niche, for as we shall see the two do not behave in the same way.)

Graphical representations of the niche, as that in Figure 5.1, frequently, though not invariably, display the characteristic bell-shaped curve typical of normal distributions. Such normal distribution may in fact be expected at least in consideration of the *fundamental* niche, for both individuals and populations. Individuals within any species will not all operate at exactly the same point on the resource continuum. Because of variations within the species or within a population in various physical characteristics (let us say, for example, bill-length in insectivorous birds) there will be variation between individuals in the optimum position selected along the resource continuum of prey size in Figure 5.1. Such physical characteristics within a population are normally distributed; the related variation in individual's niche-use will thus also show a normal distribution, and population niche will show the characteristic bell-shape. Similarly, any one individual within the population may also be expected to show a normal distribution of niche-use. While any individual may have an optimum point on the resource continuum (related to the precise length of its bill) it will of course tend to operate over a certain range around this optimum (rather than only taking insects of exactly 0.337 mm every time!); the extent of its operation around that optimum may also be expected to show a normal distribution and as a result, niche form will show for the individual, too, the characteristic normal curve. Note, however, that the *realised* niche may *not* necessarily show such a clear normal distribution. Interaction with other organisms through predation or competition may well skew and distort this ideal niche curve. (See for example Roughgarden, 1974; Southwood, 1978.)

5.3 Factors Affecting the Niche and its Parameters

Any niche, of individual or population, may thus be defined in one resource dimension in terms of its position and spread, with regard to that particular resource continuum, and the extent of use of different portions of that distribution. Various factors combine to determine the shape of this niche, combine to influence niche width and position. As already noted, the position and range of the fundamental niche are determined by physiological characteristics of the organism itself in relation to abiotic features of the environment. Thus the temperature range and actual temperature of fresh water which may be occupied by, let us say, a platyhelminth flatworm is determined exclusively by its physiological tolerances. Individuals within the population will show normal variation in temperature tolerance and the extent of use of each part of the temperature gradient overall by the population will thus show a normal distribution over its full range. However, in practice the organism may never occupy that full niche. Other factors – largely biotic – may restrict the range of temperatures used, and shift the mean of the actual realised niche expressed. Introduction to another species of flatworm into our freshwater system, offering competition to the first, may radically affect the actual niche occupied (Figure 5.2). In more general terms there are a number of factors which thus determine the breadth and position of the realised niche within this overall potential which may be summarised as predation, distribution and predictability of resources, and perhaps most importantly, inter- and intra-specific competition. Each of these and their likely effects in translation of fundamental into realised niche will be considered below.

5.3a Competition

As illustrated in Figure 5.2, interspecific competition can have a profound effect in restricting niche width and in determining actual position of the realised niche of any organism. If there are a large number of species trying to exploit the same resource there will be a strong tendency for each to specialise to utilise one particular portion of that resource *not* utilised by the other competing species. In other words there will be a tendency for each to restrict their own niche until the species occupies only that part of it which does not overlap with the niches of competing species. This will have the effect of compressing the realised niche (i.e. narrowing the niche). In addition unless competition is so to speak equal on both sides, the niche will not necessarily 'shrink evenly'. Intense competition from other species on one side or another may result in 'uneven' shrinkage and thus, effectively, a shift in niche position (Figure 5.3). Such competition may also result in extremes, in a 'skewing' of the normal distribution of niche form and resource use. If the organism is pressed by extreme competition to the limits even of the fundamental niche, it can clearly 'withdraw' no further; under such circumstances — if it remains within the community at all — the extent of occupation of the end of the niche most under competition may well be reduced, while that at the opposite end may be increased in compensation. Interspecific competition in general, thus acts to 'compress' or reduce niches.

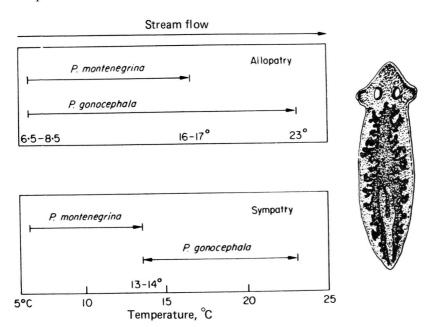

Figure 5.2: Distributions of Two Platyhelminth Flatworms, *Planaria montenegrina* and *P. gonocephala*, Along Temperature Gradients in Streams When They Occur Separately (above) and Together (below). Each species is restricted to a smaller range of thermal conditions when in the presence of the other. Source: From Beauchamp and Ullyett, 1932, after Miller, 1967 and Pianka, 1976.

By contrast, intraspecific competition tends to cause expansion of the niche. Where competition with other species is not severe, intraspecific competition will force individuals to try and exploit those parts of their environment where competition with their fellows is reduced: there will be a tendency for individuals to diversify. The species as a whole thus becomes more generalised and its niche broadens. Note, however, that while the niche occupied by the species as a whole is broadened by intraspecific competition, the individual organisms of the population may still occupy very restricted niches within this. And here it becomes clear how important is the distinction drawn earlier between the niche of an individual and that of the population to which it belongs. Competition *of whatever sort*, causes specialisation by the individual towards using that part of a set of resources where least competition with others may be experienced. Inter-specific competition causes a specialisation of both individuals and of the species as a whole; intra-specific competition causes the specialisation of *individuals* alone; the resulting diversification actually broadens the niche of the species as a whole. This is why, as emphasised on page 107 the different scales of individual or population niche must be distinguished, for as noted, they respond differently to given pressures. Thus interspecific competition causes a narrowing of the niche of both individuals and populations. Unpredictability of resources, as we shall see, results in an expansion of both individual and population niches, and predation, too, causes a change for both in the same direction. By contrast intraspecific competition narrows the niche of the individual, but broadens that of the population.

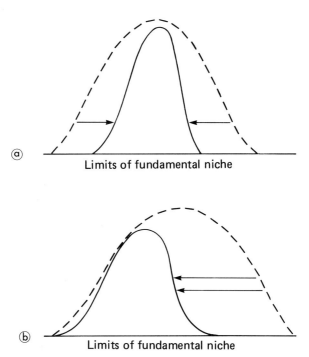

Figure 5.3: Effects of Interspecific Competition in the Realised Niche. In (a) uniform competition over all parts of the resource continuum results in an even shrinkage of the niche. In (b) intense competition on only one side of the niche causes both restriction of the niche and displacement.

5.3b Predation

Predation — and by the same token, parasitism — may also have a profound effect on niche form, once again restricting an organism or population to a limited part of the entire fundamental niche. Predation, like interspecific com-petition, acts to restrict the position of both individual and population niches, and to reduce overall niche breadth. A uniform reduction of niche breadth will result from the fact that, at the limits of its fundamental niche, the organism will already be operating under suboptimal conditions and as such is presumably more susceptible to predation at these extremes. A directional shift in niche position may also be expected along at least *some* niche dimensions in that an organism may be expected to withdraw as far as is possible from any zone of habitat overlap with a potential predator. A clear example of these effects of predation on niche width may be drawn from the studies of Paine (1974) on the community structure of intertidal regions in California; here the starfish *Pisaster ochraceus* through its predatory activities is seen to restrict the niche position of the mussel *Mytilus californianus*. The fundamental niche of *Mytilus* would permit it in theory to colonise the whole shore; the niche realised in response to the predatory activities of *Pisaster* is restricted to one small portion of this only. Paine's evidence for such conclusions is that when *Pisaster* is removed from the system, *Mytilus* rapidly expands its distribution to occupy virtually the whole shoreline.

5.3c Stability and Predictability of Resources

Intraspecific competition acts to broaden the population niche; both inter-specific competition and predation pressures act to restrict the niche realised. The actual niche width finally expressed is a product of these interacting effects. But there is one further factor which may be involved in the final shaping of the niche: predictability of resources used. In simple terms, no organism can afford to become too much of a specialist in an unstable environment where conditions and the availability of resources may be unpredictable and changeable; here is a clear case for remaining a generalist and preserving a broad niche. In more general terms the final niche width expressed by any organism must be in part a function of resource predictability. The effects of this may be viewed, rather than in itself as expanding or contracting the niche as instead providing a lower limit to niche width, providing an ultimate 'stop' below which other factors, of competition or predation, may not further reduce the niche.

There must in effect be a basic minimum size for a realised niche, below which the niche is no longer tenable: a basic minimum width to which an organism or population can reduce while still remaining viable. That 'minimum niche' must in fact be determined at that level where its resultant resource utilisation curve encloses sufficient 'area'/amount of resource to maintain the organism or population as a viable unit. And this 'basic minimum', or lower limit will vary in relation to resource predictability. The niche can be restricted to its *absolute* minimum size only if resources are entirely predictable and wide-spread. As soon as there is any uncertainty in the continued availability of any set of resources, the niche *must* be maintained considerably wider, in order to include sufficient 'area of resources' that even under the worst conditions the resources available to it *still* remain sufficient to support that niche and its occupants. In summary, the effect of resource predictability in itself is not to contract or to cause expansion of the niche. Rather considerations of such predictability of resources set the limits to which the niche can be restricted by other pressures — competition or predation.

5.4 Niche Separation

The opposing forces of inter- and intraspecific competition obviously play a powerful part in shaping the realised niche. Beyond this, competition can have no further effect on niche width, but as a consequence of its action in fashioning the individual niches to begin with, it has already had a secondary effect: that the roles of the various organisms within the community are sharply distinct. Each niche is clearly separated from all others in the phenomenon of *ecological* or *niche separation*.

If a number of species occupy a particular environment, the resources tend to be divided up between them in such a way that all available resources may be utilised fully and the maximum number of species supported, but with interspecific competition kept to an absolute minimum. One of the classic studies of ecological separation was carried out by Lamprey (1963) at the Tarangire Nature Reserve in Tanzania. Here 14 major species of herbivore were found to occupy a region where there were only three basic types of habitat: open grassland, open woodland and thick woodland. Further, there were eight species in particular which appeared to occupy almost exactly the same habitat — an observation confirmed by daily countings over four years of the numbers of animals sighted in particular areas. These species were eland, buffalo, giraffe, impala, warthog, rhino, elephant and waterbuck. They spent 5 to 10 per cent of their time in open grassland, 75 per cent in open woodland and 10 to 20 per cent in thick woodland. Yet the various species did not appear to conflict; Lamprey established that they managed to divide up the available niches between them (i) by eating different species of plants, (ii) by eating different parts of the same plants, (iii) by feeding at different heights above the ground, (iv) by occupying the same area at different times of day or different seasons, (v) by using different areas in any one particular season. Within just one of these strategies (eating different parts of the same plants) Bell (1970) has examined four species in greater detail. On the short-grass plains of the Serengeti National Park four animal species were all apparently feeding on the same species of grass without any signs of competition. The animals (zebra, *Equus burchelli*, topi, *Damaliscus korrigum*, wildebeest, *Connochaetes taurinus* and Thomson's gazelle, *Gazella thomsoni*) occupied the area in strict succession. First the zebra ate the upper parts of the stems: these are low in protein but the zebra's digestive system is adapted to cope with low quality feed by having a high rate of throughput. Then the topi ate the lower stem (which contains more protein than the upper part). The wildebeest eats the leaves, while the Thomson's gazelle eat the new green shoots which spring up within a day or two of the plant being cropped back by the other animals. Thomson's gazelle also eat the dicotyledonous plants which grow at ground level, which would otherwise increase and exclude the taller monocotyledonous grasses to the detriment of the other ungulates in the succession. A similar evaluation of niche separation amongst browsing species in East Africa has been provided by Leuthold (1978), and there are many many other examples of such separations in the literature for a whole host of different organisms.

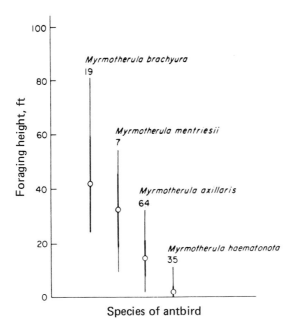

Figure 5.4: Foraging Heights of Four Species of Sympatric Antbirds (*Myrmotherula*). Means are represented by open circles, standard deviations by thickened vertical bars. Source: From MacArthur, 1972, after data of Terborgh.

These examples and the principles underlying them seem neat and aesthetically satisfying for they emphasise the sharp boundaries of the niche and also support the contention that 'two or more species with the same or similar ecologies cannot coexist in the same community' (a maxim attributed to Gause (1934), although the concept was originally formulated considerably before that time, by Lotka (1925) and Volterra (1926)) by suggesting that, as a result, presumably, of competitive interactions in the past, niches are distinct and non-overlapping. Schoener (1974) suggests that the limits on niche similarity of coexisting species should in fact result in fairly *regular* spacing of species in 'niche space', and a number of sets of species that differ primarily along a single niche dimension do indeed appear to be separated by rather constant amounts (e.g. Orians and Horn, 1969; MacArthur, 1972). A particularly clear example is derived from Terborgh's observations on the separation of foraging heights of various sympatric species of antbirds (*Myrmotherula*) (from MacArthur, 1972; Figure 5.4).

However, as usual, nothing is as clearcut as we might like to imagine and there are a number of situations where one can observe considerable overlap of niches; indeed closer examination of natural communities suggests that this is the more general rule rather than the exception.

5.5 Niche Overlap

In a study of the feeding ecology of several species of diving ducks wintering together off the coasts of Sweden, Nilsson (1969) concluded that there was considerable overlap in the food species taken. The proportions of certain food organisms in the diet of four lake-dwelling flatworms (*Polycelis tenuis, P. nigra, Dugesia polychroa* and *Dendrocoelum lacteum*) was determined, by serological means, by Reynoldson and Davies (1970). Again, there proved to be considerable overlap between the species. Such examples do not necessarily invalidate Gause's theorem, however. For the idea that two species with the same or similar ecologies cannot coexist (and that one will thus be excluded or will evolve away), is based, as we will see later, on the assumption that the species are actually in direct competition. In the examples above, the resource in question (food) is superabundant and intense competition for any resource can only exist, by definition, if the resource in question is actually or potentially limiting. In general, where niche overlap does occur as described this is indeed only when the resource in question is superabundant. As soon as resources become limiting, each species withdraws from the zone of overlap and ecological separation is re-established. Thus, in our 'food niche' examples above, each species has one particular food type or set of food types characteristic to itself: its 'food refuge' (Reynoldson and Davies, 1970). The overlap is not total, but more as in Figure 5.5. Amongst the lake-dwelling triclads studied by Reynoldson and Davies, the food niches of *Polycelis, Dugesia* and *Dendrocoelum* overlap, but *Polycelis* species feed to a greater extent on oligochaetes, *Dugesia* on gastropods and *Dendrocoelum* on *Asellus*; within the oligochaete refuge, *P. nigra* has a specific refuge in Naididae, *P. tenuis* in Lumbricidae (Reynoldson and Davies (1970), revised Reynoldson and Bellamy (1970)).

Even here the extent of dietary difference between each species and all the others averages less than 46.2% of total food at the most competitive season, when food becomes shortest; and there are many further examples where organisms appear to co-occur even in a potentially competitive situation with even greater overlap between their niches. MacArthur's study of a guild of five similar species of warbler (*Dendroica*) found together as insectivores in New England provides a classic example (MacArthur, 1972). This being the case, how much niche overlap is permissible? MacArthur and Levins (1964) examined the relative efficiency of exploitation of a range of resources by a single generalist species and by two comparative specialists. Clearly that point at which a generalist becomes more efficient at exploiting a combined niche than two specialists might be considered the point beyond which niche overlap becomes impossible (for a single generalist could then exploit the double niche more efficiently and would therefore outcompete either specialist). If we follow through their arguments therefore we can reach some conclusion as to the maximum tolerable degree of niche overlap.

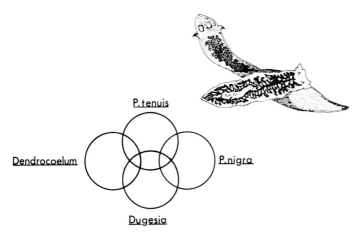

Figure 5.5: Niche Overlap Amongst Lake-dwelling Flatworms.

Suppose there is some resource which varies over a simple continuum (e.g. as Figure 5.1, insects as prey, varying in size). A large number of species may exploit different parts of this continuum; each will have an optimum point (e.g. optimum size of insect, related perhaps to a bird's beak size). Its total area of action will of course be over a range around this optimum, due to variability within the species, and, as we have already determined, each species' exploitation of the resource will be in the form of a normal distribution about its optimum point (Figure 5.1). Thus niche relationships among potentially competing species may be visualised and are conventionally represented as a series of bell-shaped resource utilisation curves along the resource continuum (Figure 5.6). Each separate distribution may be defined in terms of its mean and standard deviation (Figure 5.1). MacArthur and Levins consider the efficiency of exploitation of a pair of overlapping resources by a single 'Jack-of-all-trades' and by a pair of specialists. The degree of overlap between the resources influences the outcome. If we start with two resource distributions which are very similar, so that an animal which is good at exploiting one of them is at least moderately good at exploiting the other, we can draw two curves representing harvesting efficiency with respect to resource parameter (in our example, insect size) for animals specialising in each resource, i.e. for using the two resource units separately (solid curves in Figure 5.7). If our environment is a fine-grained mixture, in equal proportions, of the two resource units a line plotted at half height indicates the harvesting efficiency of each specialist on the mixture (dotted line).

The efficiency of a generalist operating over the whole resource and with an optimum between those of the two specialists is also indicated, and in such an environment may be seen to be greater than that of either specialist (and, it may be shown, than both specialists in combination). Suppose the resources are less similar — sufficiently separate that the (dotted) curve for the mixed resource now has two peaks separated by a trough (Figure 5.8). The generalist (J) now lies in the trough and it is clear that in this case the two specialists are superior, even on the mixed resource, to the generalist. It may be calculated that the breakpoint — that equilibrium point where the trough appears and disappears and generalist and specialists are equally efficient at exploiting the 'double niche' — occurs when the separate resource distributions overlap within two standard deviations of their means. As we have noted, these same deliberations may also be taken to suggest that there is a limiting similarity possible between the resource utilisation distributions of two coexisting species, and that at a certain degree of overlap exclusion must occur. Clearly, this too must occur at the point where one generalist becomes more efficient than two specialists, i.e. if niche overlap is within two standard deviations of the 'niche mean'.

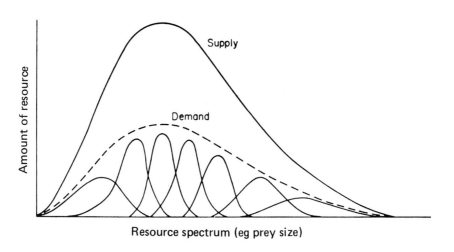

Resource spectrum (eg prey size)

Figure 5.6: Niche Relationships Among Potentially Competing Species are Often Visualised and Modelled with Bell-shaped Resource Utilisation Curves.
The uppermost bell-shaped curve represents the supply of resources along a single resource continuum, such as prey size. The vertical axis measures the amount of resource available or used over some time interval. The lower small curves are intended to represent seven hypothetical species in a community, with those species that exploit the 'tails' of the resource spectrum using a broader range of resources (that is, they have broader niches) because their resources are less abundant. The sum of the component species utilisation curves, shown by a dashed line, reflects the total use or the overall demand along the resource gradient. Pressures leading to the avoidance of interspecific competition should result in a relatively constant ratio of demand/supply along the resource continuum, as shown. Source: From Pianka, 1976.

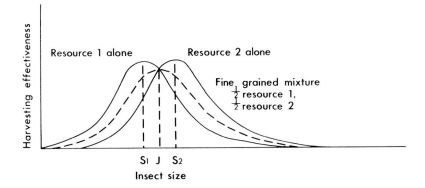

Figure 5.7: Harvesting Efficiency of Two Specialists and a Single Generalist on a Simple Resource Continuum. The graph shows the efficiency of each specialist in its own resource niche (S_1 and S_2: solid curves) together with efficiency of the same specialists and a 'Jack-of-all-trades' (J) in exploiting a mixture of the two resource types (dotted curve). Source: After MacArthur and Connell, 1966.

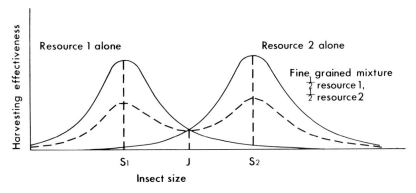

Figure 5.8: As Figure 5.7, Except that in this Case the Two Specialists are Now Quite Dissimilar in Optimum Resource Position. In a mixed environment of this type, J is now inferior. Source: After MacArthur and Connell, 1966.

Such considerations are based purely on analyses of the relative efficiency of exploitation of a given set of (unlimited) resources. In a later consideration, MacArthur and Levins (1967) have calculated the limiting similarity/overlap between species due to competitive exclusion when resources may be presumed to be limiting; they calculate a limiting similarity between the niches of 54 per cent: a point beyond which competition would lead to exclusion of one or other species. More recent models have refined this analysis still further (May and MacArthur, 1972; May, 1975a; Fenchel and Christiansen, 1976) while the extent of overlap to be observed in practice may be measured by a variety of derived formulae (Section 5.6).

Whatever the maximum possible overlap in theory, in the real community overlap will rarely extend to this theoretical potential. Real overlap will always be less than the maximum theoretically possible, because of the effects of competition within the community (see Section 6.4). Even if potential competition between the actual species primarily concerned is taken into account in calculation of overlap (as MacArthur and Levins, 1967) effects of 'diffuse' competition from other members of the community may still have a further (unpredictable) effect on permissible overlap (Section 6.6). Maximum possible overlap, as calculated here, should in theory remain a constant; maximum tolerable overlap in the real community has been shown to depend on the number of species in the community, and the pattern of species packing (Pianka, 1973, and Pianka *et al.*, 1979).

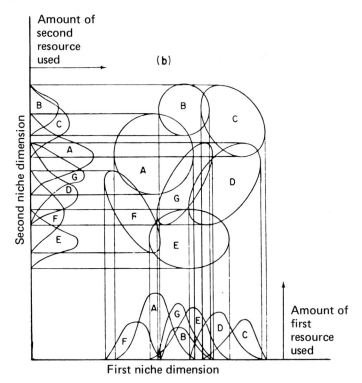

Figure 5.9: Diagrammatic Representation of Resource Utilisation of Seven Hypothetical Species Along Two Niche Dimensions, Showing That Pairs with Substantial or Complete Overlap Along One Dimension Can Avoid Competition by Niche Separation Along Another Dimension. Source: From Pianka, 1976.

Such analyses in any case refer to highly special cases where competing species differ only in their use of some single resource continuum. In practice, real organisms differ in their use of just one resource relatively infrequently however and if we consider two or more dimensions of the niche, we find that pairs of species may show moderate overlap along all dimensions — overlap far in excess of acceptable overlap for one dimension alone; indeed, in such a situation, pairs of species may have substantial or even complete overlap along one common dimension and still avoid competition by some degree of niche separation along another dimension (Figure 5.9, from Pianka, 1976). (This is, after all, only putting into general terms the evidence from studies of ecological separation like those of Lamprey (1963) and Leuthold (1978) quoted in Section 5.4, in which, while overlap in the food niche might be quite considerable, separation in the overall niche was ensured by differences in habitat use.) Ideally, therefore, an analysis of resource utilisation and niche separation/overlap should consider and quantify separation along all dimensions of the niche. In such a case it is possible to show that, provided niche dimensions are independent, overall multi-dimensional utilisation (or *full* niche) is the product of the separate unidimensional functions. Estimates of niche parameters (niche width, niche overlap) along each separate dimension can simply be multiplied together to produce the full niche picture (May, 1976; Pianka, 1976). If niche dimensions are not entirely independent, such a calculation leads to an overestimation of the actual degree of overlap between two niches (Figure 5.10 from Pianka, 1976), while if niche dimensions are totally interrelated a more accurate estimate derives as the arithmetic mean of the component niche parameters (May, 1975a).

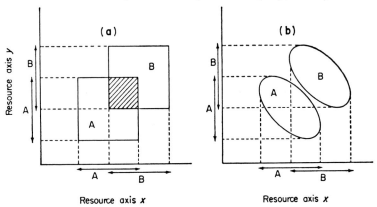

Figure 5.10: Two Possible Cases for the Use of Two Resource Dimensions (Assumed to be Constant Within Boxes or Ellipses) for Two Hypothetical Species, A and B. Although unidimensional projections are identical in both case *a* and *b*, true multidimensional overlap is zero in case *b*. Case *a* illustrates truly independent niche dimensions, with any point along resource axis *x* being equally likely along the entire length of resource axis *y*; under such circumstances, niche dimensions as orthogonal and unidimensional projections accurately reflect multidimensional conditions. However, when niche dimensions are partially dependent upon one another (case *b*), unidimensional projections always result in overestimates of true niche overlap. Source: From Pianka, 1976.

Using such a theory, Leuthold has in fact attempted to quantify the degree of overlap occurring between the four species of browsing ungulates of his study in the Tsavo National Park in Kenya (as above; Leuthold, 1978), calculating niche overlap along each of three dimensions (food plant type, browsing level, habitat preference) as

$$0_{jk} = \Sigma p_{ij}p_{ik} / \sqrt{\Sigma p_{ij}^2 \Sigma p_{ik}^2} \qquad (5.1)$$

(after Pianka, 1973), where

0_{jk} = mutual overlap between the jth and kth species, and p_{ij}, p_{ik} are the proportions of the ith resource utilised by species j and k respectively. (The index can assume values between zero and unity, see Section 5.6). He has shown that, while there is considerable overlap between various species pairs in each separate dimension, calculation of the overlap of the three dimensions together restores separation to a great extent (Table 5.1). Leuthold points out that the residual niche-overlap observed is likely to be overestimated: both due to the calculation of the 'combined overlap' as the *product* of the separate unidimensional overlaps, while the niche dimensions measured were almost certainly not totally independent, and also due to the fact that he has considered only three dimensions of an n-dimensional niche. Other, more recent examples have examined multi-dimensional niche overlap among, for example, spiders (Turner and Polis, 1979) and estuarine fishes (Thorman, 1982) using a variety of different indices. Examination of all these examples suggests that, under natural conditions whatever the degree of overlap on any single resource dimension, in the multidimensional situation niche separation is effective and virtually complete. In practice some overlap even in the 'full niche' is probably permissible: although it is impossible to determine just how extensive this may be. Nonetheless it is clear from the proliferation of studies stimulated by the developing theory of niche structure, that organisms frequently show considerable ecological overlap in one dimension and that considerable multi-dimensional overlap is not uncommon.

Table 5.1: Ecological Separation among Browsing Ungulates. Indices of overlap (0) in the three niche dimensions considered, and combined overlap among pairs of the four herbivore species are calculated in green and dry seasons. The index of combined overlap is the product of the indices in the three dimensions.

Species pair	Season	Niche dimension			Combined overlap
		Food	Habitat	Browsing level	
Kudu/gerenuk	Green	0.97	—	1.0	—
	Dry	0.99	0.56	1.0	0.55
Kudu/giraffe	Green	0.98	—	0.67	—
	Dry	0.98	0.97	0.32	0.31
Gerenuk/giraffe	Green	0.95	0.68	0.67	0.43
	Dry	0.999	0.50	0.32	0.16
Kudu/rhino	Green	0.50	—	0.78	—
	Dry	0.61	0.93	0.86	0.49
Gerenuk/rhino	Green	0.38	—	0.78	—
	Dry	0.48	0.69	0.86	0.28
Giraffe/rhino	Green	0.39	—	0.21	—
	Dry	0.47	0.89	0.16	0.07

Source: From Leuthold, 1978.

5.6 Measures of Niche Width, Separation and Overlap

The study of niche form and niche relationships has become one of the most explosive areas in community ecology in the past few years. More and more studies are devoted to analyses of niche width, separation or overlap within communities in one or many dimensions. It is in large part this proliferation of studies which has heightened our appreciation of the extent to which niche overlap may occur in real communities. A variety of measures have been proposed so that niche widths, and overlaps may be quantified. While it is *not* our intention to review all such indices here (an able critique is offered by Abrams, 1980 or Slobodkichoff and Schulz, 1980), it may be valuable to present here, briefly, those measures in commonest use.

5.6a Niche Width

The concept of niche width or niche breadth refers essentially to the 'diversity' of resource use shown by any one organism or group of organisms. Those whose resource use is restricted to a small portion of the available resource spectrum are considered to have narrow niches, those which exploit a relatively diverse set of resources within the resource continuum are defined with broad niches. As various workers have sought measures of niche width which they might apply, most have therefore turned to ready-made measures of diversity, originally developed as indices for the diversity of species within a community (Chapter 12) and have applied such indices instead to the diversity of resources used by any organism.

Simple indices of niche breadth are thus based upon such species diversity measures and may be calculated as

$$B = \frac{1}{\sum\limits_{1}^{j} p_{ia}^{2}} \quad \text{(after Simpson, 1949)} \tag{5.2}$$

$$\text{or } B = - \sum\limits_{i}^{j} p_{ia} \log p_{ia} \quad \text{(Shannon-Wiener index see page 322)} \tag{5.3}$$

In each case, p_{ia} measures the proportion of individuals of species i associated with resource j (or the proportion of total resource use by individual i which is made up from resource j). Colwell and Futuyma (1971) have criticised these simple diversity measures on a number of grounds — essentially that the measures are highly dependent upon the number of resources considered. They note that 'there are two situations in which one might wish to compare niche metrics: among species within a community, and between species of different communities. In both cases the paramount difficulty is standardisation of the procedure so that measurements are comparable for different species [and different communities]' [our addition]. Colwell and Futuyma develop further measures of niche breadth, which are reconsidered and refined by Hurlbert (1978), but these become rather complex for general use. Other measures are suggested by Feinsinger et al. (1981) or Thorman (1982).

5.6b Niche Overlap/Similarity

In the same way that measures of niche breadth are all primarily derived from various indices of diversity, the various measures of niche overlap are usually based on some analysis of resource partitioning. Thus most are functions of $\sum_{a=i}^{a=n} p_{ia}$ (as above page 120): that partition of the ath resource which is used by the ith species or alternatively *the number of individuals of species i using the ath resource partition*. The measures of niche overlap most commonly used are:

$$0_{ij} = 1 - \tfrac{1}{2} \sum_{a=1}^{n} \left| p_{ia} - p_{ja} \right| \quad \text{(Schoener, 1968)} \tag{5.4}$$

$$0_{ij} = \sum_{a=1}^{n} p_{ia} p_{ja} \Big/ \sqrt{\sum_{a}^{n} (\Sigma p_{ia})^2 (\Sigma p_{ja})^2} \quad \text{(Pianka, 1973)} \tag{5.5}$$

$$0_{ij} = \sum_{a=1}^{n} p_{ia} p_{ja} \Big/ \sum_{a=1}^{n} p_{ia}^2 \quad \text{(Levins, 1968; MacArthur, 1972)} \tag{5.6}$$

Calculated as percentage similarity, overlap may also be assessed (as Goodall, 1973) as

$$0_{ij} = Ps_{ij} = \Sigma_j \min (p_{ia}, p_{ja}) \tag{5.7}$$

Other measures have since been developed and are reviewed by Hurlbert (1978), Abrams (1980) and Lawlor (1980).

Each of these indices measures overlap along one resource dimension only. As noted above, multidimensional overlap may be derived as the arithmetic sum of separate overlaps if resource dimensions are interdependent, the product, if they are independent. Frequently, of course, it is very difficult to determine whether or not the separate dimensions are or are not dependent – or to what extent. Slobodkichoff and Schulz (1980) offer a valuable test for dependence and independence of resource dimensions used.

The different indices obviously each have difference strengths and weaknesses. May (1975a) points out for example, that the overlap matrix produced for Levins' index is not symmetrical: overlap of i upon j does not equal overlap of j upon i. Although this has several disadvantages, it has the signal advantage that it is sensitive to the number of individuals involved in each species. Slobodkichoff and Schulz further note that the different measures embrace to different degrees, competitive events. They stress a clear distinction between the concepts of resource use overlap and competitive pressure resulting from niche overlap, and note that the same distinction should be recognised in the different indices. They suggest that Pianka's (1973) overlap index represents resource use overlap (true niche overlap) while Levins' (1968) index (with its sensitivity to numbers of individuals) is more an index related to competitive pressure resulting from niche overlap.

5.7 Niche Relationships and Community Structure

Clearly the various characteristics of the niche, the limits to similarity and overlap between 'adjacent' niches, the phenomenon of niche separation, all have profound consequences in the structuring of communities. An understanding of the fundamentals of niche structure may allow us to draw a number of conclusions about the way in which communities may be organised (Chapter 2). This indeed is one of the main hopes stimulating all the recent work in the niche.

Schoener (1974), as we have already noted (Section 5.4), has suggested that restrictions of niche structure are such that at least along a single resource dimension organisms should be regularly spaced within the community; from different premises DeVita (1979) draws the same conclusion. Pianka (1981) points out that such overdispersion of species might be expected (from considerations of potential competition) even in multidimensional resource space, with each species minimising its interactions with all others. Other constraints of niche composition or limits to overlap and separation have further implications for community structure: and our understanding of the fundamental principles of niche relationships is potentially very powerful in offering an analysis of community design. But, how does our theoretical conception of niche structure relate to the real world? To what extent is this regular spacing of organisms in resource space a real phenomenon? How many, and which, niche dimensions are actually important in separating species? Are individual *niches* even the basic unit involved in community structure, or are organisms associated into clusters of related niches of functionally similar species ('guilds'), with organisation of resource space undertaken with respect to these guilds as the basic unit rather than the individual niches? (Pianka, 1980). There is certainly considerable evidence that such clustering of species does occur (Section 2.9). But in such case are these guilds merely a result of built-in design constraints upon consumer species, does the apparent clustering simply reflect natural gaps in resource space, or can guild structure evolve even when resources are continuously distributed as a means of reducing diffuse competition? And what factors are involved in the spacing of individual species *within* the guild? (Pianka, 1980). In brief, our theoretical knowledge of community design is not yet strong enough, our understanding is not as yet sufficiently complete for this approach to be used in a practical way in community analysis. There are still many problems unresolved. Perhaps the biggest problem, apart from that of translation of the theory itself into parameters which can be measured in real communities, is the converse: that of measuring the appropriate characteristics or real communities which can be used to advance the theory, and answer some of those questions we have catalogued here.

Attempts to define some of the 'rules' of community design have concentrated on studies of the separation or overlap of niches within real communities (e.g. MacArthur, 1970; May, 1974; Yoshiyama and Roughgarden, 1977; Rappoldt and Hogeweg, 1980), extending this to an analysis of what is referred to as 'species packing' within communities. But it is extremely difficult to synthesise the conclusions of all this work, and the implications for community design: the literature itself is confused and conflicting. Different authors appear to work from different premises, and have worked in markedly different communities which may well not obey the same 'rules'. (Thus, in an analysis of changes in niche relationships with increasing species diversity in communities of desert lizards, Pianka (1973) found evidence of decreased niche overlap with increasing species number. Cody (1974) using methods similar to those of Pianka, in an analysis of bird communities in scrub habitats, came to the diametrically opposite conclusions: finding that average overlap between 'adjacent niches' within a community increased with increasing diversity.) In addition, 'packing' is variously defined by different authors in terms of the number of species that can be accommodated per unit volume of resource space (or in a single dimension, per unit distance of a resource continuum) (e.g. Roughgarden, 1974; Rappoldt and Hogeweg, 1980), as the closeness of packing of resource utilisation distributions for non-overlapping species (Pianka, 1975), the extent of overlap observed in resource utilisation within the community or between adjacent pairs of species within the community (MacArthur, 1970; Roughgarden and Feldman, 1975). While all these do indeed represent facets of the way species are packed into a community it is true, all equally represent rather *different* facets. The number of species which can be accommodated is a measure of community fill; the pattern in which these species are arranged within niche space — their actual spatial relationship to one another — is an aspect of packing design: while the closeness of packing of non-overlapping species, or the degree of overlap within the community are *derived* functions of the extent of community-fill, the packing design and the type of organism involved. Nor are these separate elements necessarily related to each other in any simple way. Failure to distinguish these different aspects, or assumption that they show necessary correlation, leads to a great deal of confusion. Thus extent of resource overlap for example (the measure most commonly referred to as 'species packing' (e.g. Roughgarden and Feldman, 1975)) is *not* as has sometimes been implied a valid index of community fill at all, indeed it is not necessarily related to community fill in any way. For while it is the case that increased 'fill' of a community may well be reflected by an increase in niche overlap in many instances (as Cody, 1974), an increase in the number of species within a given resource space in other circumstances may equally well be accommodated without such increased overlap but by restriction of niche widths. (Thorman (1982) has shown a decrease in niche widths amongst guilds of estuarine fish with increasing species number.) Yet again, a greater number of species may be accommodated merely by expansion of the actual resource spread exploited (as Pianka, 1973: this is why increased species number was accompanied by decreased niche overlap in his study of desert lizards).

Measurement of the extent of niche-overlap within the community is no more and no less than a measurement of niche overlap. Likewise closeness of packing of non-overlapping niches is precisely that and no more. Extension beyond this to rules of community organisation is fraught with complexity.

5.8 Parallel Niches

So far we have been considering the characteristics and attributes of our 'multi-dimensional hypervolume' within its community, within a theoretical community. Returning to the real world, so to speak, what may be learnt from an examination of equivalent niches in parallel communities? First of all, can two niches in different communities be considered the same (so that, for example, there is always an insectivorous, earthworm-eating small mammal in any suitable community) or is the niche defined on the spot by the pressures and limitations of the particular community involved (i.e. it would be possible to find an insectivorous, earthworm-eating small mammal if there are vacancies in these roles, but it is equally likely that one might find one which was insectivorous only and another, earthworm-eating only dividing up the niche between them)? Clearly, from our considerations of what a niche is, it must be defined by the circumstances of the particular community in which it occurs, shaped by interspecific competition, predation and so on. Thus it would be misleading to define general classes of niche. However, there does appear to be, despite this, some evidence of a surprising constancy in niche structure as we noted in Chapter 2. Cody, for example, has shown that, in the separate ecosystems of the Californian chaparral and the Chilean matorall, the resources available to insectivorous birds are divided up in almost exactly the same way, albeit between totally different sets of species (Cody, 1974). Further, on a third continent, in South African macchia, equivalents of exactly the same classes of insectivorous birds are found again (Cody, 1975). Similar parallels are shown by Fuentes (1976) among assemblies of Californian and Chilean lizards (see above, Section 2.10).

Where such parallels do occur there are remarkable similarities in the morphology and physiology of the organisms exploiting the equivalent niches (Figure 2.8), showing that the ecological pressures of being exactly moulded to a particular way of life are an extraordinarily potent, and *constant*, force. Karr and James (1975) have indeed used morphological similarities in order to establish equivalence of niches in different communities in their own analysis of the constancy in resource partitioning suggested by Cody and others, and there are countless examples of such parallel adaptation. In East and Southern Africa, the hyena and aardwolf (the striped hyena, *Hyaena hyaena* and the aardwolf *Proteles cristatus*) show striking convergence (although this has, more recently, been claimed as mimicry (Greene, 1977)); the hawks, falcons and owls all show many common features to a common way of life, and between the marsupial fauna of the Australian continent and placental mammals, there are many further parallels. One of the most exactly replicated series is that of the marsupial mole (*Notoryctes typhlops*), European mole (*Talpa europaea*), and the various North American moles (e.g. *Scalopus aquaticus*, the Eastern mole). In each continent independently a mammal has developed to fill the burrowing, insectivorous, earthworm-eating role we defined in the previous paragraph. In each case, the 'strictures of the job' have produced a mammal with a torpedo-like body shape, non-existent neck, heavy, spade-like 'hands' and long, sensitive, naked snout. In each the eyes are no longer the primary sense organs; in each, a feature otherwise unknown in mammals, the fur is rooted in such a way that it may lie equally easily brushed forwards or backwards: a remarkable adaptation for free backward and forward movement in a restricted tunnel.

6 Interspecific Competition and Community Structure

6.1 Introduction and Definitions of Competition

Much of our previous discussion has relied, whether implicitly or explicitly, on considerations of ecological competition. In the last chapter particularly, such consideration became more and more explicit as we introduced ideas of both inter- and intraspecific competition and their role in shaping the niche, niche position, niche breadth and degree of permissible overlap. Many of these principles of ecological competition are self-explanatory — almost second nature — which is why it has been possible to use them so far without stopping to consider them in real detail. However, having now established something of the context, we must now turn to a more rigorous examination of competition in its own right, as one of the major biotic factors which link organisms into a community.

As soon as two organisms start to exploit the same resource — even within the permissible limits of niche overlap — they begin to influence each other in terms of competition. One may interfere with the other's ready access to the resource, or may use up some of the resource in prior exploitation, leaving less available to the 'second comer'. In either case, the presence of the competitor influences, deleteriously, the other's free use of a resource. Such competition may be defined as 'the active seeking after, or utilisation of a given resource by two or more individuals of the same, or different species, the resource itself being actually or potentially limiting' (e.g. Milne, 1961; Miller, 1967). As usual, in order to be sufficiently rigorous, to qualify bald and oversimplistic statements, the definition becomes unmanageably clumsy. However, the points that any single definition of ecological competition has to cover may be examined separately.

Competition for a given resource may be between individuals of the same species (*intraspecific* competition) or different species (*interspecific* competition). In either case, such competition may be direct, through actual direct interactions of competitors seeking the same resource and thus interfering with each other's uninterrupted access to and use of that resource (*interference competition*: an example here might be allelopathy between competing plant species); alternatively, the competition may be indirect: one individual affects the other's uninterrupted use of the resource through prior exploitation (*exploitation*

127

competition). In this last case, although the competitors do not meet and directly restrict each other's use of the resource, by occupying the resource separately (perhaps at different times?) each nonetheless affects the other's use of that resource by diminishing the total amount available. Often, of course, a competitive interaction will contain elements of both interference and exploitation.

Competition through exploitation requires that the resources competed for be limited in supply; (if there's more than enough for everyone, they can't be competing for it). Recognition of this extra restriction for exploitation competition has generally led to its insertion within any definition of competition as a whole: that the resource competed for should be 'actually or potentially limited in supply'. Yet clearly *interference* between organisms for use of a resource may occur, and have a real effect upon those organisms, whether or not the resource in question is limited in supply. Indeed organisms may interfere even when they are seeking to exploit totally different resources. In this sense, however, the interference becomes an interaction in its own right, quite separate from the idea of competition. When organisms interfere with each other's free use of resources, but those resources are not common to the organisms concerned, they are hardly *competing*. Pianka (1981) has in fact suggested that some potential for exploitation competition is in practice a prerequisite for interference *competition*. It is thus general practice to restrict the term competition – whether interference or exploitation – to those situations where organisms interact in their use of a shared resource which is, at least, potentially, limited in supply.

This competitive 'seeking after or utilisation of shared resources' has profound effects on the competitors. Competition can be viewed in terms of our simplistic scheme of biotic relationships (Table 2.1) as a double-negative interaction: an interaction where both participants are disadvantaged by the relationship. Its immediate effect on interacting populations is in suppression of population growth and vitality: a depression of population growth and development. (This in turn may have even more significant repercussions, as we will discuss later in the chapter.) Note, however, that this does not necessarily mean that as soon as we observe a situation in which two species co-occur – and in that co-occurrence are seen to inhibit each other's abundance – we can at once attribute this to competition: for in fact suppression of population growth can occur for a variety of reasons, not solely competition; as we have just discussed, interference effects may occur quite outside our definition of competition. Holt (1977) makes a useful distinction between true competition as we have defined it here and 'apparent' competition, stressing that while in both cases interactions result in mutual inhibition of population abundance, true competition requires this effect to be mediated through the interactions due to sharing a limited resource.

Both inter- and intraspecific competiton, as we have seen, play an important part in defining each organism's niche within the community, and thus in the shaping of the community as a whole. By the same token, intraspecific competition increases diversity/variation within a population ('broadens the niche') and may ultimately lead to the formation of new species and adaptive radiations. Further, both intra- and interspecific competition may be invoked as forces potentially responsible for regulating the numbers of animal populations. All these facets of the varied effects of competition are, however, dealt with in other chapters, and here we will restrict ourselves to considerations of the actual mechanisms and short-term implications of competition, and more particularly of the role of interspecific competition and its effects within the community.

6.2 Interspecific Competition

In such a context, let us re-establish what interspecific competition is believed to do. To begin with if two or more species with sufficiently similar ecologies find themselves together, each will have a direct, depressing effect on the other's free use of the available resources. This will be translated into a depressive effect on population growth and development in the competitors and in effect restricts both rate of growth and ultimate ceiling reached. Ultimately, intense competition may result in the exclusion of one or other species: one may change its ecology – either as a rapid adaptation through changes in the behaviour of individuals, or, in the longer term, through evolutionary change within the species – or one or other species may become locally extinct. Clearly both these 'solutions' are different aspects of the same overall effect, in the sense that a change in ecology is just as much a form of 'exclusion' from the contested niche as is actual extinction within the niche. Either produces a marked effect upon the structure of the community.

6.3 The Mechanics of Competition

Perhaps the best way to explain these various effects is to look at the way in which competition affects populations, using a simple model. In effect, what is happening in a competitive situation is that the normal pattern of population growth is being damped to a greater or lesser degree by the competition.

In an unlimited environment, population growth will follow a simple geometric progression as the full natural rate of increase of the population is expressed. If in one generation some imaginary population doubles in size, in the next generation that double population will double again – and so on. This sort of increase leads to a pattern of population growth as shown in Figure 6.1 for a population of pheasants (*Phasianus colchicus*) introduced for the first time to a new environment: Protection Island, Washington (Einarsen, 1945). The growth of such a population can be described by a differential equation

$$\frac{dN}{dt} = r_m N \qquad (6.1)$$

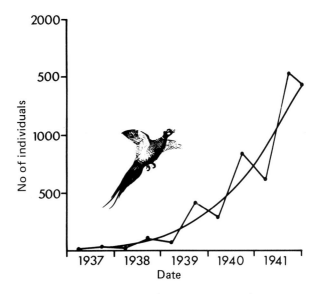

Figure 6.1: Exponential Growth of Pheasant (*Phasianus colchicus*) Populations on Protection Island, Washington. Smoothed curve connects average numbers at the end of each year. Source: After Einarsen, 1945.

where $\frac{dN}{dt}$, the rate of change of number with time, equals the product of the natural rate of increase, r_m, and population size at any instant, N. Such growth cannot, however, be maintained indefinitely. As resources become limiting, intraspecific competition has a dampening effect on this free growth. We may incorporate a second component in our equation to represent this dampening effect of intraspecific competition.

$$\frac{dN}{dt} = r_m N(1 - \frac{N}{K}) \tag{6.2}$$

when N is again instantaneous population size, r_m is intrinsic rate of increase and K is a measure of the total population the environment can support at balance: the *carrying-capacity* of the environment for that species. (It is easy to justify this interpretation of K, by considering what happens to our theoretical population above as the ratio of N to K alters. Where N>K the population will decrease; where N<K the population will increase. Where N = K the population will remain stable: thus K represents the stable population size imposed by the environment.) This new equation produces a population growth curve of the familiar sigmoid form of Figure 6.2, where, after a period of rapid increase, the population levels off to a plateau: a situation far more commonly experienced in nature. (Note: the full complexities of population growth and development will be discussed in full in later chapters. At this stage, we introduce only the barest essentials, and the simplest 'models', sufficient merely to allow us to develop our

theme of interspecific competition. For fuller treatment of the characteristics of population growth, the reader is referred to Chapter 7.)

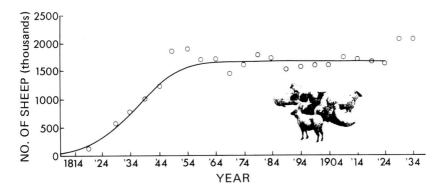

Figure 6.2: Typical Sigmoid Curve for Population Growth. The graph plots the development of sheep populations in Tasmania. Source: From Davidson, 1938 after Odum, 1959.

In exactly the same way, it is possible to incorporate into our equation a further dampening function to represent the effects of interspecific competition from a second species (species 2) on the growth rate of the first (species 1). In its simplest form (and these models are rather oversimplistic) this term is calculated to include an effect from the size of the population of the competing species (N_2) and a 'competition coefficient' ($\alpha_{1, 2}$) summarising the effect of inter- relative to intraspecific competition on slowing the growth rate of species N_1. This, α, is likely to depend on the degree to which exploitation of resources by species 2 overlaps with that of species 1 (hence $\alpha_{1, 2}$: the effect of species 2 on 1). Our equation for the population growth of species 1 now becomes:

$$\frac{dN_1}{dt} = r_1N_1 \left(1 - \frac{N_1}{K_1}\right) - r_1N_1 \left(\alpha_{1, 2}\frac{N_2}{K_1}\right) \text{ or} \tag{6.3}$$

$$\frac{dN_1}{dt} = r_1N_1 \left(1 - \frac{N_1}{K_1} - \alpha_{1, 2}\frac{N_2}{K_1}\right) \tag{6.4}$$

K_1 appears in the denominator throughout since we are interested in the effects of both intra- and interspecific competition on the resources used by species 1.

The growth of species 2 can equally be described as

$$\frac{dN_2}{dt} = r_2N_2 \left(1 - \frac{N_2}{K_2} - \alpha_{2, 1}\frac{N_1}{K_2}\right) \tag{6.5}$$

where the term $\alpha_{2, 1}\frac{N_1}{K_2}$ summarises the competitive effect of species 1 on species 2. The strength of competition is reflected in the degree to which growth is res- tricted, and, if great enough, may ultimately cause extinction (Figures 6.3, 6.4).

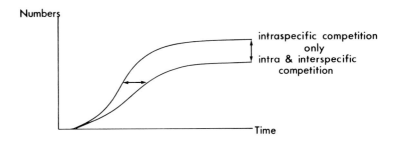

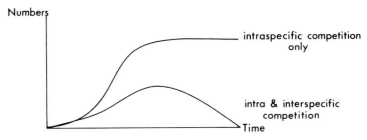

Figure 6.3: Effects of Introducing an Element of Interspecific Competition into the Population Equations of Page 131. Population growth is demonstrated here for a theoretical population but compare with Figure 6.7.

Figure 6.4: As Figure 6.3, but Strength of Competition Has Increased Such as to Cause Extinction.

The theoretical analysis developed here upon a logistic model for population growth was first suggested by Lotka (1925) and Volterra (1926) upon whose work these competition equations are based. It should be emphasised that it is possible to derive similar equations based upon other mathematical premises. The Lotka-Volterra equations greatly oversimplify the process of interspecific competition and a number of alternative models – or refinements of the simple Lotka-Volterra equations as presented here, have been proposed (e.g. Taylor and Taylor, 1979; Hanski, 1981; reviews by MacArthur, 1972; May, 1976). However the basic conclusions do not differ significantly from those presented here and the simple differential equations we have used provide a straightforward conceptual framework for analysis of the biological events.

The competition coefficient α may be calculated for *real* populations to quantify the degree of competition experienced for some resource. It may be derived from theoretical considerations for stable populations, but clearly such a figure depends to some extent on the assumptions implicit in the particular model used. Alternatively, α may be determined empirically from experimental manipulation of co-occurring populations – discovering the precise effect of increasing or decreasing the numbers of each population or of adjusting the amount of available resource (see Section 6.9). May (1975a) notes that an α

matrix can be drawn up for any pair of potentially competing species, with an α value corresponding to each resource dimension of common usage.

Clearly our theoretical competition equations can be extended to accommodate competition, both interference and exploitation, from more than one competing species. Additional elements of the same type may be added to extend to a three-, four-, five-, or multi-species complex. Here, however, for simplicity, we will restrict our considerations to a two-species case. Note, in addition that these same basic equations can be modified to include the effects of other interactions on population growth: positive effects from mutualistic partners can be included by altering the sign before the appropriate $\alpha\dfrac{N}{K}$ element of a pair of equations. Parasitic effects can be represented by altering the sign to $+$ in one case $-$ in the other, and so on. In fact, the paired *signs* of the various biotic interactions summarised in Table 2.1 correspond precisely to the mathematical value of the α element in these equations which would represent their effect on the two interacting populations.

6.4 Niche Overlap and Competition

Such analyses show clear parallels to those employed in the practical study of niche overlap (Sections 5.5 and 5.6); indeed in a number of studies measures of niche overlap and competition coefficients are treated as equivalent: niche overlap indices are taken as direct estimates of competition coefficients for use in theoretical analyses (e.g. Schoener, 1968; Levins, 1968; Pianka, 1969 etc.). Yet, although clearly intimately *related*, the two are *not* analogous. And, however tempting it may be, equating overlap with competition can be dubious and misleading (Colwell and Futuyma, 1971). As Pianka (1981) notes 'although niche overlap is clearly a pre-requisite to exploitative competition, overlap need not necessarily lead to competition unless resources are in short supply'.

Maximum permissible niche overlap is calculated for the most part on the basis of niches remaining separated (or at least on the basis that two specialists will be able to maintain a more efficient exploitation of the resources than a single generalist). Limits to coexistence through competition may occur long before this. Niche overlap up to a maximum of 100 per cent is possible in niche overlap theory along one single resource dimension, so long as the species involved are sufficiently separated along some other dimension of the niche.

Yet exploitation competition along this one resource dimension would have been sufficient to cause exclusion of one or other species long before this. In brief niche overlap theory does not, in general, take into account the finite nature of the resources involved or relative number of individuals in each species, merely considering the efficiency of their exploitation (although, as we have noted above (page 123) different overlap indices take different account of the effects of competition). Ecological competition considers the result of such overlap when resources are in short supply. (For further consideration see Slobodkichoff and Schulz, 1980; Pianka, 1981).

6.5 The Effects of Interspecific Competition Within the Community: Exclusion and Coexistence

Our theoretical analysis of the effects of competition shows us that it is in theory possible for competition to have such a depressive effect upon a population as to cause its extinction, if the species does not shift its ecological 'tastes' in advance of this. Clearly if this is the case, this has profound implications for the shaping of the structure of communities and for their species composition. Indeed it was mathematical examinations of the phenomenon of competition by Lotka (1925) and Volterra (1926) whose equations we have just examined that led to the formulation of the idea, that 'two species with the same or similar ecologies cannot coexist in the same environment' (page 113). But what evidence have we that such events actually occur in practice?

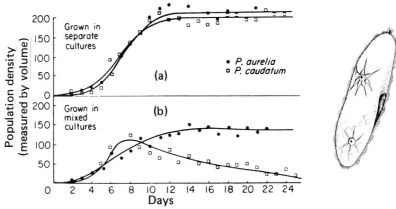

Figure 6.5: Growth of *Paramecium* Populations (*P. caudatum, P. aurelia*) in Separate Culture (a) and in Mixed Culture (b). Source: After Gause, 1934.

The mathematical conclusions of Lotka and Volterra were first enshrined into the empirical theorem as which they are now widely quoted, by Gause (1934). Gause's original statement was based on the results of certain laboratory experiments with *Paramecium*. He showed (Gause, 1934) that, in separate culture, the species *Paramecium aurelia* and *P. caudatum* each showed a pattern of population growth with time as in Figure 6.5a. Yet, if the two species were cultured together – although under otherwise identical conditions – populations of *P. caudatum* were suppressed and unable to establish themselves (Figure 6.5b). These same effects were later demonstrated for flour beetles (*Tribolium*) by Park (1948, 1954, 1962). In separate culture, populations of *Tribolium castaneum* and *T. confusum* each showed characteristic patterns of growth (Figure 6.6a). In competitive, mixed culture, one species excluded the other (Figure 6.6b). But the experiments of Park and his colleagues showed that the outcome of the competitive interaction was not always the same – that, in fact, while under hot, moist conditions *T. castaneum* excluded *T. confusum*, under colder, more arid conditions, *T. confusum* consistently outcompeted *T. castaneum* (as in our example Figure 6.6b). Further experiments established that the outcome of any interspecific competitive interaction depends on (a) initial population densities, (Neyman *et al.*, 1956), (b) environmental conditions (Park, 1954) and (c) the genetic constitution of the strains of competing species (Park *et al.*, 1964). Indeed by varying these parameters it is possible to produce a reversal of exclusion, or even to create a situation in which two potentially competitive species may in fact coexist.

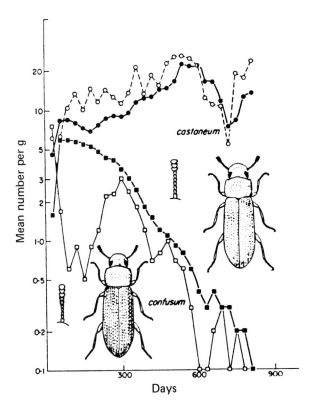

Figure 6.6: Growth of Populations of the Flour Beetles *Tribolium castaneum* and *T. confusum* in Separate and in Mixed Culture. Source: From Park, 1948, after Varley, Gradwell & Hassell, 1973.

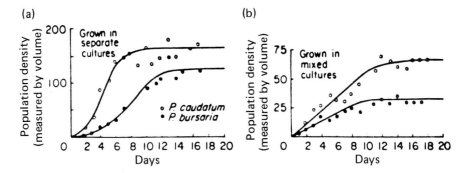

Figure 6.7: Coexistence of *Paramecium caudatum* and *P. bursaria* in Mixed Culture (b).
Population curves for the two species in separate culture are shown in (a).

While the most frequently quoted of Gause's experiments with *Paramecium* demonstrate clear exclusion, in another of his original experiments he was in fact able to show coexistence of two species: *Paramecium caudatum* and *P. bursaria* (Figure 6.7b). The population curves of these two species in separate culture are, however, markedly different (Figure 6.7a) and Gause concluded that they could coexist because their ecologies 'were not sufficiently similar to result in exclusion'. More recent experiments have, however, been able to demonstrate sustained coexistence under certain conditions of species which are known, under other conditions, to be fully exclusive. Ayala (1970) working with competitive cultures of *Drosophila* has shown that at temperatures of approximately 23°C the two species *Drosophila pseudo obscura* and *D. serrata* are able to coexist. Yet at temperatures of 25°C or above *D. serrata* outcompetes *D. pseudo obscura* and at temperatures of 22°C or below, *D. pseudo obscura* has competitive advantage (convincing demonstration that each *can* potentially outcompete the other and, as shown by Park, that the outcome is dependent on various environmental factors).

Competitive advantage varies with environmental conditions, and, in changing, may reverse the outcome of experimental exclusions or even permit coexistence. In Ayala's experiments coexistence occurs at a temperature where neither fly species is at an optimum and where therefore their respective competitive abilities are very weak. Coexistence might equally be possible in a patchy environment: an environment spatially or temporally heterogeneous, where rapid alternation of environmental conditions may result in corresponding alternation of competitive advantage of the various species. Hutchinson (1953) has suggested that the role of regular environmental fluctuations in permitting the coexistence of ecologically similar forms depends on the relationship between the periodicity of environmental fluctuations and the generation time of the organism (Figure 6.8). If the generation time of the organisms is very short in relation to the period of environmental fluctuation, competition can occur in one direction over several generations within one phase of the environmental cycle and thus be able to run its course to elimination of one or other species before environmental conditions change. If the generation time is too long relative to the environmental change any one individual must cope in its lifetime with all environmental conditions; the better-adapted species overall may still outcompete the lesser. However, if the relationship between environmental periodicity and generation time is intermediate, the environment may alternately favour different species for periods of a few generations and both may coexist.

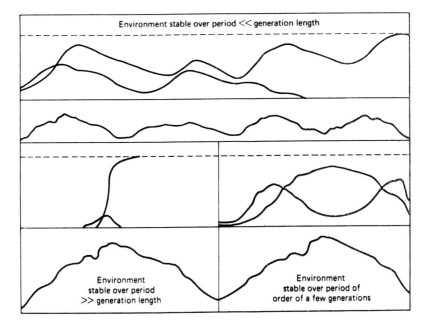

Figure 6.8: Ideal Course of Competition Between Two Species as Regulated by the Relation Between Generation Length and the Period over which the Environment may be Taken to be Stable. Source: From Hutchinson, 1953 after Collier *et al.*, 1973.

Other possible situations permitting coexistence can be conceived when competition may favour each of two species at different stages of the life cycle etc. etc.

All these findings may be simplified, and explained by reference to the mathematical rationalisation already introduced. The outcome of a competitive interaction depends on the relative strengths of the competition coefficients, α, in our equations:

$$\frac{dN_1}{dt} = r_1N_1 \left(1 - \frac{N_1}{K_1} - \alpha_{1,2}\frac{N_2}{K_1}\right) \text{ and}$$

$$\frac{dN_2}{dt} = r_2 N_2 \left(1 - \frac{N_2}{K_2} - \alpha_{2,1} \frac{N_1}{K_2}\right)$$

The relative value of the two elements $\alpha_{1,2} \frac{N_2}{K_1}$ and $\alpha_{2,1} \frac{N_1}{K_2}$ is what determines the outcome of a competitive interaction: and since the values of the two elements are clearly dependent to some extent on extraneous conditions, we can see where there may be different outcomes in different circumstances — exclusion of one or other species, or coexistence.

The actual conditions for exclusion or coexistence in each case can be derived by solution of the differential equations above. If species 1 is to outcompete species 2, $\alpha_{1,2}$ must be relatively small, while $\alpha_{2,1}$ is correspondingly large; if species 2 is to exclude species 1, then $\alpha_{2,1} \ll \alpha_{1,2}$. For coexistence both $\alpha_{2,1}$ and $\alpha_{1,2}$ must be very small, so that the effects of interspecific competition add very little to those of intraspecific competition. In terms of our equations, species 1 and 2 coexist in terms of our equations if $\frac{dN_1}{dt} \geqslant 0$ and $\frac{dN_2}{dt} \geqslant 0$. For this to be the case

$$\frac{N_1}{K_1} + \alpha_{1,2} \frac{N_2}{K_1} \leqslant 1 \tag{6.6}$$

and $$\frac{N_2}{K_2} + \alpha_{2,1} \frac{N_1}{K_2} \leqslant 1 \tag{6.7}$$

which means the species may coexist if

$$\alpha_{1,2} \leqslant \frac{K_1 - N_1}{N_2} \tag{6.8}$$

$$\alpha_{2,1} \leqslant \frac{K_2 - N_2}{N_1} \tag{6.9}$$

All this may be represented visually by drawing population trajectories for the two populations and seeing where they reach equilibrium. These trajectories represent for each species an envelope of the changes in population number with respect to carrying capacity, and to population sizes of their own and the competing species. In effect, they are a graphical representation of the solution of our competition equations, for all values of N_1 and N_2. (Thus for given K the value of N_1 is derived for all values of N_2, and vice versa.) The trajectories, when superimposed, show zones where populations of N_1, N_2 may increase or must decrease. The slope of each trajectory is a function of the competition coefficient α. If the competition coefficients are adjusted correctly, then the population trajectories for two species can reach stable equilibrium as in Figure 6.9. When species 2 is rare it can increase in number even when species 1 has reached its maximum population size. Conversely, when species 1 is rare, it can increase even in the presence of species 2.

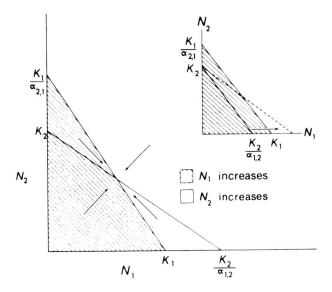

Figure 6.9: Conditions for the Stable Coexistence of Two Competing Species. The population trajectories in Figure 6.9 demonstrate that the point of intersection of the equilibrium lines represents stable coexistence: all trajectories converge upon that point. When species 2 is rare, it can increase in number even when species 1 has reached its maximum population size. Conversely, when species 1 is rare, it can always increase in the presence of species 2. We can conclude that two species will coexist when competitive interactions between them are weak, specifically when each species is more strongly limited by intraspecific competition than by interspecific competition.

Both graphical and algebraic analyses lead to the same conclusion, that the two species can coexist when, for whatever reasons, the competitive interactions are weak — specifically when each species is more strongly limited by intraspecific competition than by interspecific competition. Coexistence arising from rapid reversal of the relative competitive advantage of two species can also be visualised and explained with our equations, by incorporating into them a changeable α coefficient.

While coexistence might occur between potential competitors, Ayala (1970) found in his experiments with *Drosophila pseudo obscura* and *D. serrata*, that when they were in mixed culture, neither population behaved exactly as it did when reared separately. Both productivity (measured as adults emerging per week) and total population size of the mixed culture was considerably lower than that of either species in isolation (Table 6.1). To find an explanation for this curious result, we must return to our earlier considerations of the two separate components of a competitive interaction: interference and exploitation. On grounds of exploitation competition alone, one would expect to find mixed cultures producing total numbers of flies equivalent to the total numbers produced in a single species culture — always assuming the two species to be equal in their production of individuals from a unit portion of food. Even if the species were markedly different in this regard, one would expect to find a total population reflecting the proportion of the total resource used by each population, combined in each case with the production of individuals to be expected for that species, per unit of the resource. In practice, much lower populations are recorded. Such a result must imply that some degree of competition by interference is still experienced by each population, insufficient to cause exclusion of one or other species, but enough to impair their ability to make the fullest use of the resources available.

Table 6.1: Outcome of Competition Experiments in which *Drosophila pseudo obscura* Coexists with *D. serrata*

		Productivity (adults emerging per week)	Adult population size
Species raised separately			
*D. pseudo obscura***		269	664
D. serrata		537	1,251
Species raised together			
D. pseudo obscura		133	278
D. serrata		104	252
	Total	237	530

**Arrowhead chromosome arrangement. Similar experiments with the Chiricahua chromosome arrangement had nearly identical outcomes.
Source: From Ayala, 1970.

The degree of interference is bound to depend on the population sizes of the competing populations: this must mean that the competition coefficients of our competition equations should be variable with respect to population size. Under conditions of pure exploitation competition, the coefficients would be constant in this respect; indeed in Ayala's experiments where environmental conditions are constant throughout, and coexistence results merely from reduced competitive abilities of both species in a constant but suboptimal environment (rather than from reversal of competitive advantage under changing environmental conditions), coefficients of competition derived from exploitation alone would be absolutely constant.

The population trajectories illustrated in Figure 6.9, derive from our competition equations by assigning each species a constant α; these trajectories thus represent in practice, the development of populations under the influence of exploitation competition alone. The actual population size in a mixed culture at equilibrium, according to Ayala's findings, lies well below the intersect of these 'exploitation' trajectories (Figure 6.10). If however, we incorporate an element of interference competition into our considerations by substituting a variable α, derived as some function of population sizes, new population trajectories arise which are concave in form and whose intersection is indeed well below that equilibrium population size that might be predicted from pure exploitation competition equations with constant α (Figure 6.11).

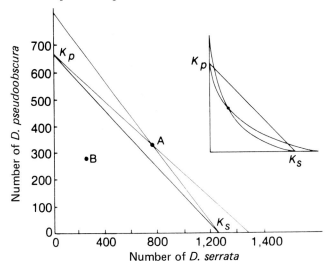

Figure 6.10: Comparison Between a Prediction of the Logistic Competition Model for Coexisting Species (A) and the Outcome of an Experiment with Coexisting Species of *Drosophila* (B). When coexistence occurs the model predicts that the equilibrium population (point A) must lie outside a line connecting the two carrying capacities (K_p and K_s). Ayala's result is indicated by point B. A modification of the graphical model to incorporate an element of interference competition is shown to the right (Figure 6.11).

All these various experiments have been based in the laboratory and may be criticised as such. Yet it is already clear from our deliberations in Chapter 5 that, even in the field, organisms are unlikely to compete to exclusion since, in a multidimensional situation, the degree of overlap in any one niche dimension can be much greater than that acceptable in the one dimension alone. In overall conclusion therefore, Gause's hypothesis, which we have set out to evaluate, is grossly oversimplified. In the field, competition coexistence is perfectly feasible — even between organisms which engage in comparatively intense competition for some single resource — as long as niches are sufficiently separate along some other dimension(s). In the laboratory, under simple, controlled conditions where niche dimensions can be reduced to a minimum, competitive exclusion of species will occur if competition is powerful. But if competition is weak (and the population restricted more by intraspecific competition than by interspecific competition) or highly variable in a heterogeneous habitat, coexistence can still be maintained.

6.6 Diffuse Competition and Indirect Competitive Effects

The traditional approach to interspecific competitive interactions is somewhat restricted: only the direct effects of each species on any other target species are considered. However the effects of such primary interaction must be compounded by contributory effects from many of the other species of the community which do not appear to be directly involved.

The fact that maximum observed niche overlap is characteristically less than is theoretically possible (even when the limits to niche overlap are calculated on the basis of competitive exclusion between the overlapping species (MacArthur and Levins, 1967)) has been attributed to the additional effects of 'diffuse competition' (MacArthur, 1972) from the rest of the community in which the organisms in question occur — effects not accounted for in the basic model which considers only the direct competition of the primary protagonists (Section 5.5).

Such interactions are still directly competitive ones: however, indirect interactions, not specifically competitive, must also occur (Levine, 1976). Thus, two species with non-overlapping diets which are preyed upon by a common predator may nonetheless have a net negative effect on each other's population density: as one becomes less abundant, the predator turns its attentions more to the alternative prey (Pianka, 1981). Similarly two predatory species with little or no dietary overlap may influence each other indirectly, if their prey species compete: an increase in either predator population, resulting in a decrease in its particular prey, reduces competition for the other prey species and hence enhances conditions for the other predator (Vandermeer, 1980). Such effects, competitive or mutualistic, are however better regarded under the umbrella of Holt's 'apparent' competitive effects.

6.7 Competition as a Selection Pressure Promoting Change

Sharing a resource, even if it does not necessarily lead to a species' exclusion, still reduces the potential of each competing species. Thus natural selection, as we have already suggested, will favour any development which tends to avoid interspecific competition of this sort and reduces competitive overlap. Ultimately, each species will tend towards a position where competition from *all* surrounding members of its community is reduced to the minimum possible. (We have already discussed this idea in our consideration of determinants of the 'realised' from the 'fundamental' niche.) If there is room in the community, we may observe, in the limit, total separation of ecologies (unless the need to use some particular resource is so critical as to outweigh the disadvantages of competition).

In the natural community niche space is sufficiently well-packed and niche positions in most cases sufficiently well-established, that there is little room for further change, and continued competition will tend merely to preserve the status quo. The fact that competition *is* having an effect however, is apparent from considerations of the end result, and also from various artificial laboratory experiments.

Prolonged coexistence of laboratory cultures may be shown to be accompanied by actual genetic changes in competitive ability of the co-occurring species. (Here is another factor in our list of possible mechanisms to permit coexistence: relative competitive ability of the competing species continually changing and reversing with *evolutionary* change.) Pimentel *et al.* (1965) demonstrated clear genetic changes in competitive ability in the housefly *Musca domestica* and the blowfly *Phoenicia* (=*Lucilia*) *sericata*. Their experiments were based on the hypothesis that in a situation in which one member of a competing pair of species is abundant and the other is rare, the strongest selection pressures on the abundant species will derive from intraspecific competition, while selection should act on the rarer species most strongly through interspecific competition. Improvement in the interspecific competitive ability of the rare species should thus occur more rapidly than that of the abundant species. Eventually, the rare species should improve in competitive ability to the extent that it is able to reverse the abundance relationship, or at least reach a stage where it itself is more powerfully influenced by intraspecific events. We have already considered the importance of the relative strengths of inter- and intraspecific competition in permitting coexistence in the first place (page 139: interpretation of Figure 6.9); Pimentel and his co-workers continue to consider the likely genetic changes arising from prolonged coexistence under such conditions, suggesting that if competitive differences between two species are slight, this feedback process may allow a rather more dynamic coexistence than we have so far considered, as the species alternate in relative genetic competitive ability.

Table 6.2: Results of Competition Tests Between Housefly and Blowfly Populations Involving Individuals of Wild Stock and Individuals From a Mixed-Species Population Cage Maintained for 38 Weeks with High Densities of Houseflies and Low Densities of Blowflies.

| | | Contests won by | |
	Week	Housefly	Blowfly
Wild housefly X wild blowfly	0	2	2
Wild housefly X wild blowfly	38	4	1
Exp. housefly X wild blowfly	38	3	2
Wild housefly X exp. blowfly	38	0	5
Exp. housefly X exp. blowfly	38	0	5

Source: Data from Pimentel *et al.*, 1965.

The competitive ability of wild flies was compared by Pimentel *et al.* with that of experimental flies which had experienced competition with each other over many generations in mixed populations with relatively high densities of houseflies and low densities of blowflies. The competitive abilities of the flies were tested by placing small groups of the two species in a competition cage and allowing outcompetition to exclusion. The results (Table 6.2) show quite clearly that flies which have experienced some degree of continued interspecific competition are more successful in competitive encounters with the other species than naive flies (experimental blowflies vs wild blowflies), whereas for that species which had been most strongly influenced by intraspecific competition throughout (experimental houseflies) little change in interspecific competitive ability from 'wild-type' is observed. These results confirm some genetic change during continued competition and offer at least a mechanism whereby a theory of coexistence by evolutionary alternation of competitive ability as envisaged by Pimentel *et al.* may operate. Further, as we have noted, they offer clear evidence for competition as a powerful and potent selection pressure.

Over a longer time scale, such genetic changes may lead to major changes in a species' ecology, in physiology or morphology, to establish ecological separations of the sort we have discussed earlier. (Where continued interspecific competition is experienced, such changes as are reported by Pimentel *et al.* may clearly be unidirectional.) Indeed it is axiomatic that competition must produce such changes: we have argued all along that the phenomena of ecological separation, niche spacing along a single dimension, niche breadth, are all the result of interspecific competition. But this is something that is in fact very difficult to prove. In looking at the end result: two species with differing ecologies, how can one *prove* that this is as the result of real or potential competition in the past? Nonetheless there are, if we are cautious in our interpretation, a number of examples available very suggestive of evolutionary changes as a result of interspecific competition.

6.8 Niche Shifts and Evolutionary Change due to Competition

There are, for example a number of studies in the literature comparing the ecologies of pairs of species when found in sympatry and in allopatry. A study of the distribution of two species of flatworms (*Planaria montenegrina* and *P. gonocephala*) along temperature gradients in freshwater streams has shown that each species is restricted to a much smaller range of thermal conditions when they occur sympatrically than when they occur separately (Beauchamp and Ullyett, 1932, from Pianka, 1976) – an example already introduced in Chapter 5 (Figure 5.2). Huey *et al.* (1974) have established for a species of subterranean skink (*Typhlosaurus lineatus*) in the Kalahari desert that the proportion of larger prey items in the diet is far higher when the animal is in sympatry with another species (*T. gariepensis*) than when it occurs in isolation. Both species feed almost exclusively on termites – largely on the same few species; in sympatry *T. lineatus* concentrates far more on the larger castes and larger species.

Niche shifts of this kind may of course occur relatively rapidly. But if maintained over a long period, such shifts may lead to other adjustments, in morphology or physiology, to adapt the species more closely to its new lifestyle. For the *Typhlosaurus* skinks studied by Huey *et al.*, snout-vent lengths of *T. lineatus* increase abruptly at the boundary of its sympatry with *T. gariepensis*; moreover, heads become proportionately larger in sympatry (Figure 6.12). Both these changes are correlated with the dietary shift in sympatry towards a larger size of insects (Huey *et al.*, 1974). Such an observation can be presented in more general terms.

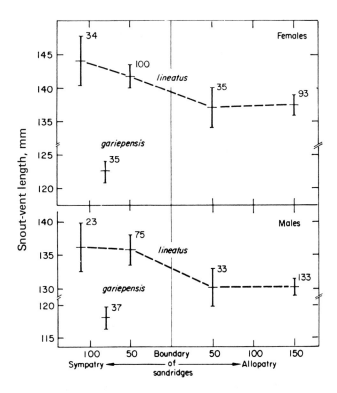

Figure 6.12: Step Clines in Mean Snout-vent Lengths of Subterranean Skinks (*Typhlosaurus lineatus*) Associated with the Presence of a Smaller Congeneric Species (*T. gariepensis*). Head proportions also change with proportionately larger heads occurring in sympatry. Source: Huey *et al.*, 1974, after Pianka, 1976.

When two species in a community are competing for some set of limited resources, each can be expected to evolve in response to the presence of the other. If the extent of competition is determined in part by the measure of some quantitative character, natural selection would be expected to lead to increased and persistent differences in the distribution of that character in the two species (Slatkin, 1980; after Brown and Wilson, 1956). Perhaps the best known example of this type of 'character displacement' (Brown and Wilson, 1956) is in the apparent change in bill length in two species of nuthatch (*Sitta*) where they

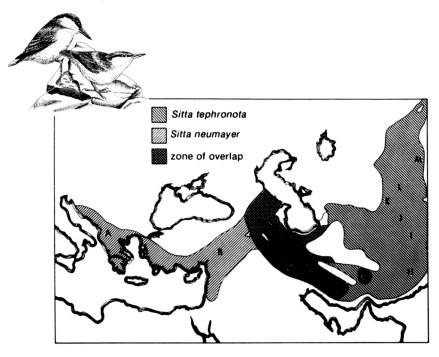

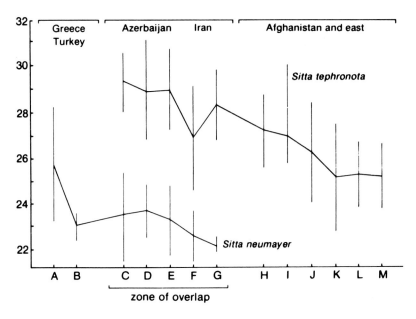

Figure 6.13: Geographical Ranges and Character Displacement in the Bill-length of the Western Rock Nuthatch (*Sitta neumayer*) and Eastern Rock Nuthatch (*S. tephronota*). Vertical lines represent the range of beak lengths in each local population. Source: After Ricklefs, 1971; from data of Vaurie, 1951.

occur in sympatry (Vaurie, 1951). The Western Rock Nuthatch (*Sitta neumayer*) extends through Greece, Turkey and Iran into Afghanistan; the range of the Eastern Rock Nuthatch (*Sitta tephronota*), although it extends much further to the East, overlaps with that of *S. neumayer* in Afghanistan. In allopatry, the range of beak sizes (and thus, it is presumed, the size range of insect prey) of the two species overlaps considerably; in sympatry, the bill lengths of each change so that their size ranges become quite distinct (Figure 6.13). (We should perhaps note however that a more recent analysis of the material suggests that this particular example — however often it has been quoted — is not perhaps a good one: that the bill sizes of each species even in allopatry show a marked clinal change over their whole range and that as a result, differences between beak size in allopatry and when in sympatry with a potential competitor could be explained purely in terms of a continuation of this clinal change and would thus have occurred whether the competitor was present or not (Grant, 1972).)

Hutchinson (1959) lists this and other examples of character displacement (Table 6.3) and notes that the magnitude of character shift seems remarkably constant. He calculates that the ratios of mouthpart sizes among all the coexisting congeneric species cited, whether insects, birds or mammals, range only from about 1.2-1.4 (Hutchinson, 1959), a constancy borne out by many additional studies since (e.g. Schoener, 1965; Grant, 1968; Diamond, 1973; Fenchel, 1975; Pulliam, 1975a). This suggests a minimum required distance between two species in resource space before competition is reduced to an acceptable level and presents an interesting counterpart to Schoener's, theoretically based, suggestion that species differing primarily in their use of a single resource dimension (as congeneric species are very likely to do) may be expected to be relatively evenly distributed in niche space, and to show a minimum niche separation of between 1.1 and 1.4 x niche breadth (Figure 5.2 and Schoener, 1974. But see the somewhat irreverent critique offered by Horn and May 1977.) The principle suggests ecological and evolutionary adjustment in the sympatric species in response to competition. (And it is after all, exactly these same considerations that we invoke for competition *within* species, leading to adaptive radiations and speciation.)

However, as noted by MacArthur (1972) and Slatkin (1980), such character displacement is by no means an inevitable consequence of competition based on the value of some phenotypic character. There are some conditions that may lead, as above, to extinction of one or other species; other conditions yet that may actually lead to *convergence* despite the competition. (The common selection pressures of being well adapted to exploitation of a common resource would tend to produce a convergent response in the two species irrespective of competition. Indeed competition itself might prove one of the most important of such selection pressures!) In an extremely able review, Slatkin (1980) concludes that two sets of conditions may obtain which *will* lead to significant character displacement. Where species are of approximately equal abundance at equilibrium, character displacement will occur if there is some difference in the set of limiting resources of the two species. In such a situation, competition will 'encourage' each species to emphasise their particular exclusive 'refuge',

Table 6.3: Mean Character Displacement in Measurable Trophic Structures in Mammals (Skull) and Birds (Culmen).

	Locality and measurement when sympatric	Locality and measurement when allopatric	Ratio when sympatric
Mustela nivalis	Britain; skull ♂39.3 ♀33.6 mm	(*boccamela*) S. France, Italy ♂42.9 ♀34.7 mm	♂100 : 128
		(*iberica*) Spain, Portugal ♂40.4 ♀36.0	♀100 : 134
M. erminea	Britain; skull ♂50.4 ♀45.0	(*hibernica*) Ireland ♂46.0 ♀41.9	100 : 109
Apodemus sylvaticus	Britain; skull 24.8	unnamed races on Channel Islands 25.6-26.7	
A. flavicollis	Britain; skull 27.0		
Sitta tephronota	Iran; culmen 29.0	races east of overlap 25.5	100 : 124
S. neumayer	Iran; culmen 23.5	races west of overlap 26.0	
Geospiza fortis	Indefatigable Isl.; culmen 12.0	Daphne Isl. 10.5	100 : 143
G. fuliginosa	Indefatigable Isl.; culmen 8.4	Crossman Isl. 9.3	
Camarhynchus parvulus	James Isl.; culmen 7.0	N. Albemarle Isl. 7.0	James 100 : 140 : 180
	Indefatigable Isl.; culmen 7.5	Chatham Isl. 8.0	100 : 129
C. psittacula	S. Albemarle Isl.; culmen 7.3	Abington Isl. 10.1	Indefatigable 100 : 128 : 162
	James Isl.; culmen 9.8	Bindloe Isl. 10.5	100 : 127
	Indefatigable Isl.; culmen 9.6		
C. pallidus	S. Albemarle Isl.; culmen 8.5		
	James Isl.; culmen 12.6	N. Albemarle Isl. 11.7	S. Albemarle 100 : 116 : 153
	Indefatigable Isl.; culmen 12.1	Chatham Isl. 10.8	100 : 132
	S. Albemarle Isl.; culmen 11.2		
			Mean ratio 100 : 128

Source: Data for *Mustela* from Miller (1912); *Apodemus* from Cranbrook (1957); *Sitta* from Brown and Wilson (1956) after Vaurie; Galapagos finches from Lack (1947); from Hutchinson, 1959.

adapting so that it becomes in practice their major resource. In these circumstances, postulates Slatkin, displacement will be much larger than initial differences between resource spectra; small, possibly unmeasurable differences between species could lead to significant displacements. The other set of conditions which may lead to displacement is when the resource distribution in competition is skewed in some way. In these circumstances, no difference between resource spectra of the two species is required for displacement, but in resultant equilibrium, the relative abundances of the two species will be very different.

Despite the intuitive appeal of the concept of character displacement as an explanation for differences between sympatric species – or differences within the same species where in sympatry and allopatry – it is not necessarily the only possible explanation. As pointed out by Grant (1972) and others, alternative explanations can be put forward, achieving the same end result, with no need for evolutionary change. Such authors argue that the same phenomenon could result purely from demographic processes. Those species, or individuals within a species, that differ from those already present in a community, can successfully invade such a community without leading to extinction of a similar species. Those that do not differ, cannot. Thus differences between sympatric species could arise purely because only those individuals within each species that (by normal population variation) do differ from the other species, may invade and persist. Such an argument has the merit that demographic changes can occur over a shorter time-scale than evolutionary ones (although they may later be 'fixed' genetically). By contrast, while essentially similar species *could* evolve displacement in response to each other's presence, it is also possible that one species would be excluded before significant genetic changes had had time to develop.

Whatever the mediating process in any particular set of circumstances, demographic or evolutionary, such examples suggest very powerfully some influence of interspecific competition. Yet despite the cogency of these (and many other) examples our conclusions are still in practice only deductive: we cannot *prove* that the niche-shifts or character-shifts are due to competition (except in rare experimental cases, e.g. Crowell and Pimm, 1976; Pimm, 1983). We are merely presented with a *fait accompli*. Indeed we cannot prove that interspecific competition has *ever*, actually or potentially occurred outside the laboratory. Various authors have, at different times, claimed that interspecific competition cannot and does not exist in the field. Natural selection causes organisms to evolve for their maximum 'fitness' or efficiency. Maximum efficiency must involve avoidance of interspecific competition: thus competition may never occur. In answer to such claims there have been a number of attempts to demonstrate that interspecific competition does have an effect on natural populations and is indeed a potent selection pressure. We should perhaps end this chapter then, as maybe we should have begun, by reviewing some of these studies in our own bid to demonstrate the existence of interspecific competition in natural systems.

6.9 Interspecific Competition in Natural Systems

Reynoldson and Bellamy (1970) established a set of criteria for establishing the existence of competition in the field. These are now widely accepted, and may be presented here.

1. The comparative distribution and/or relative abundance of the two potentially competing species should be amenable to explanation based on competition.
2. It is necessary to show that the competing species are utilising a common resource which may provide the basis of competition.
3. There should be evidence from the performance of the particular species populations in the field that intraspecific competition is occurring. This may relate to fecundity, survival, growth rate of individuals or some other appropriate parameter. This criterion assumes that if persistent interspecific competition is occurring then intraspecific competition must also be taking place.
4. Both the resource which is being competed for and the population should be manipulated separately in the field with predictable results based on the hypothesis that competition is occurring. It is insufficient to manipulate only the absolute amounts of a resource since its availability may be altered irrespective of the competition process. For example, many populations are likely to respond to an increase in food whether or not competition is occurring because the same amount (or more) may be obtained with less expenditure of energy.
5. Events following the introduction or removal (or reduction) of a competing species should be consistent with the competition hypothesis. This criterion differs from criterion 4 since it concerns interspecific events only. Criteria 4 and 5 are clearly the most crucial to observe: results of criterion 5 enable the empirical determination of the competition coefficients α, as noted in Section 6.3.

Reynoldson and Bellamy apply their own criteria in a study of competition between *Polycelis tenuis* and *Polycelis nigra* (two of the species of the flatworm assemblage introduced in Section 5.3). An opportunity was taken to follow the effects of competition between these two species, when *P. tenuis* colonised a lake previously occupied only by *P. nigra* (Reynoldson and Bellamy, 1970). The study showed that under these conditions, *Polycelis nigra* (the original species of the lake) was reduced to become a relatively small proportion (10 per cent) of the total triclad population; since the total *Polycelis* population remained more or less constant, there must have been actual replacement of *P. nigra* by *P. tenuis*.

Such natural experiments are rare, however, and as a result most studies of the influence of interspecific competition in natural systems have involved experimental manipulations of these systems. Thus Pontin (1969) dug up nest mounds of the wood ant *Lasius flavus* and repositioned them near mounds of a congeneric species *L. niger*. He found that, under such circumstances, queen production was considerably suppressed: a result at least suggestive of competitive interaction. (Pontin further showed that where *L. flavus* nests were placed close to other *L. flavus* nests, creating conditions which might promote *intra*-specific competition between the adjacent colonies, queen production in the nests was suppressed still further, satisfying Reynoldson and Bellamy's third criterion. The effects of intraspecific competition were in fact some three times as strong as those of interspecific competition: in this case competition may be presumed to be complete for every dimension of the niche.)

More recently Hairston (1980) has demonstrated for two sympatric species of terrestrial salamanders that experimental removal of *Ptethodon jordani* from marked plots in the Great Smoky Mountains and Balsam Mountains of North Carolina, resulted in a statistically significant increase in abundance of the sympatric *P. glutinosus*. Removal of *P. glutinosus* (the less abundant species) did not result in an equivalent increase in population numbers of *P. jordani*, but *did* increase markedly the proportion of young within the population.

In another field experiment on competition between two sympatric species of starfish *Pisaster ochraceus* and *Lepasterias hexactis*, Menge (1972, 1979) removed all the individuals of *Pisaster* from one small island reef and added them to a similar reef elsewhere (effectively doubling their density on the second site); a third undisturbed reef nearby acted as a control. Menge showed that, while the size of the control animals remained unchanged, the average weight of individual *Lepasterias* increased significantly with removal of *Pisaster* and decreased with its addition. Again this is highly suggestive of some competitive interaction, a conclusion supported by the further observation that, on undisturbed reefs, standing crop of *Lepasterias* (biomass per m^2) is inversely correlated with that of *Pisaster* (Menge, 1972).

Pisaster ochraceus is involved in yet another example which may be adduced as evidence for interspecific competition. This starfish is the dominant predator in an intertidal system of some 16 species of invertebrates; many of these species overlap broadly in feeding ecology but seem able to coexist without evidence of competition. However, when *Pisaster* is artificially removed from the system, the species number within the prey community falls dramatically with the total exclusion of seven species, to leave a new community of only eight species (Paine, 1966). This can surely be attributable only to interspecific competition within the prey community: competition which was suppressed in the presence of the dominant predator. The activities of the predatory starfish must normally be such as to maintain the populations of all its prey species well below their maximum, so that, as a result, with superabundant resources, potential competitors may coexist. With the removal of *Pisaster*, however, prey species could increase in numbers, resources became limited and interspecific competition resulted in the exclusion of many of the original species.

Examples such as these, whatever their ideological flaws, are surely hard to refute in their powerful implications of the force of interspecific competition within ecological communities.

7 Population Structure and Analysis

In our studies of community structure and function so far, it is clear that many of the effects observed are the results of factors affecting the dynamics of the component populations. In the next few chapters we will examine in more detail the dynamics of single species populations and populations of interacting species.

7.1 What is Population Ecology?

Populations of animals and plants exhibit their own special dynamic behaviour through the action of changing rates of emigration (in animals), immigration, birth and death. It is the quantification and explanation of the numerical changes resulting from these four factors which concern the population ecologist. The old ecological phrase 'the balance of nature', although not easily definable, at least reminds us of the way early population ecologists were stimulated by the apparent persistence of populations over long periods of time. It often seems that such persistence, despite large fluctuations, cannot be explained if only random processes of colonisation and extinction are operating. Instead, populations often seem to fluctuate below a level at which the habitat would be destroyed and above a level where extinction is a serious danger (Figures 7.1, 7.2 and 7.3). The challenge of population ecology is the quantification of the rates of immigration, emigration, birth and death in order to explain what is influencing the timing and magnitude of the fluctuations, the mean level around which the population changes occur and the role of factors which prevent over-exploitation and extinction. This desire to explain is often stimulated by the hope of manipulation; it is no coincidence that much population ecology theory has been derived from work on insects; in two recent surveys of regulation in animal populations (Podoler and Rogers, 1975; Stubbs, 1977), 70 per cent of the examples were insect pests. The pragmatic advantage given by a short generation time, providing data quickly, coupled with their frequent role as pests, have led to population ecology being dominated by studies on insects. For similar reasons, weed studies are frequent in plant population ecology. The aim of predicting and reducing the amplitude of pest population fluctuations or lowering the mean level around which the fluctuations occur can lead to attempts to lessen pest effects through crop manipulation, rational insecticide use,

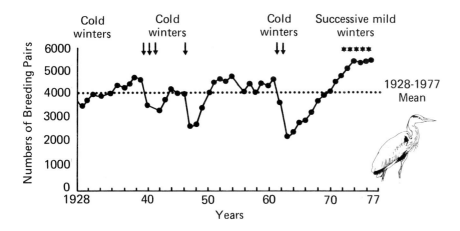

Figure 7.1: Grey Heron Population Levels in England and Wales, 1928-1977. Source: From British Trust for Ornithology Annual Census of Heronries.

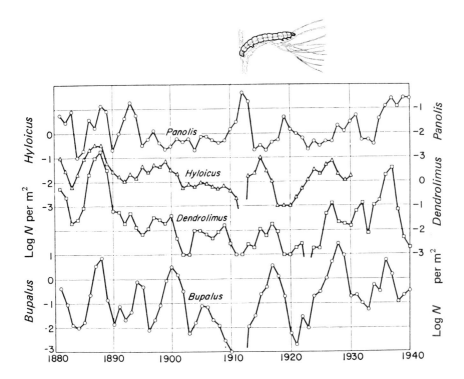

Figure 7.2: The Results of 60 Successive Census Counts of Moth Pupae or, in the Case of *Dendrolimus*, of Hibernating Larvae per m² of Pine Forest Floor at Letzlingen, Germany. Source: From Varley, Gradwell and Hassell, 1973.

maximisation of indigenous natural enemies' effects or the introduction of biological control agents. Population harvesting, too, needs basic information on population properties. The International Whaling Commission and some commercial fisheries use models of populations to set catch levels.

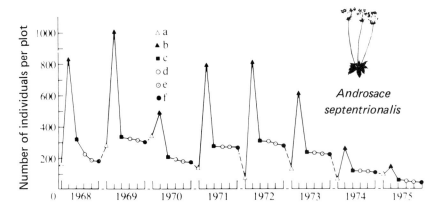

Figure 7.3: The Population Dynamics of *Androsace septentrionalis* Over an 8-Year Period. (a) Beginning of germination; (b) maximum germination; (c) end of seedling phase; (d) period of vegetative growth; (e) flowering; (f) fruiting. Source: From Symonides, 1979 and Silvertown, 1982.

One can approach the understanding of population dynamics in an inductive or deductive way; in the former, simple models of parts of the population process can be constructed, often with rather restrictive assumptions and omissions, in the hope that real populations, albeit often laboratory ones, may be better understood. A deductive approach begins with real field population events, often recorded over many generations, and through careful recording of births and deaths, followed by analysis and/or simulation, attempts to identify the main restraints and disturbing factors acting on the population. We will begin with the simple, almost classical, laboratory-orientated induction approach to single-species population growth; although it has obvious limitations, there are several reasons why it is not just of academic interest.

7.2 Theoretical Population Growth

We cannot discuss and develop a theoretical model for single-species population growth without classifying our organism ecologically in some way first. We need to treat organisms with overlapping generations, such as bacteria, protozoa, birds, mammals, most higher plants and some insects, differently from those with discrete generations. The latter include such insect groups as butterflies and moths in temperate regions, where all the caterpillars present at one time arise from a cohort of eggs laid by females more or less in phase and in the same generation; the next cohort of eggs to appear originates from new adults, the parents of which have by this time died or at least ceased ovipositing.

7.2a Overlapping Generations

Overlapping generations are dealt with using differential equations in which the rate of change of the number of organisms, dN/dt, is related to the number of organisms at time t, $N(t)$. The simplest (and most unrealistic) form which this relationship can take is:

$$\frac{dN}{dt} = r_m N \tag{6.1}$$

which, if the population at time t_0 is N_0, can be solved from:

$$N_t = N_0 \exp (r_m t) \tag{7.1}$$

r_m is called the intrinsic rate of natural increase (Birch, 1948) and is birth rate − death rate under fixed conditions; when it is positive, growth is exponential (geometric, Figure 7.4); when negative, an exponential decline in numbers occurs. In natural logarithms, this is written:

$$l_n N_t = \ln N_0 + r_m t \tag{7.2}$$

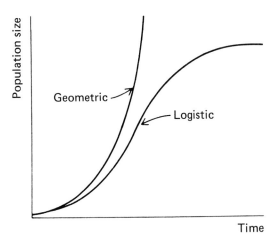

Figure 7.4: Population Growth. Geometric growth in an unlimited environment and logistic growth in a limited environment.

In logarithms to the base 10:

$$\log N_t = \log N_0 + r_m t \, (\log e) \tag{7.3}$$

Using equation (7.1) with e at 2.718, a population starting at 100 at t_0 and $r_m = 0.5$, $N_t = 100.e^{0.5} = 100 \times 1.6487 = 165$.

This type of growth in an unlimited environment, and its derivatives, is of interest for five main reasons, in spite of its simplicity. It is of *historical interest* because of the belief of Malthus in 1797 that populations (including that of man) increase in this geometric way (e.g. 2, 4, 8, 16, 32 etc.) while food supplies increase arithmetically (e.g. 2, 4, 6, 8, 10, 12, etc.). This spelt doom for mankind in Malthus' view and both Darwin and Wallace were influenced by this idea in their development of the theory of natural selection.

To calculate r_m we need information on reproductive delay, distribution of progeny in time, length of life etc. (Southwood, 1978). Having calculated it, we have a convenient theoretical way of *comparing organisms' reproductive potential*, either between species or for one species under different environmental conditions; this is the second advantage of the above simple approach to population growth. r_m has been used in this way by Dixon and Wratten (1971) to compare winged and wingless morphs of the black bean aphid, *Aphis fabae*, to show that the winged emigrant is at a reproductive disadvantage compared with the wingless adult. In studies of host-plant resistance to insect pests, r_m has also been used to rank plant and pest species. In a study of the performance of three aphid species on 22 vetch relatives of broad beans (*Vicia*), r_m ranged from 0 to 0.30, giving pointers for future resistant plant breeding programmes (Birch and Wratten, 1983). The assumptions about r_m are true only for short periods of time for most organisms because the environment available to them is not unlimited: it has a carrying capacity defining the maximum number of animals which can live there without destroying it. Darwin realised that animals' reproductive potential was rarely fully achieved; although we saw in Figure 6.1 that pheasant populations on Protection Island showed exponential growth for the first four years after introduction, this obviously cannot be continued indefinitely, because, as mentioned earlier, numbers eventually reach an approximate balance with their environment. What happens is that as a population approaches its carrying capacity (K), the actual rate of increase slows down, i.e. a negative feedback process operates whereby population growth rate responds to its own density. When the *rate* of population change decreases with increasing density (or *proportional* mortality increases) we call this process *density dependence*. This idea is of fundamental importance in population ecology and will be expanded later. The equation incorporating the effects of the environment's carrying capacity is:

$$\frac{dN}{dt} = r_m N \left(1 - \frac{N}{K}\right)$$
(6.2)

This can be re-written, in the integral form, as:

$$N = \frac{K}{1 + a \exp(-r_m t)}$$
(7.4)

where a is a constant defining the point of inflection of the curve which is produced (Figure 7.5). (For an example of this type of growth in a real population, see Figure 6.2.) This integral form of the equation is known as the logistic equation (or the Verlhurst-Pearl equation) and the curve it describes is the logistic curve. The way it *demonstrates density dependence* (by implication in this case through the action of intraspecific competition) is the third justification for this theoretical approach. K can vary through time, of course, and the extent to which a population follows it will depend on the population's response time $(\frac{1}{r_m})$ (Figure 7.6).

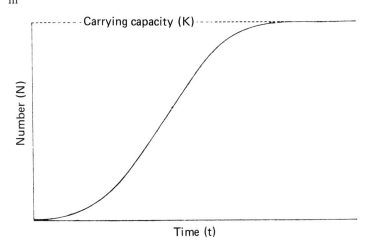

Figure 7.5: The 'Logistic Curve'.

The fourth reason for this approach is that the logistic equation, when applied to interacting populations, forms the basis of well-known models of *interspecific competition*. These have been dealt with in Chapter 6. The fifth use of the geometric logistic growth models is that they provide a framework for an *ecological classification of organisms* according to the main selection pressures to which they are thought to have been subjected during their evolution. This will be expanded in Chapter 10; we can simply say here that an organism which is 'r-selected' is an exploiter of short-duration habitats and has evolved to maximise r_m; staying below the carrying capacity (K) and the operation of density-dependent regulatory processes are of low importance in its strategy. A K-selected organism, however, has a relatively low reproductive rate, stays below its carrying capacity and density dependence is likely to be important in its population processes.

Equations 6.2 and 7.4 can be 'improved' in that they show the action of density dependence instantaneously. There is bound to be a delay in most real examples, attributable to a generation of the organism or to environmental recovery. A model which allows for this is:

$$dN/dt = r_m N \left[K - N(t-T)/K \right] \qquad (7.5)$$

where T is the time lag.

This equation can generate a range of population behaviours which vary in character according to the length of the feedback delay (T) in relation to the system's natural response time $1/r_m$ (Table 7.1).

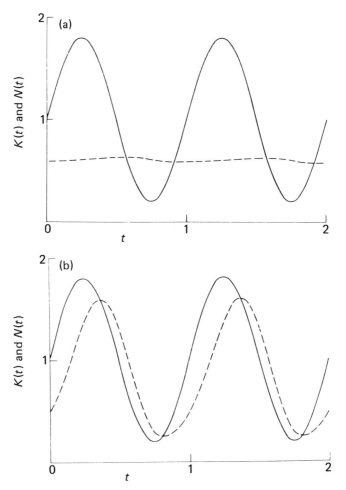

Figure 7.6: (a) Illustrating the Way a Population [N(t), Dashed Line] Varies if it Obeys a Logistic Equation with a Time-Dependent Carrying Capacity [K(t), Solid Line]. This figure is for the case when the population's natural response time (1/r_m) is long compared to the periodicity in the environment (T), i.e., for relatively small r_m. Specifically, the figure is for $r_m = 0.2$, $T = 1$, $K_0 = 1$, $K_1 = 0.8$; i.e., $rT = 0.2$. (b) As for (a) but in the opposite limit when the natural response time is short (relatively large r_m) so that the population tracks environmental variations. The details are as for (a) except that $r_m = 10$, whence $rT = 10$. Source: From May, 1981.

Table 7.1: Properties of Limit Cycle* Solutions of Equation 7.5

$\dfrac{1}{r}$	$N(\text{max})/N(\text{min})$	Cycle period/T
1.57, or less	1.00	—
1.6	2.56	4.03
1.7	5.76	4.09
1.8	11.6	4.18
1.9	22.2	4.29
2.0	42.3	4.40
2.1	84.1	4.54
2.2	178	4.71
2.3	408	4.90
2.4	1,040	5.11
2.5	2,930	5.36

*See Figure 13.1.
Source: From May, 1981.

There are situations when, perhaps surprisingly, apparently good visual fits to the geometric and logistic curves seem to occur in real populations. Some classical laboratory experiments seem to demonstrate this with stored products – insects or protozoa (Figures 6.1, 6.2, 6.5 and 6.6). However, graphical agreement does not necessarily mean that a particular equation is the only appropriate description of the data – there is evidence that population age-structure, for instance, may be an important factor in some early data sets but the logistic equation cannot deal with this (see Varley *et al.*, 1973).

Even where data seem to be well-fitted by the logistic model they are usually laboratory derived with constant food; field population graphs are a visual mess in comparison but the incorporation of more 'noise' does not necessarily invalidate the underlying processes. Those with time-delays can perhaps best be summarised as 'minimally realistic' (May, 1981) but we are still left with the situation of populations with non-overlapping generations.

7.2b Discrete Generations

Models for this type of population are often framed in difference equations. In these, time is a discrete variable and the general form is:

$$N_{t+1} = F(N_t) \tag{7.6}$$

The difference equation equivalent of equation (7.1) is:

$$N_{t+1} = \lambda N_t \tag{7.7}$$

so that the number at time $t = \lambda^t N_0$ (7.8)

λ (lambda) is termed the finite rate of increase and is the number of times the population multiplies itself/generation ($1n\lambda = r$ which is the difference equation equivalent to the intrinsic rate of increase in a continuously growing population). Equation (7.7) generates geometric growth, as does equation (7.1). There is a large range of more complex non-linear functions which can also turn equation (7.7) into one with density-dependent reduction in population growth rate. These functions may stay in the domain of single-species populations or may describe the effect of predators or parasitoids (Chapter 9). A whole range of single-species population growth patterns can be generated even by one of the many forms of equation (7.6) simply by varying the value of r and the starting population. Equation (7.9), for instance, can generate the range of patterns shown in Figure 13.1.

$$N_{t+1} = N \exp [r(1 - N/K)] \qquad (7.9)$$

These patterns are theoretical, however, so an attempt to fit field data to one or more of the forms of equation (7.5) may be of greater relevance. This was done by Hassell *et al.* (1976) and a graphical summary of the types of patterns generated is given in Figure 13.2. The implication is that natural populations show some stability while laboratory ones tend towards cyclic or chaotic behaviour.

The charitable view of the models discussed so far is that they describe the behaviour of populations uninfluenced by competitors, predators or by the vagaries of the abiotic environment; when we study real populations the usual absence of such patterns as shown in Figures 7.5 and 13.1 could be said to reflect the complication of the situation, rather than negating the explanation of the population's intrinsic behaviour. The uncharitable view is that these ideas are too simplistic and that as all populations interact with others within their community, it is only through a study of real field populations that we will approach a thorough knowledge of population behaviour. Whatever our view, once we begin to study a real population, with the complexities of age-structure, intra- and interspecific competition, predation and parasitism, migration etc., our methods must change. There seem to be four main approaches to the study of population processes and there is considerable debate about the value and limitations of each method. The approaches are: (1) the analytic (life table) and related methods; (2) the theoretical (usually difference equation model) approach; (3) the synthetic (simulation modelling) approach; and (4) survey and experimental manipulation of field populations. They all share the aim of identifying causes of variation in reproductive performance or of varying mortality with a view to explaining fluctuations and identifying factors, if any, which regulate, i.e. impose an upper and/or lower limit on a population through negative feedback.

In this chapter, we will deal with the two most contrasting methods, the analytical and synthetic approaches. The other methods will be dealt with in the framework of predation (Chapter 9).

7.3 The Analytic (Life Table) Approach

A life table is a population budget. In it, survival of a population through time is recorded and, in one type, causes of mortality of each stage are identified.

7.3a Time-specific life tables

The alternative names for this type of life table give a clue to its properties; they are sometimes called static, current or vertical life tables. This is because they deal with animal or plant populations in which generation overlap occurs and the data they contain are in essence a frequency distribution of the ages of individuals in a cross-section of the population taken at a specific time. On the assumptions that the population size is constant and that the age-structure is also constant, these tables give a general picture of survivorship with age. They are general, however, and although they show the pattern of survival and mortality with age, they do not enable the causes of mortality to be identified, nor the extent of density dependence or regulation to be quantified. They do not form the basis of more detailed population models but do have the advantage of enabling us broadly to classify organisms according to their reproductive and survival strategies (see Chapter 10). The construction and practical use of these tables are well described in Dempster (1975) and Wratten and Fry (1980) so an outline only will be given here.

We begin by deciding on what is known as the pivotal age: this is the age interval into which we segregate our population data. For man, this may be one, five or even 10 years, for insects, perhaps days or weeks while for bacteria it could be hours. It is traditional to begin with 1000 individuals so the raw data are corrected to this figure. We have the option of either ageing individuals in a sample of the living population or of collecting data from dead organisms. In practical terms, both methods depend on the acquisition of a random sample of organisms. The method based on collecting data from corpses, skulls etc. is particularly open to sampling bias. A hypothetical life table is given in Table 7.2 and the symbols in it are as below:

Table 7.2: A Hypothetical Time-specific Life Table

x	l_x	$\log l_x$	d_x	q_x	e_x	L_x	T_x
1	1000	3.0	550	550	1.21	725	1210
2	450	2.7	250	556	1.08	325	485
3	200	2.3	150	750	0.80	125	160
4	50	1.7	40	800	0.70	30	35
5	10	1.0	10	1000	0.50	5	5

x = pivotal age class
l_x = no./1000 surviving at the start of the age interval
d_x = no./1000 dying during the age interval
q_x = no. dying/1000 alive at the start of the age interval e.g. at the start of age class 2, $q_x = 250/450$ x 1000

L_X = the average no. of individuals between age x and age (x + 1) i.e.
$(l_X + l_X + 1)/2$

T_X = the total no. of individuals aged x and beyond i.e. all the mid-points added up from the bottom first

e_X = the life expectation, in units of x, for individuals attaining age x i.e. T_X/l_X

We can obtain a visual representation of survival by plotting l_x against x but it is usual to plot $\log_{10} l_x$ against x because this shows how survival *rate* changes. In the example in Table 7.3, for instance, although a different number of organisms die in successive age intervals, the fact that the same proportion dies in each one is shown by the fact that the logs decline by the same amount at each stage (0.3) and therefore that the log line is straight (Figure 7.7).

Table 7.3: Survivorship Data Showing a Constant Death Rate

l_x	$\log l_x$	d_x
1000	3.0	300
500	2.7	250
250	2.4	125
125	2.1	63
63	1.8	

Examples of each of the three major types of survivorship curve can be found among plant and animal populations but we have to be sure that the data were collected rigorously before we accept them completely; when we see a type 2 curve for instance, like that shown in Figure 7.8a for the lapwing, it suggests a constant survivorship. However, if the data were obtained from the ringing (banding) of free-flying birds, nestlings would be omitted; nestling mortality is high in most birds, so the true shape would probably be between types 2 and 3. The low early loss shown by a type 1 curve is typical of mammals (largely K-selected) which exhibit parental care; a true type 2 curve would represent organisms which die before they are physiologically old and weak while a type 3 curve typifies many *r*-selected organisms such as insects, fish, parasites, annual plants etc.

Sometimes when time-specific life tables are constructed from field-collected data, the curves are not smooth but may include one or more prominent inflexions. This normally means that the requirement of constant age structure has been broken but, when this does happen, extra ecological information is often obtained. For instance, Figure 7.9 shows a survivorship curve for mussels (*Mytilus edulis*) collected from an English rocky shore. The prominent 'lump' in the curve could be traced back to a year of high 'spat' settling on the rocky substrate; this high settling rate followed severe winter storms which provided settling space by removing many mussels from the colony.

Apart from the disadvantages of the average statement about survival which time-specific life tables provide, accurate ageing of individual plants or animals

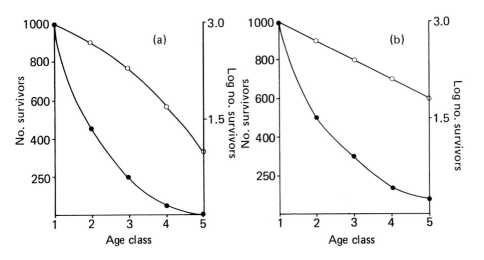

Figure 7.7: Survivorship Curves on Arithmetical and Log Axes for the Data in (a) Table 7.2 and (b) Table 7.3.

is often difficult or at best, tedious. Even when it is possible, it is frequently the case that this reveals an age structure which is not constant, possibly coupled with a non-static population growth rate. These properties will lead to interesting ecological information in their own right but time-specific life tables cannot deal with them.

Where different age-classes can readily be recognised within a population and each age-class may have characteristic and different rates of fecundity and mortality, it is clearly desirable to preserve such detail of information, and make use of it in our model for population analysis, rather than merely submerging these age-differences in a 'population average'. Indeed, as we have noted, unless the age-structure of the population is immutable, no such constant population average can be calculated for fecundity or survivorship. In such a context it is possible to analyse the dynamics of the population using matrix algebra. The approach is still essentially one of life table analysis, but permits some added sophistication, adapting to populations with changing age-structure and incorporating age-specific values of fecundity and mortality.

The use of matrix algebra for population analysis in this way was pioneered by Leslie (1945, 1948). The concept is a simple one. In any closed population, or a population sufficiently large that immigration and emigration are insignificant in their contribution to population change, the only factors affecting population size and structure are birth and death. If we thus divide a population up into a number of distinct age-classes, we can define the dynamics of that population in terms of a series of fecundities and survivorships ascribed to each of these classes. These may be conveniently summarised in a matrix (Table 7.4).

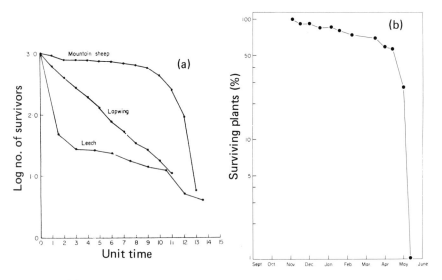

Figure 7.8: (a) Survivorship Curves for the Mountain Sheep (Murie, 1944), the Lapwing (Lack, 1943) and the Leech, *Glossiphonia* (Mann, 1957). (From Dempster, 1975). (b) Survivorship Curve for the Chickweed (*Cerastium atrovirens*) on Fixed Dunes at Aberffraw, North Wales. Source: From Mack, 1976 and Silvertown, 1982.

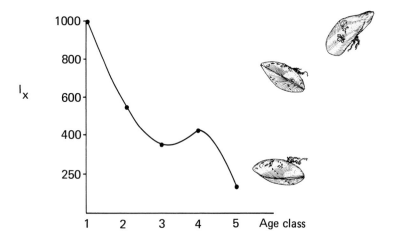

Figure 7.9: A Survivorship Curve for a Population of Mussels (*Mytilus edulis*) Which did not Exhibit a Constant Age Structure.

Since all new-born animals are recruited into age-class 0, the fecundities of all age-classes of the population, $(f_0, f_1, f_2, f_3, f_4 \ldots f_n)$, are entered against the input into age class 0 at time $t + 1$ (row 0).

By definition however, survival is measured as a probability of surviving from one age class to the next. Probabilities of survival $(P_0, P_1, P_2 \ldots P_n)$ are thus

Table 7.4: The Structure of a Population Matrix

Age classes at time t					Age classes at time $t + 1$
0	1	2	3	4	
f_0	f_1	f_2	f_3	f_4	0
P_0					1
	P_1				2
		P_2			3
			P_3		4

entered in the matrix as a diagonal: P_0 represents the probability of surviving from age-class 0 at time t, to age-class 1 at time $t + 1$, P_1 survival from age-class 1 to 2 and so on.

Actual population structure at any one time (t) (the number of organisms in a given age-class) may be represented by a vector

$$N_0$$
$$N_1$$
$$N_2$$
$$N_3$$
$$N_4$$

In biological terms, succeeding generations result from the action of the population parameters of fecundity and survivorship (summarised in the matrix) upon the existing population as represented in the vector. Mathematically, the population at time $t + 1$ results from multiplication of the column vector through the Leslie matrix. Therefore:

Population at time $t + 1$

Age class $0 = N_0^t$ (at time t) $f_0 + N_1^t f_1 + N_2^t f_2$

$$1 = N_0^t \, P_0$$
$$2 = N_1^t \, P_1$$
$$3 = N_2^t \, P_2$$
$$4 = N_3^t \, P_3$$

In our analysis here, the population has been divided up in terms of age-classes. In fact the different categories recognised in the analysis need not necessarily be age-related: they must be defined merely so that whatever is contained within the class shares the same fecundity and survivorship rate. The model can thus be refined to consider separate sexes within different age-classes (permitting recognition of differential production of males or females at birth, or differential post-natal survival of different sexes) or to accommodate other subtleties.

Fecundities and survivorships may be entered as variable functions, rather than constants, so that some form of density-dependent change may be built in to the model. In addition it has been suggested (Pollard, 1973) that it is also possible to include a mechanism to account for the population effects of immigration and emigration as well as birth and death.

Leslie matrix models thus have considerable flexibility. They can accommodate populations with changeable age-structure and they can preserve age-specific differences in fecundity and mortality of the species studied. To date, such matrixes have been applied primarily to analysis of vertebrate populations (although there is no reason why they should not be adopted for other organisms). Thus Pollard (1973) has modelled changes in human populations with Leslie matrix techniques; Beddington (1974) has used Leslie matrix algebra to analyse the effects of different harvesting regimes on red deer populations, and Putman (1983a) has also developed models to assess the effects of exploitation on fallow deer. Indeed in most cases where Leslie matrix methods have been used, it has been in this latter context, in applied analysis of the effects on populations of exploitation or harvest. Leslie matrix algebra has been employed in this context most notably amongst whales and marine fish.

Although matrix algebra can deal with overlapping or non-overlapping populations, there remains another method (restricted to populations with discrete generations) which has been used very frequently in recent years; it is the construction of age-specific life tables and their subsequent analysis (Section 7.3b).

7.3b Age-specific Life Tables and k-factor Analysis

If we envisage a population of a uni-voltine egg-laying insect such as a butterfly in a temperate climate, the application of the time-specific life table approach can easily be seen to be impossible. Such an insect may lay eggs over a relatively short period in the spring; these hatch into larvae which may moult three or four times as they grow, and then pupate. A 'cross-section' of the population during the larval stages, for instance, cannot give a complete frequency distribution for individuals' ages as eggs, pupae and adults are not present. The approach of an age-specific life table is to follow a cohort of individuals, or samples from them, from the ovipositing adults to the next generation of adults, recording the density of each stage and, wherever possible, the extent and causes of losses. An egg cohort is not a batch of eggs but rather all the eggs laid by the females of a particular generation. In practice, important consideration of sampling theory must be applied at this stage, but the reader is referred to Elliott (1977), Southwood (1978) and Wratten and Fry (1980) for the practical details.

Of the many published age-specific life tables, that of Harcourt (1971) is summarised in Table 7.5. This author worked on the Colorado beetle (*Leptinotarsa decemlineata*) in Ontario, Canada. Here it has one generation/year, passing

Table 7.5: Six Years' Life Table Data for the Colorado Beetle

log D	1.			4.	
4.07	0.10	k_1	Not deposited	3.79	0.14
3.97	0.02	k_2	Infertile	3.65	0.03
3.95	0.02	k_3	Rainfall	3.62	0.00
3.93	0.07	k_4	Cannibalism	3.62	0.12
3.86	0.02	k_5	Predation	3.50	0.01
3.84	0.00	k_6	Rainfall	3.49	0.08
3.84	0.34	k_7	Starvation	3.41	0.10 k_7
3.50	0.00	k_8	Parasites	3.31	0.02
3.50	−0.02	k_9	Summer adults	3.29	0.01
3.52	2.32	k_{10}	Emigration	3.28	1.42
1.20	0.05	k_{11}	Frost	1.86	0.52
1.15	2.92	K		1.34	2.45 K

	2.			5.	
3.86	0.06			4.05	0.08
3.80	0.02			3.97	0.03
3.78	0.02			3.94	0.00
3.76	0.08			3.94	0.10
3.68	0.05			3.84	0.02
3.63	0.09			3.82	0.10
3.54	0.15	k_7		3.72	0.19 k_7
3.39	0.01			3.53	0.02
3.38	0.00			3.51	−0.01
3.38	2.20			3.52	3.04
1.18	0.14			0.48	0.14
1.04	2.82	K		0.34	3.71 K

	3.			6.	
3.75	0.08			3.05	0.13
3.67	0.03			2.92	0.04
3.64	0.00			2.88	0.00
3.64	0.09			2.88	0.13
3.55	0.06			2.75	0.06
3.49	0.20			2.69	0.01
3.29	0.09	k_7		2.68	0.01 k_7
3.20	0.02			2.68	0.06
3.18	−0.03			2.62	0.00
3.21	1.95			2.62	0.00
1.26	0.18			2.62	0.07
1.08	2.67	K		2.55	0.51 K

Source: From Harcourt, 1971.

the winter as hibernating adults in the soil; these emerge in the spring and lay their eggs on the potato foliage. The four larval instars feed near the top of the plant and, when fully grown, pupate in cells in the soil. Summer adults emerge from these cells, feed for a month and then hibernate. The usual questions which work of this type poses are:

(1) What are the main causes of fluctuations in the animal's numbers, i.e. disturbing factors?

(2) What factors, if any, act in a density-dependent way, with the potential to regulate?

A third question, sometimes asked, is whether the main population processes operating on the species differ between area.

Harcourt sampled the developmental stages of the beetle on a regular basis and the data in Table 7.5 show the decline in numbers through the generation based on the total numbers on 96 'hills' on each of which grew a cluster of stems. A cursory glance at most life tables usually gives a good idea of where most mortality occurs, especially if each reduction in numbers is expressed as a percentage reduction in the initial egg number. In Table 7.5, it is clear that in five of the six years of the study, very large losses from the population took place through emigration of the adults before autumn hibernation. This is not strictly a mortality, as their fate is unknown, but as far as the study population is concerned, this emigration is a major loss. Identifying the largest percentage is a useful preliminary but we need to carry out further analysis in order to be able to ask questions more relevant to the dynamics of the study population. In many studies, k-factor analysis (Varley and Gradwell, 1960) is the next stage. This approach begins by converting the successive mortalities into 'log killing powers'. This is simply the subtraction of the log numbers of the organism after the mortality acts from the log numbers before it acts. In Table 7.5, the first mortality in the first generation studied is 'eggs not deposited'; this is the extent to which females did not lay their full egg complement because of low temperatures and bad weather. The k-value for this can be calculated as:

$$\log 11{,}799 - \log 9{,}268 = 4.07 - 3.97 = 0.10$$

$$\text{or} \quad \log \frac{11{,}799}{9{,}268} - \log 1.27 = 0.10$$

The main advantages of expressing mortalities in this way are (1) the successive mortalities are easily calculated, and can be added up to give total generation mortality (K); if we had expressed the loss of eggs as a percentage of the preceding number, addition of these percentages would have had no meaning; (2) the k-value is a proportional measure, because the greater the proportional loss of organisms between sampling occasions, the higher the k-value. The numbers dying may vary markedly but if the proportional mortality is unchanged, so is the k-value,

$$\text{i.e.} \quad \log \frac{200}{100} = 0.3 = 50 \text{ per cent death}$$

$$\log \frac{700}{350} = 0.3 = 50 \text{ per cent death}$$

As one of the aims of life-table construction and analysis is the detection of density dependence (a proportional increase in mortality with density), the fact that k-values are proportional measures is obviously an advantage. Having calculated the k-values, the next stage is often to compare visually the way individual k-values change between generations with changes in overall genera-tion mortality. If one k-value clearly parallels changes in K, then we can call this the key factor causing population change. The Colorado beetle key factor is clearly k_{10} (adult emigration), not because it is the largest mortality but because it most closely follows changes in K (Figure 7.10). Therefore, with the limitation that we have only six generations' data, it appears that emigration of adults before autumn hibernation is the main factor causing between-generation changes in the beetles' numbers. There are three situations, however, when visual comparison may not be sufficient. These situations are: (1) one clear key factor may not be obvious; (2) we may wish to compare the effects of the same k-value in different places; (3) we may wish to rank the mortalities in order of their importance in contributing to population change.

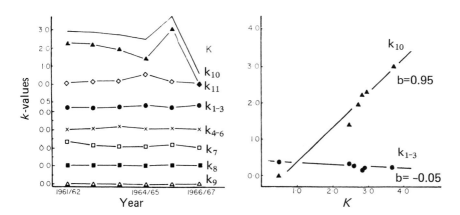

Figure 7.10: k-factor Analysis of the Mortalities Acting on the Colorado Beetle, *Leptino-tarsa*. k_{10}= adult emigration. Source: Data from Harcourt, 1971; after Podoler and Rogers, 1975.

An example of a published k-factor analysis which did not yield an obvious key factor through visual comparison of trends in k-values is the study of a grassland bug, *Leptoterna dolobrata*, (McNeill, 1973) (Figure 7.11). The relative importance of the k-values in causing population change is also difficult to determine in this case. An important pair of studies which enable us to compare one species in two places are those of Embree (1966) and Varley and Gradwell (1968) on the winter moth (*Operophtera brumata*) in Canada and England, respectively. This comparison has important implications for the theoretical basis of biological control (see Chapter 9) but in the context of k-factor analysis, it is not easy to compare visually the role of k_1 (disappearance of eggs and young larvae) in two places (Figure 7.12a, b).

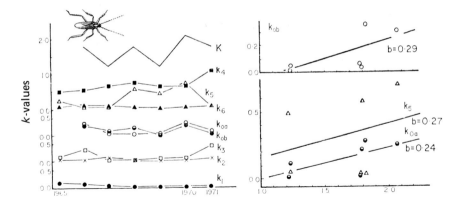

Figure 7.11: *k*-factor Analysis of the Mortalities Acting on the Grass Bug *Leptoterna*. Source: From McNeill, 1973; after Podoler and Rogers, 1975.

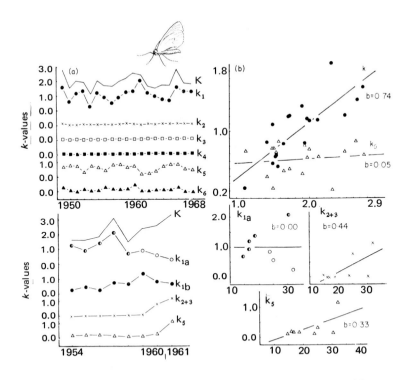

Figure 7.12: *k*-factor Analysis of the Mortalities Acting on the Winter Moth (*Operophtera*). Upper: England. Lower: Canada. k, = winter disappearance. k_{1a} = mortality of eggs and young larvae before (●) and after (○) introduction of parasites. Source: From Varley and Gradwell, 1968 and Embree, 1966; after Podoler and Rogers, 1975.

In these situations, a useful method, proposed by Podoler and Rogers (1975), is to plot each k-value in turn against total generation mortality (k) and then to calculate the regression coefficient. The mortality which gives the slope (b) with the highest vlaue is by definition the key factor; the maximum value for the relationship is around 1.0. When this is done for the Colorado beetle data in Figure 7.10, it confirms the dominant role of k_{10} which was already visually obvious. However, in the case of the grass bug, it is apparent that the absence of a single obvious key factor is confirmed but we can see that three of the eight mortalities contribute significantly to population change; in this case, k_{oa} is variation in potential fecundity, k_{ob} is failure to lay the full egg complement and k_s is loss of late-instar nymphs other than mortality through parasitism. This regression approach also helps to clarify the complicated winter moth example. In Canada, the key factor was k_{2+3} (larval mortality from introduced parasites), a slope of 0.44, while in England the key factor was k_1, the disappearance of eggs and young larvae (b = 0.74). There was a suggestion, however, that before the introduction of the parasites from Europe, the same key factor as that in England was operating (Figure 7.12b - filled circles).

In these examples, we should be reluctant to call any one mortality the key factor if its slope is not markedly different from the rest, or if all slopes are very low. Sometimes, however, it is acceptable to identify two key factors as in Figure 7.13, if their slopes are reasonably steep and they differ little.

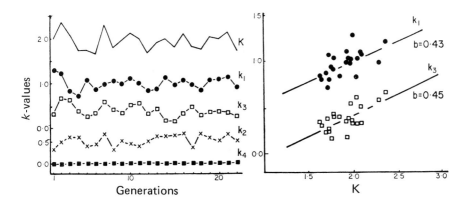

Figure 7.13: k-factor Analysis of the Mortalities Acting on the Mediterranean Flour Moth (*Anagasta*). Source: From Hassell and Huffaker, 1969; after Podoler and Rogers, 1975.

In a statistical sense, the above regression technique is 'illegal' because two requirements of regression are that the axes should be independent and the independent variable free from error. These conditions are not obeyed here so we should restrict our use of these graphs to the identification, more clearly than in the visual comparison method, of the mortality(ies) which contribute most to changes in K. We will return to the restrictions of regression analysis later in assessing density dependence in k-factor analysis.

The second major question concerning the dynamics of a study population remains: do any factors act in a density-dependent way and do they show regulatory potential? We can use the same examples as before but this time we need to evaluate responses to density rather than correlations with overall generation mortality. As a first step, we plot each k-value against the log of the density on which it acts; in Table 7.5 the six values of k_7 (larval starvation) for example, are plotted against the log of the density before starvation begins; 0.34 from generation one is plotted against 3.84, 0.15 from generation two is plotted against 3.54 and so on. Any positive relationship suggests that the mortality is acting in a density-dependent way, as is suggested by Figure 7.14.

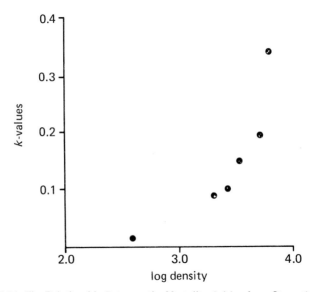

Figure 7.14: The Relationship Between the Mortality Arising from Starvation in Colorado Beetle Larvae and log Larval Density (Table 7.5) Source: data from Harcourt, 1971.

Before we can assess the contribution to population regulation, if any, of our suspected density-dependent factor, statistical problems remain, similar to those in the use of regression in detecting the key factor. As each k-value is often calculated from the log of the density on which it acts, the two measures are not statistically independent so regression analysis, with associated significance tests, is not valid. Also, simple regression analysis assumes linearity, which may not always apply, and that the independent variable is free from error. As the density values are based on population samples around which there is (quantifiable) variation, there is error around the independent variable. The best way of overcoming these problems is the 'two-way regression test' of Varley and Gradwell (1968). If both the regression of log initial density on log final density ($\log N_t$ on $\log N_{t+1}$) and the reverse give slopes which are significantly diffeent from b = 1 and are on the same side of this line the existence of density dependence may be accepted. In the case of the Colorado beetle, the fact that k_7 was in fact density dependent seems to be an enigma, because k_7 is the key factor. We saw that key factors tend to disturb populations (cause fluctuations) while density-dependent factors have the potential to regulate. Here we have the apparent paradox of a density-dependent key factor. So, having satisfied the statistical criteria for the detection of density dependence, we are left with the more difficult problem of assessing the relevance of any density-dependent factors which we identify in the regulation of the numbers of the study organism. In the process, we should be able to clarify the apparent inconsistency of a density-dependent key factor. It is obvious that a mortality which increases its proportional effects slowly in relation to population growth rate is unlikely to have a regulatory effect. A prey population increasing between generations as follows: 10, 50, 250, 1,250, 6,500 etc., is unlikely to be influenced by the following consumption/generation by a predator: 0 (0 per cent), 3 (6 per cent), 18 (7 per cent), 95 (8 per cent), 600 (9 per cent), even though the mortality is acting in a density-dependent way and may have been demonstrated as such in a k-factor analysis. We would call such a result weak density dependence and it would contribute little to the restriction of population fluctuations. Sometimes, however, a density-dependent factor can be too strong to regulate a population disturbance, inducing a tendency for the population to fluctuate away from its mean level rather than to return to it (regulation); the density-dependent key factor is therefore in the former, over-compensating, category. To see the mechanics of such a process, we need to incorporate the results of k-factor analysis in a simple model. We will begin with a simple, hypothetical model and then see how the process is extended for a real life table and its analysis. Here and in the next two chapters it will be helpful to refer to Figure 7.15 which shows a range of graphical ways that mortality can be related to density.

In Figure 7.16, a range of hypothetical linear density-dependent responses is shown. Without a knowledge of the fecundity of, and other mortality agents acting on, the study organism, we are not able to say which, if any, are too weak or too strong to effect regulation, apart from line B. This relationship, with a slope of one, and, if acting as the only density-dependent factor, would bring about perfect compensation regulation (see below). We need, therefore, to incorporate density-dependent responses of different strengths into a population model. The model below is crudely simplistic but makes the point clearly. We will take a population with a fixed growth rate/generation of 5 (i.e. λ of 5 in equation 7.7). With no mortality acting on it, this would increase geometrically. We can easily calculate the consequences for stability or fluctuations of the incorporation of the density-dependent relationships of the different 'strengths' in Figure 7.16 in our model. As the k-values are in logs, our calculations in the model can be in logs too, which simplifies the arithmetic; log 5 = 0.7. Note that there is a threshold density in Figure 7.16 below which no density-dependent mortality occurs; this is realistic in that predation or mortality from intraspecific competition, for example, would not be expected to operate at very low densities.

Figure 7.15: Types of Density Relationships and Different Graphical Representations

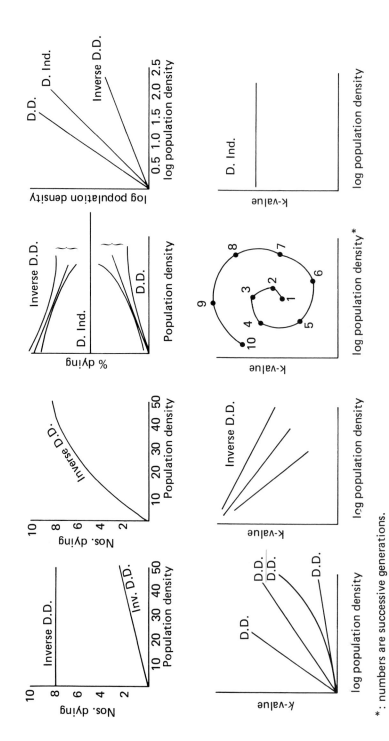

*: numbers are successive generations.

D.D. = Density dependent, D. Ind. = Density independent.

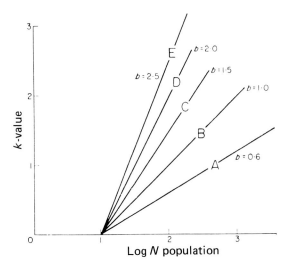

Figure 7.16: A Range of Linear Density-Dependent Relationships of Different Strengths (Source: From Varley *et al.*, 1973)

The basic formula for the numbers (N) in generation 2 is:

$N_2 = (N_1 -$ appropriate density-dependent mortality$) \times 5$

In logs, $\log N_2 = \log N_1 - k + 0.7$ (7.10)

If we start with a population of two organisms and use line B, the calculation for generation 2 is:

$\log N_2 = \log 2 - 0 + 0.7 = 0.3 + 0.7 = 1.0$ (log 10)   (7.11)

We calculate or read off the appropriate k-value from Figure 7.16; in this case, $k = 0$. Generation 2 therefore has a population of ten. These organisms also multiply without restraint because no density-dependent mortality operates at this level either, so:

$\log N_3 = \log 10 - 0 + 0.7 = 1.7$ (log 50)   (7.12)

At this point, density-dependent mortality does enter the calculation, and we have:

$\log N_4 = \log 50 - 0.7 + 0.7 = 1.7 - 0.7 + 0.7 = 1.7$ (log 50)   (7.13)

In this simple way, we have achieved perfect population stability by the third generation (similar to line B in Figure 7.17); mortality exactly *compensates* for reproduction. This result would occur with any value of λ and any starting population, although the level at which the stable population occurred would differ.

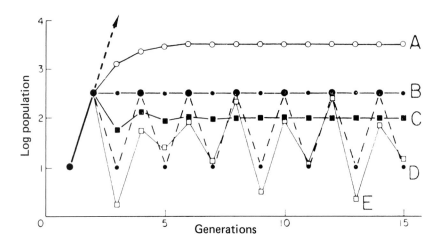

Figure 7.17: The consequences of the Curves A-E in Figure 7.16 on a Population Which has the Initial Size of 10 and in Which log F = 1.5.
A, gradual attainment of stability; B, attainment of stability in one generation; C, dampened alternations leading to stability; D, alternations with alternate generations of equal size, and E, irregular sized alternations. (Source: From Varley *et al.*, 1973)

With the same starting number (2) and the same value for λ (5), the consequences of the other density-dependent relationships in Figure 7.16 are more complex. If we take a 'strong' relationship (line E), we can produce a population graph with violent fluctuations i.e.

$$\log N_2 = \log 2 - k + 0.7 = 1.0 \tag{7.11}$$

$$\log N_3 = \log 10 - k + 0.7 = 1.7 \tag{7.12}$$

$$\log N_4 = \log 50 - 1.6 + 0.7 = 0.8 \text{ etc.} \tag{7.14}$$

By including a sufficiently strong density-dependent relationship we can even cause our population to fall to extinction; other combinations can give stable oscillations or oscillations of declining amplitude. It is now clear how a key factor, such as that operating on the Colorado beetle population, can be strongly density dependent; it could be weakly density dependent too although in most published life table analyses there is no significant relationship between the key factor *k*-value and the log of the density on which it acts; a truly random mortality with respect to the population density which it affects. Note also that in calculating the population graphs in Figure 7.17, those relationships in which stability was brought about also demonstrate how a regulatory agent reduces its proportional effects as density declines. In this way, it can contribute to the avoidance of extinction as well as to the avoidance of over population. Figure 7.17 used F = 32 however.

Table 7.6: One Year's Life Table for the Winter Moth

	Percentage of previous stage killed	Number killed per m^2	No. alive per m^2	Log no. alive per m^2	*k*-value
Adult stage ♀♀ climbing — year x			4.39		
Egg stage ♀♀ x 150			658	2.82	
Larval stage					0.84=k_1
Full grown larvae	85.3	561.6	**96.4**	1.98	0.03=k_2
Attacked by *Cyzenis*	6.7	**6.2**	90.2	1.95	0.01=k_3
Attacked by other parasites	2.3	**2.6**	87.6	1.94	
Infected by *Microsporidian*	4.5	**4.6**	83.0	1.92	0.02=k_4
Pupal stage					0.47=k_5
Killed by predators	66.1	**54.6**	28.4	1.45	
Killed by *Cratichneumon*	46.3	**13.4**	15.0	1.18	0.27=k_6
Adult stage ♀♀ climbing trees — year x+1			7.5		

The figures in heavy type are those actually measured. The rest of the life table is derived from these. Source: Modified from Varley *et al.*, 1973.

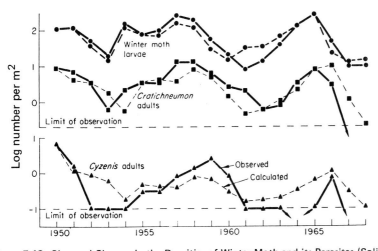

Figure 7.18: Observed Changes in the Densities of Winter Moth and its Parasites (Solid Lines) Plotted Beside the Densities of These Species as Calculated by a Mathematical Model (Broken Lines). Source: From Varley *et al.*, 1973.

A model arising from a k-factor analysis of a real population is obviously more complicated than that above, but the principles of its construction are the same. A well-publicised model based on a k-factor analysis is that of the winter moth *Operophtera brumata*. Details can be found in Varley *et al.* (1973) but its essence is summarised in Table 7.6 and Figure 7.18. Agreement between prediction and observed events is very good, which is a vindication of this particular k-factor analysis. Models based on k-factor analyses can go wrong, however, as we will see when we return to the winter moth in the predation chapter (Chapter 9). To understand why this may happen, the limitations of this approach to population ecology need to be discussed.

7.3c Pros and Cons of k-*Factor Analysis and Resulting Models*

The apparent clarity with which k-factor analysis identifies and quantifies the effects of mortality agents in population processes is belied by the fact that many published studies of animal and plant populations are not based on this method. We have seen one major restriction on its use, that of the requirement for a population with discrete generations, but a cursory examination of modern ecological journals will reveal many population studies which do not use k-factor techniques including some of organisms with non-overlapping generations. Why is this and what are the alternatives?

k-factor analysis is based on age-specific life tables and no matter how carefully an age-specific life table is constructed, the following problems are likely to remain to some degree.

(1) Some mortalities overlap. k-factor analysis assumes that mortalities act successively. When, for example, an insect is dissected or reared in the laboratory to detect the presence of a parasite, we have no way of knowing whether that parasite would actually have killed the larva or whether, instead, a predator would have consumed the larva and the parasite it contained. Mortality attributable to predation is likely to be underestimated in these circumstances and that for parasitism may be overestimated. A similar situation could arise in a plant life table in which seeds may be killed by fungus (which could be shown by laboratory storage) and/or eaten by grazers.

(2) The true succession of mortalities may not be known. Although it is obvious that mortality of early developmental stages must act before that of late ones, problems may arise when a series of different mortalities acting on one developmental stage have to be ordered, without accurate knowledge of the order in which they act in the field. A very thorough k-factor analysis is that carried out by Dempster (in Dempster, 1975) on the cinnabar moth (*Tyria jacobaeae*) feeding on ragwort (*Senecio jacobaea*) in a heathland habitat. The mortalities acting on the moth's larvae were predation, starvation, parasitism, and in Dempster's life table they were taken as operating in succession in the above order. Dempster states, however, that 'This (order) is not strictly true, since although starvation

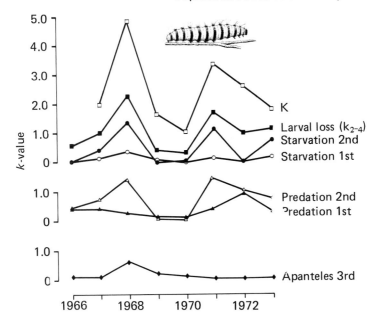

Figure 7.19: The Impact of *k*-Factors for Starvation, Predation and Parasitism, on Larval Mortality in the Cinnabar Moth. Dempster's order of the action of larval mortalities was starvation (2), predation (1) and parasitism (3). The graph shows the new *k*-values for starvation and predation when their order of action is reversed. □, K; ■, larval loss; ○, starvation 1st; ●, starvation 2nd; ▲, predation 1st; △, predation 2nd; ◆, parasitism. Source: Modified from Dempster, 1975.

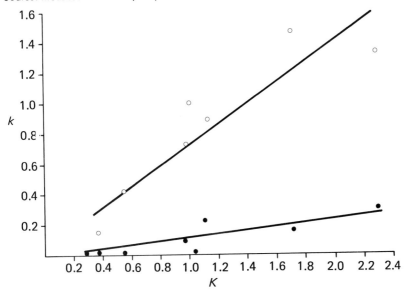

Figure 7.20: Podoler and Rogers' (1975) Test for the Key Factor in the Recalculation of *k*-values for Cinnabar Moth Larval Starvation (●) and Predation (○). Mortality order changed in data from Dempster (1971). Lines fitted by eye.

affects mainly older instars, it can cause high death rates among very young larvae'. In Figure 7.19, larval k-values have been calculated for the above mortalities in Dempster's order and then for predation and starvation in reverse order. Parasitism is kept as the third mortality in the series, so its k-values are unchanged. However, predation and starvation k-values change up to seven-fold. More importantly, the visual comparison of the k-values with generation K takes on a different pattern (Figure 7.19) and the regression of k on K no longer suggests that starvation is the key factor but that predation is (Figure 7.20). There are probably good phenological and ecological reasons why such a crude reversal of cinnabar moth mortalities may be unfair, but Figures 7.19 and 7.20 do show that there is a danger that the major ecological conclusions of k-factor analysis may, under some circumstances, be methodological artefacts. The well-known life table for the winter moth, constructed by Varley and Gradwell (1968) also makes similar assumptions about the order of mortalities. It assumes that parasitism acts after predation. Varley, Gradwell and Hassell (1973) state that, if this order is reversed, '. . . in most population models (derived from the life table) these altered assumptions produce identical effects'. As the aim of many life tables is to produce a population model of the study organism, as we saw on page 173, a check on the implications of changing the order of mortalities, where there is some doubt about the order, would be useful. What Varley, Gradwell and Hassell mean by 'most population models', however, is not clear; what did the others predict . . .?

(3) The mortality categories may be so wide that they contain and therefore hide others. The key factor in the winter moth life table was a mortality called 'winter disappearance' (k_1). This was a very wide mortality category operating between the count of wingless females in midwinter and the count of fully grown larvae in the following May. Dempster identified 11 mortalities for the equivalent part of the cinnabar moth's life cycle, for instance; they were:

> variations in adult fecundity
> adult mortality and dispersal
> egg infertility
> egg failure to hatch
> other causes of egg mortality
> starvation of larval instars I and II
> predation + unknown losses of instars I and II
> starvation of instars III and IV
> predation etc. of instars III and IV
> parasitism of larval instar V
> starvation etc. of larval instar V.

Varley, Gradwell and Hassell (1973) discounted (without presentation of the relevant data) predation on adult females, egg mortality and larval mortality. They considered that the larval mortality that did occur was mainly a result of young larvae dispersing from those trees whose buds opened late and therefore provided no food to the newly hatched caterpillars. There remains the possibility, however, that k_1 may have included other mortalities which, for all we know, may have exhibited strong relationships with K or with egg density, the latter possibility indicating a hitherto unsuspected density-dependent factor.

(4) Once the study has begun, incorporation of newly recognised mortalities is usually impossible. A parasite or pathogen may remain undetected for the first few studied generations, only to be recognised once the study is under way. Retrospective incorporation of its earlier effects is impossible so the approach depends heavily on a sound knowledge of the organism's life history, and mortality agents, at its inception.

(5) The relationship between k and K, or k and log density, may not be linear. The two-way regression test for density dependence and Podoler and Rogers' graphical test to identify the key factor both assume linearity in the data. Figure 7.21 shows a plot of k on K which looks markedly curvilinear; as a result, this mortality gave a low linear regression coefficient and another mortality with a higher b-value was identified as the key factor. A type of mortality which often shows curvilinearity when plotted as a k-value against the log of the density on which it acts is that resulting from intraspecific competition (see Chapter 8). In Figure 7.22, intraspecific competition for food leading to larval starvation and death gives a clear curvilinear response to larval density. This is because there is a threshold density of approx. 1000 larvae/kg food plant below which significant competition does not occur. The two-way regression test carried out on the data in Figure 7.22 did not demonstrate density dependence but this test's dependence on linearity creates doubt about whether or not density dependence really did operate in this case.

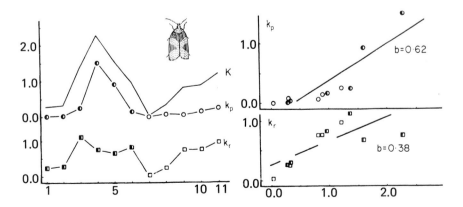

Figure 7.21: k-factor Analysis of the Mortalities Acting on the Black-headed Budworm, *Acleris*. Source: From Morris, 1959; after Podoler and Rogers, 1975.

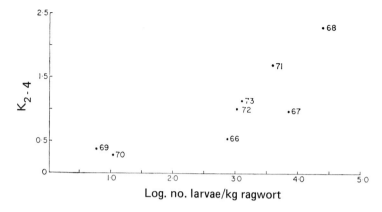

Figure 7.22: The Relationship Between Larval Mortality (k_{2-4}) and Larval Density in the Cinnabar Moth. Source: From Dempster, 1975.

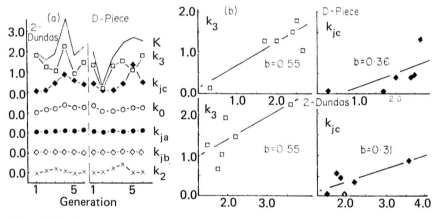

Figure 7.23: k-factor Analysis of the Mortalities Acting on the West Indian Cane Fly, *Saccarosydne*, in Two Areas. Source: From Metcalfe, 1972; after Podoler and Rogers, 1975.

(6) A large number of generations must be studied for reliability. As k-factor analysis depends heavily on analyses in which each point on the graph represents a generation's data, reliability demands adequate data, especially as in the plots of k on K, orthodox statistical tests cannot be used (see p. 172). There are examples in which established relationships seem rather tenuous; in Figure 7.23, there are large gaps in the density data and sometimes only one point seems to confer linearity on the relationship, so again detailed ecological conclusions may not be merited from the data as they stand.

(7) Major population processes may change during the course of a long-term study. As plots of k on log density or k on K are based on the complete study, which in the case of the winter moth was many years, they are in a way an 'average' description of the relationships and in this sense remind us of the averaging process involved in the construction of time-specific life tables. However, it does not follow that the same key or density-dependent factors will operate throughout a study; they may change between generations within a year or different factors may operate at different mean population levels. Most k-factor analyses cannot detect such changes and the next section, concerning the simulation of population events, contains an example where such changes did occur in a population during its study.

(8) The data are always incomplete because mortalities acting on the main mortality agents are not quantified. This argument was advanced by Gilbert *et al.* (1977) about Dempster's cinnabar moth life table. They pointed out that the work '... cannot be used to illustrate the parasite-host (i.e. parasite-moth) relationship because the level of hyperparasitism was not recorded'. The logical extension of this argument, however, is that the mortality factors acting on all biotic mortality agents of the study organism must be quantified. As the aim of most life tables is deliberately much more circumscribed than this, the argument is, in most cases, irrelevant.

For the above reasons, as well as its reliance on organisms with discrete generations, there are many population studies in which k-factor analysis is not used. A completely different approach, of increasing use in the last decade, is that of simulation modelling.

7.4 Simulation of Population Events

This synthetic approach to the study of population dynamics involves building up a picture of population changes by assembling separately studied component processes in a detailed model. This model's predictions are then compared with field events. They may not match well at first so further experimentation or changes in the ecological assumptions are made in an effort to fit the model better to reality. Once the match is good and the model has been satisfactorily validated conclusions can begin to be drawn about the roles of the different processes, just as was done following the calculation of k-values from an age-specific life table.

We will use as a case study the lime aphid (*Eucallipterus tiliae*) system modelled by Barlow and Dixon (1980). This is a fairly simple ecological system, comprising two herbivores and their natural enemies on lime (*Tilia*). If the procedures below appear complex, it would be wise to consider how more complex they may have to be if a system such as aphids on wheat were chosen, where 100 plus natural enemies may prey on seven migratory aphid species feeding on a fast-growing and variable wheat plant.

The main components of the lime aphid system are shown in Figure 7.24. Each of the boxes contains its own subsystem comprising interrelated variables. The subsystem for the aphid itself is shown in Figure 7.25. The work leading to

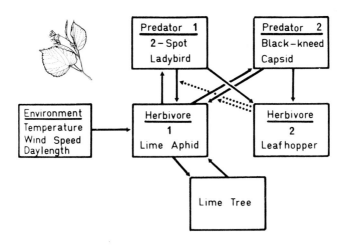

Figure 7.24: Simple Diagram of the Lime Aphid Population System, Showing the Main Components and Interactions. Source: From Barlow and Dixon, 1980.

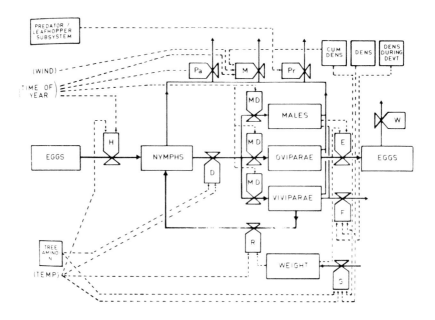

Figure 7.25: Detailed Relational Diagram of the Lime Aphid Subsystem. Rectangles denote state variables, squares subsidiary variables, brackets driving variables and valve symbols factors affecting rates of flow. →a flow of energy or matter; −−→ a link denoting an effect of one factor on another; H, egg-hatching rate; D, development rate; MD, morph determination; E, rate of egg-laying; W, winter mortality rate; F. emigration rate of alates; G, growth rate; R, reproductive rate; P_a, rate of parasitism; P_r, predation rate; M, mortality rate from other causes. Source: From Barlow and Dixon, 1980.

the model began by summarising population data accumulated over eight years. Although the individual years' population changes seemed very different (Figure 7.26) there were clear patterns within and between years and it is these which form the 'backbone' of the model.

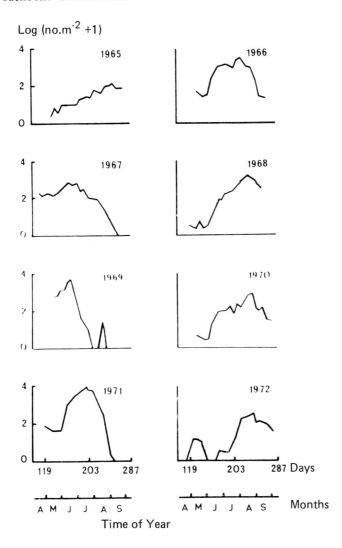

Figure 7.26: Observed Trends in Total Population Numbers of the Lime Aphid, Averaged over Two Trees, During Each Year from 1965 to 1972. Source: From Barlow and Dixon. 1980.

Eggs are laid on the twigs in the autumn by wingless females called oviparae and hatch in the spring; the individuals in the first generation are called funda-trices (sing. fundatrix). There is a clear negative relationship between the number

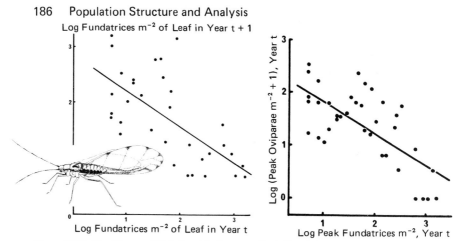

Figure 7.27: The Observed Relationship Between Peak Densities of Lime Aphid Fundatrices in Successive Years. Each point represents 1 tree in 1 year. Source: From Barlow and Dixon, 1980.

Figure 7.28: The Observed Relationship Between Peak Density of Lime Aphid Oviparae in One Year and Peak Density of Fundatrices in the Next. Source: From Barlow and Dixon, 1980.

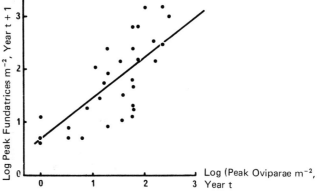

Figure 7.29: The Observed Relationship Between Peak Density of Lime Aphid Fundatrices and That of Oviparae in the Same Year. Source: From Barlow and Dixon, 1980.

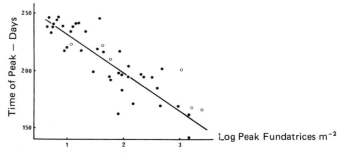

Figure 7.30: The Observed Relationship Between Peak Numbers of Lime Aphid Fundatrices and the Time of the Overall Peak Each Year. ● point from the field (1 tree in 1 year); ○ point from insectary populations. Source: From Barlow and Dixon, 1980.

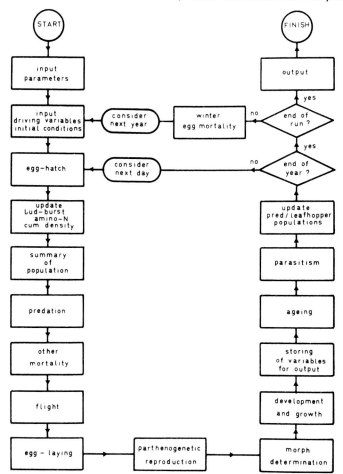

Figure 7.31: Flow Diagram of the Lime Aphid Model. Source: From Barlow and Dixon, 1980.

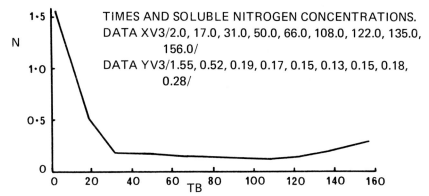

TIMES AND SOLUBLE NITROGEN CONCENTRATIONS.
DATA XV3/2.0, 17.0, 31.0, 50.0, 66.0, 108.0, 122.0, 135.0, 156.0/
DATA YV3/1.55, 0.52, 0.19, 0.17, 0.15, 0.13, 0.15, 0.18, 0.28/

Figure 7.32: Soluble Nitrogen Content of Lime Phloem (N, % Dry Weight) Throughout the Year. (TB, time in days after bud-burst). Source: From Barlow and Dixon, 1980.

of fundatrices in successive years (Figure 7.27) and between the number of autumn oviparae and the number of fundatrices in the spring of the same year (Figure 7.28). Large numbers of autumn oviparae lead to large numbers of spring fundatrices (Figure 7.29). The timing of the major population peak during the summer varied markedly (Figure 7.26) and this was related to the number of fundatrices hatching from the eggs (Figure 7.30). Many detailed relationships were incorporated, relating to aspects of the insects' growth rate, movement, predation rates etc. and to properties of the tree itself such as timing of bud-burst and leaf amino-nitrogen levels. These 'sub-routines' are parts of the boxes in the flow diagram of the model which is shown in Figure 7.31.

The operation of the model begins as below:

(1) Input of fixed parameter values, such as number of days in each month, weight of prey eaten by each predator/day, weights of each aphid and leaf-hopper instar etc., survival probabilities of the predators, leaf nitrogen levels etc. These values are 'constants' in that events in the model do not change them but their values are not necessarily fixed through the year. For instance, leaf soluble nitrogen levels seem to follow a standard pattern, being high at bud-burst, low in the summer with a small rise at senescence. Figure 7.32 shows this pattern together with the relevant part of the FORTRAN program.

(2) The next stage is to set the starting conditions on the fixed start date for the model (March 1). These will be abiotic factors such as temperature, the values for which can come either from a between-year average or from a file of daily data for the year being simulated. Levels of biotic factors such as egg numbers are also set at this stage. These levels can be chosen by the modeller or can come from the previous year's run.

(3) Individual sub-models (the components from which the overall model is built up) are now included stepwise. The first to be entered is the egg-hatch sub-routine, which gives the timing and pattern of egg hatch for the chosen year, driven by the hatching/temperature relationship in Figure 7.33. The cumulative population hatched is updated and the number hatching subtracted from the egg population and entered into the first nymphal age class, individuals of which are assumed to have a standard weight. These hatching numbers were obtained from an integration of data from laboratory experiments on hatching rates in relation to temperature (Figure 7.34), a laboratory-derived temperature threshold for egg hatch (Figure 7.35) and average field temperature values for the appropriate time of the year (Figure 7.36).

(4) Other sub-routines are incorporated in this way, the relationships comprising them being derived, as above, from laboratory and/or field observation coupled with some averaging and a few generalisations and assumptions. One sub-routine which was later to be shown to be very important in the insects' dynamics was sub-routine FLIGHT. This has four major components, shown in Figure 7.37 together with the way part of the program handled them. Once all sub-routines are written, the program is run for a particular year and its output can be compared with observed population changes (Figure 7.38) and with the empirically

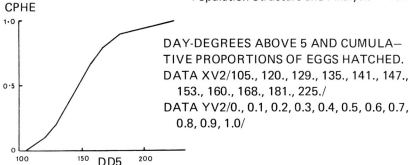

CPHE

DAY-DEGREES ABOVE 5 AND CUMULA–
TIVE PROPORTIONS OF EGGS HATCHED.
DATA XV2/105., 120., 129., 135., 141., 147.,
    153., 160., 168., 181., 225./
DATA YV2/0., 0.1, 0.2, 0.3, 0.4, 0.5, 0.6, 0.7,
    0.8, 0.9, 1.0/

Figure 7.33: Relationship Between the Cumulative Proportion of Lime Aphid Eggs Hatching (CPHE) and Summed Day-degrees Above 5°C from 1st March (DD5). Source: From Barlow and Dixon, 1980.

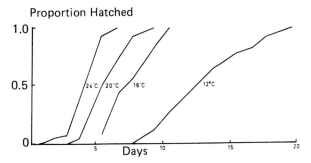

Figure 7.34: Distribution of Egg-hatching in Time for the Lime Aphid at Four Constant Temperatures. Source: From Barlow and Dixon, 1980.

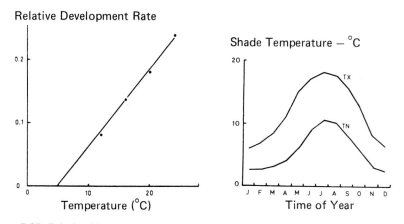

Figure 7.35: Relationship Between Relative Development Rate of Eggs of the Lime Aphid and Temperature. Source: From Barlow and Dixon, 1980.

Figure 7.36: Long-term Average Maximum (TX) and Minimum (TN) Daily Shade Temperatures Throughout the Year in the Lime Aphid Study Area. Source: From Barlow and Dixon, 1980.

established within- and between-year relationships (Figure 7.39). In Figure 7.38 there is general agreement between the model's predictions and real events but the model seems to miss some of the more detailed population changes. Barlow and Dixon (1980) considered that the main cause of disagreement was the omission from the model of some weather effects, acting on flight and mortality. If the model had been run continuously, simulating a connected series of seasons from 1965 to 1972, the above fairly small errors would have become compounded and the fit to the data would have become poorer with each year.

```
SUBROUTINE FLIGHT      (AA,AS,AM,TEMP,WIN,DAY,CUM,MM,IMAX,E)
REAL AA(30,6),AS(9,2),AM(30,2),TEMP(2,366),WIN(366)
INTEGER DAY
FMT=0.
FVT=0.
DO 1 I=1,IMAX
AGE-CLASS SKIPPED IF NO ALATES PRESENT
IF((AA(I,1)+AA(I,3)).LE.1E-6) GOTO1
ADULT COMPONENT
FA=0.005*((AS(2,1)+AS(3,1)+AS(4,1))/4.+AS(5,1''+AS(6,1)+AS(7,1)+AS(8,1))
```

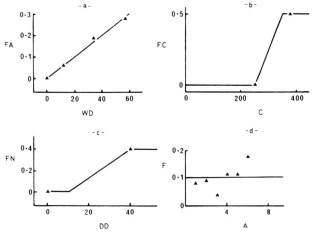

Figure 7.37: The Components of Daily Adult Lime Aphid Flight. (a) The proportion flying (FA) dependent on current weighted population density (WD, see text); (b) the proportion flying (FN) dependent on mean total density experienced during nymphal development (DD); (c) the proportion flying (FC) dependent on cumulative density (aphid-weeks, C); (d) the background level of flight (F) relative to adult age (A). Source: From Barlow and Dixon, 1980.

Having matched real events (and improved the fit by changing appropriate variables) the most interesting stage, in terms of population dynamics, is reached. In this, we can change assumptions or parameter values to find out:

(a) whether errors in the less-understood parts of the model would have major consequences e.g. if aphid fecundity, measured in a leaf clip-on cage, was 10 per cent too high, what would be the effect of reducing it to a more realistic value?

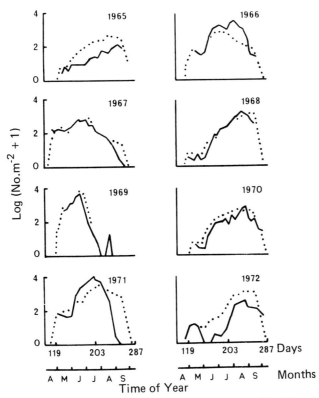

Figure 7.38: Comparison of Observed Aphid Population Trends on Two Lime Trees From 1965 to 1972 (——), and Those Generated by the Model Given the Same Weather and Initial Conditions Each Year (. . . . .). Source: From Barlow and Dixon, 1980.

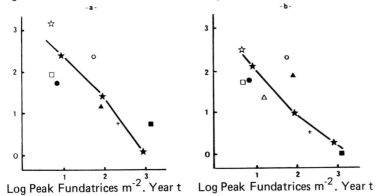

Figure 7.39: Comparison of Relationships Generated by the Lime Aphid Model (★ — ★) Between (a) Fundatrix Densities in Successive Years, and (b) Peak Densities of Fundatrices and Oviparae Within Each Year, and Those Observed in the Field on Two Trees. ●, 1965; ○, 1966; +, 1967; ☆, 1968; ■, 1969; □, 1970; ▲, 1971; †, 1972; each point is the average for that year over both trees. Source: From Barlow and Dixon, 1980.

(b) what are the main density-dependent, potentially regulating factors?
(c) what are the main disturbing factors in the population (i.e. the 'key factors').
(d) areas in which more experimental work is required.

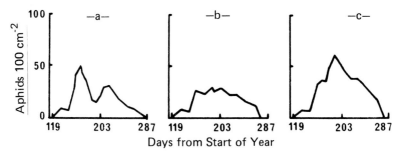

Figure 7.40: Effects of Different Nymphal Flight Components on Population Behaviour, Generated by the Lime Aphid Model, when Each One is the Only Control Process Acting. (a) The normal nymphal component. (b) The normal nymphal component with no threshold. (c) The nymphal component with no threshold and reduced slope. Densities are numbers/100 cm^2. Source: From Barlow and Dixon, 1980.

(a) How important are inaccuracies? In Figure 7.40, the consequences of changing the effects of an aphid's experience as a nymph on its subsequent flight are investigated; the extent to which the graphs differ shows how rigorous the data needed to be in the first place.

(b) Density-related processes. If we consider the year-to-year relationship between fundatrix numbers, the consequences of the removal of a factor on the slope can be investigated. Taking the negative slope as it stands and allowing it to be the only factor in the aphid's dynamics, population stability would result eventually, after about 10 generations. If we remove a factor from the model and the slope of the line changes markedly, the consequences of the changed slope for population stability will tell us something about that factor's role; this is a related approach in simulation to the investigation of k/log density relationships in the previous section. In Figure 7.41, it can be seen that removing some processes makes little difference while removal of others, such as predation and nymphal effects on subsequent flight, decreases the slope of the line, which decreases stability. We can now put each of the processes back into the model but this time, acting alone (Figure 7.42). We see that some of these factors are capable of producing a relationship not very different from the original one, which was regulatory. Predation is the closest, but shows this property only at low and intermediate density ranges, suggesting that the population can 'escape' from the restraints of a limited predation pressure; we will see in Chapter 9 that this has been demonstrated several times, both experimentally and theoretically and it is encouraging that, here, a third approach to population ecology, that of simulation, can reveal the same fundamental population process.

A final lesson from the above case study is that we should not expect to find one factor which is alone responsible for regulation or disturbance. Figure 7.42 shows how predation cannot remain the regulatory factor at higher prey density ranges and Barlow and Dixon's model showed that a hierarchical series of factors are responsible. In some years, the situation seems to be: predation at low and intermediate densities, followed by predation 'helped' by the nymphal effect on subsequent flight at higher ones. In other years, a different mechanism operates, based on an increase in flight or mortality in relation to accumulated density during the year. One of the drawbacks of k-factor analysis was that it was less likely to be able to detect the action of such hierarchical mortalities or between-year changes in the major process operating; our simulation case study has shown again that whether or not the subtleties, complexities and variations of real population events are revealed depends heavily on the methodology chosen. This fundamental point that methodology may strongly influence ecological conclusions will be emphasised again when we consider the dynamics of predation in Chapter 9. Other simulation studies which, by escaping the methodological 'straight-jacket' imposed by life-table and k-factor analyses, have shown the complexities of population processes, include those of Hughes and Gilbert (1968), Gilbert and Gutierrez (1973) and Gutierrez *et al.* (1974).

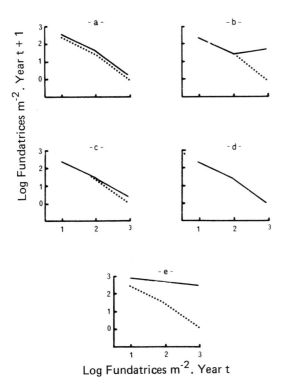

Figure 7.41: Effect on the Year-to-Year Relationship Between Lime Aphid Fundatrix Numbers of Removing Each Control Process in Turn From the Model. (a) The adult flight component; (b) the nymphal flight component; (c) changes in adult weight; (d) cumulative density effects; (e) predation. —— new relationship; . . . . . original relationship. Source: From Barlow and Dixon, 1980.

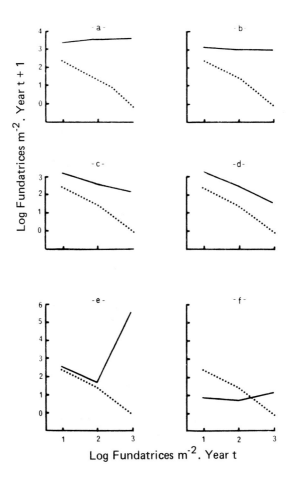

Figure 7.42: Effect on the Relationship Between Lime Aphid Fundatrix Numbers in Successive Years, of Each Control Process Acting in Isolation. (a) The adult flight component; (b) the nymphal flight component; (c) changes in adult weight; (d) cumulative density effects; (e) predation; (f) predation with aphid background mortality doubled. —— new relationship; . . . . . original relationship with all processes included. Source: From Barlow and Dixon, 1980.

This chapter would be incomplete, however, if we allowed the restrictions of different methodologies to preclude a consideration of the current state of population ecology theory. We need to consider whether or not there are patterns emerging which enable us to bring order out of chaos and produce general statements concerning which, if any, processes and agents commonly (1) regulate and (2) disturb populations and under what circumstances they may do so. This synthesis will be attempted in the next section.

7.5 Towards a General Population Theory

Population ecologists have always been interested in patterns and strategies although attempts in the past to bring order from a chaos of data have often been frustrated by a lack of the latter. In the past decade or two, however, more light and less heat seem to have been generated and some general principles do seem to have emerged. In this section, we will consider three important questions in population ecology and consider current views. These questions are: (a) are density dependence and regulation common events? (b) do populations commonly return to equilibrium levels as a result of density-dependent mortality or does density dependence operate only when environmental carrying capacity is approached? and (c) do the mortality agents which are (1) density dependent and (2) disturbing differ qualitatively according to the organism's position in an ecological classification?

7.5a The Frequency of Density Dependence

Some ecologists would feel disinclined to discuss this question because it may appear to be largely of historical interest. It is true that much of the intensity of the debate of the 1950s and 1960s, stimulated by the writings of Andrewartha and Birch (1954), has now declined and the existence of density dependence is now generally accepted. However, the fact that the debate lasted so long can be seen largely as a result of confused ecological thinking. This chapter has tried to clarify and separate different processes in population ecology and has tried to show how methodology can sometimes influence conclusions. With the wisdom of retrospect it now seems likely that a combination of methodology and confused ideas was responsible for misinterpretations in one of population ecology's greatest debates.

In essence, Andrewartha and Birch felt that expressions such as 'steady density', 'control' and regulation are 'allegorical', stemming from the 'dogma' of density dependence. They felt this largely because of an analysis of the population of thrips (small sucking insects in the Order Thysanoptera) carried out by Davidson and Andrewartha (1948). In this, a multiple regression analysis of the peaks in the thrips' population in relation to weather variables was carried out. Depending on how the peak numbers were calculated, up to 84 per cent of the variance in peak number could be 'explained' by weather. Inasmuch as a predictive equation was produced for thrip numbers then this analysis was useful ecologically, although, as always, correlation and regression do not prove causality. However, it was Andrewartha and Birch's later interpretation which caused the trouble: '... not only did we fail to find a density-dependent factor but ...' (with 84 per cent of the variance explained) '... there was no room for one.' What Davidson and Andrewartha had done, however, was effectively the equivalent of only the first stage of a k-factor analysis, i.e. identification of the key factor(s) *causing population change*. They found no density dependence because they did not look. Such confused thinking appeared even among 'supporters' of the Nicholsonian idea of density dependence. Lack (1954), in what was a classical ecology text of its time, said of the Bobwhite Quail data of Errington (1945) that the fluctuations could be accounted for by two density-dependent checks. If we substitute 'factors' for 'checks', this sentence could be interpreted as the action of over- or under-compensating density dependence (p. 176), but the combination of 'checks' and 'fluctuations' in the same sentence defies ecological understanding.

Table 7.7: Position of Sycamore Leaves Suitable and Less Suitable for the Sycamore Aphid, Relative to Other Leaves in the Canopy.

Leaf	Completely overlapping	Large overlap	Slight overlap	No leaf immediately below
Suitable	4	2	4	16
Less suitable	15	3	1	6

Source: From Dixon and McKay, 1970.

Table 7.8: Sycamore Aphid Densities Immediately Before and After a Mortality Caused by a Strong Wind.

Tree	No. of aphids/forty leaves Day 1	Day 6	% reduction
1	2548	1044	59
2	754	587	22
3	419	341	19

Source: From Dixon and McKay, 1970.

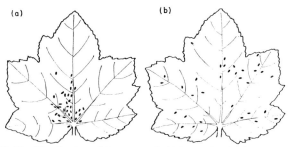

Figure 7.43: The Distribution of Sycamore Aphids on a Leaf (a) During a High Wind and (b) After the Wind had Abated. Source: After Dixon and McKay, 1970.

Abiotic variables are sometimes invoked in population ecology in a slightly different, but still confused, way, that of mistaking them as acting in a density-dependent way. It is in this area that we have to be especially careful in our choice of words but the example in Tables 7.7 and 7.8 and Figure 7.43 should clarify the situation. Figure 7.43 shows how the sycamore aphid (*Drepanosiphum platanoidis*) clusters by the main leaf veins during wind because here the insects are more protected from leaf-to-leaf brushing. That the wind acting in this way can be a major population influence is shown in Table 7.7. In Table 7.8, density-dependent mortality has occurred through the action of wind: the higher populations have lost higher proportions of individuals. However, the ecological conclusion is that at high population densities the aphids were forced, through *intraspecific competition*, to feed away from the leaf-vein protection and also on leaves subject to much brushing by others. Here, the density-dependent mechanism is biotic (competition) while the mortality *agent* is abiotic (wind). Without a biotic mechanism changing the proportion of organisms at risk it is difficult to see how abiotic factors can be involved in regulation. Unless we accept as common the action of '. . . an array of density-independent factors' (Dempster, 1975) keeping populations within limits (i.e. persistence through random (stochastic) processes only) then we are forced to accept that density-dependent regulation is a common event.

7.5b Equilibrium Levels or Carrying Capacity?

It has been fundamental to much of our discussion of populations in this chapter than when regulation occurs, populations tend to be pulled towards an equilibrium level and that this level may be more or less fixed and well below the maximum which the habitat could hold. In chapters on competition and predation, this has also been taken as the norm. However, when we leave the precision of mathematical models and look at real field populations, such as that in Figure 7.1, we could conclude that such equilibrium levels are difficult to detect. Dempster and Pollard (1981) went further and suggested that not only are they difficult to detect but, in many insect populations, do not exist. Instead, these authors suggested that fluctuations in the numbers of many insects are determined (*sic*) primarily by fluctuations in the carrying capacity of their habitats. Figures 7.44 and 7.45 were among those presented by Dempster and Pollard as evidence for this. They propose a resource-limited 'ceiling' model in which the population may fluctuate between zero and its carrying capacity and the only density dependence in the system is the restriction of growth as the population approaches the ceiling. The ceiling is determined by environmental carrying capacity and is likely to fluctuate. In the equilibrium type of model, the population fluctuates around its equilibrium and is 'pulled' towards it by the relaxation and intensification of density dependence below and above it, respectively.

It is relatively easy to seek out examples of populations which seem to fluctuate well below an apparent carrying capacity; the numbers of winter moth larvae in Figure 7.18, for instance, did not greatly exceed about 100/m² of tree canopy area. It is difficult to imagine that crude leaf number was limiting or that it varied by nearly two orders of magnitude between years. Similarly, the numbers of aphids on wheat shown in Figure 7.46 fluctuate by a factor of about 200 while presumed stem numbers hardly vary at all. Dempster and Pollard would point out, however, that no measurements were made in either of the above cases of *real* carrying capacity; winter moth larvae are washed off the

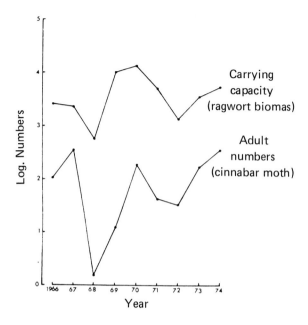

Figure 7.44: The Numbers of the Cinnabar Moth Compared with the Carrying Capacity of the Habitat in Terms of Available Food. Source: From Dempster and Pollard, 1981.

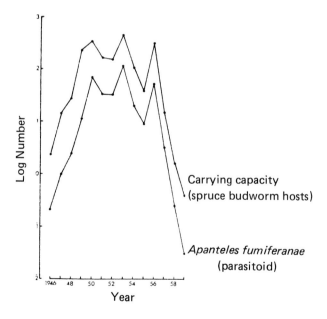

Figure 7.45: Variations in the Numbers of *Apanteles fumiferanae* Compared With Fluctuations in the Availability of Hosts. Source: Data from Miller, 1963; after Dempster and Pollard, 1981.

more exposed oak buds, so not all buds are equal, and we saw earlier how sycamore leaves are often highly unsuitable for aphids. Even in the wheat aphid example, there could be subtle physiological differences between the wheat plants which restrict aphid numbers, making crude stem counts irrelevant. Although the ceiling model probably cannot be refuted, and does not need to be in that some populations may obey it, perhaps the best set of examples supporting the equilibrium alternative are those of Beddington, Free and Lawton (1978). These workers analysed published cases of clear suppression of invertebrate numbers following the introduction of a predator or parasitoid. They calculated the ratio of the numbers after predator release (the new host equilibrium) to the numbers before (i.e. approaching the hosts' carrying capacity). For field examples, this ratio was very low, not exceeding 0.025. The extent of suppression achieved if host density dependence alone (through intraspecific competition) is incorporated into an unstable host-parasitoid model is only about 0.4 (Beddington, Free and Lawton, 1975). Even though there are field cases where resource limitation (the ceiling model) appears to operate, the biocontrol analysis above shows that low equilibrium levels achieved through natural enemies are also possible. Dempster and Pollard's point was that a quest for data which support the equilibrium model may preclude a search for those supporting resource limitation. As we have seen earlier, in population ecology, hypotheses cannot be tested unless the appropriate question is clearly asked and the appropriate data collected. Further insight into this question of equilibrium may be obtained in Section 8.3, but we should still retain an open mind on the topic of equilibria until accurate data on resources have been collected more often.

7.5c Mortality Patterns and Ecological Classifications

Throughout this chapter and elsewhere in the book we have made the distinction between factors which bring about fluctuations in populations and those which exhibit density dependence, with the potential to regulate. Some recent analyses have attempted to collate and classify the results of published studies to show which mortalities most commonly are responsible for regulation/fluctuations. Lack (1954) attempted such a synthesis, from many fewer case studies than are now available. Lack believed that most birds' numbers were regulated *via* food shortage outside the breeding season, while those of gallinaceous birds, deer and most plant-feeding insects are limited by predation. In 1964, Lack had not changed his view other than expressing some doubt about the role of predation in the regulation of gallinaceous birds' numbers. Major recent surveys of density dependence in life table data have concerned plant-feeding invertebrates and vertebrates (Caughley and Lawton, 1981) and a wider range of animals (Stubbs, 1977). Southwood (1975) analysed key factors (i.e. the causes of population change) in a similar way.

Caughley and Lawton (1981) considered that the possible regulators of phytophagous insect populations could include natural enemies, food (i.e.

intraspecific competition) and disease. There was no pattern in these mortalities in the survey; each category was well represented and some studies revealed no obvious density dependence. For grazing vertebrate populations, too, important mortalities ranged from food supply and quality, weather and predation although they have not always been analysed in such a way so that true density dependence could be detected. Intraspecific competition does seem to be very important in vertebrate population regulation, however, as we will see in Chapter 8.

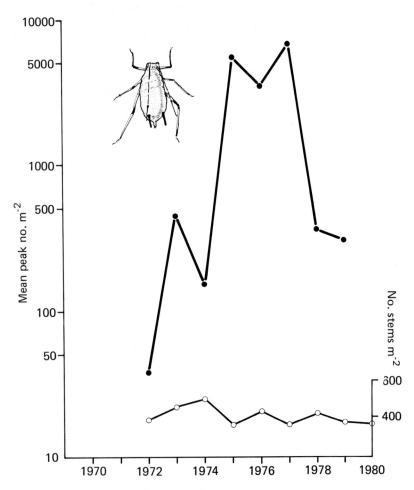

Figure 7.46: The Numbers of the Aphid *Sitobion avenae* on Wheat in S. England over a 9-year Period, with Hypothetical Numbers of Stems/m^2 Over the Same Period. Source: From Vickerman *et al.*, 1983.

Stubbs' analysis used the habitat of the surveyed animals as a framework and expressed its likely influence by placing the organisms on the *r-K* continuum (see p. 272). She divided the density-dependent mortalities into four categories, according to where in the life cycle they act: adult loss, reduced fecundity, parasitism and predation, loss of immature stages. For temporary-habitat animals (*r*-strategists), 86 per cent of the density-dependent mortality acted on the young stages, while for *K*-strategists this figure was only 15 per cent. Instead, parasitism and predation (30 per cent) and reduced fecundity (35 per cent) were more frequent in these animals (Table 7.9). Stubbs also investigated the 'strength' of the mortality in relation to its type. *b*-values for plots of *k* on log density, like those used in Section 7.3(b), showed that parasitism, predation and disease were compensating or undercompensating while intraspecific mortalities (acting) largely on the *r*-strategists) tended to overcompensate, tending towards scramble competition (Chapter 8). Where *K*-selected animals exhibited clear intraspecific competition, values of *b* were less than 1.2, i.e. only weakly overcompensating. *r*-strategists showed a switch from weak density dependence at low densities to strong, overcompensating density dependence at high population levels; this allows them to use limited-duration resources to the full, then suffer high 'mortality' which, for many *r*-strategists, takes the form of a high emigration rate (Chapter 8).

Table 7.9: The Incidence of Density-Dependent Mortalities at Various Times in the Life-cycle, Divided up into *K*- and *r*-Strategists

	K (no.)	Total (%)	*r* (no.)	Total (%)
Adult loss	5	16.7	1	4.8
Reduced fecundity	9	30.0	1	4.8
Parasitism/predation	8	26.7	1	4.8
Young/larval loss	8	26.7	18	85.7
Total	30		21	

Source: From Stubbs, 1977.

We should be surprised if Lack's (1954) analysis, based on taxonomic groupings, was robust enough to be of general use but Stubbs' paper, based on a spectrum of animals' strategy, does obey ecological commonsense. There will always be exceptions in such analyses, however, not least because of the difficulties of placing organisms on the *r*-*K* continuum. It shows us, however, that competition and natural enemies are important regulatory driving forces in population ecology and the next two chapters give them the detailed consideration they deserve.

8 Competition and Population Stability

8.1 Introduction: Inter- and Intraspecific Competition and Population Stability

All biotic interactions are potentially regulatory; the section at the end of the last chapter attempted to show which did so most commonly. This chapter will consider in more detail the effects of competition. Intraspecific effects are more commonly considered important in relation to populations while interspecific competition is most implicated in community events. However, here we will discuss the population effects of both.

The effect of *interspecific competition* on populations is usually destabilising – leading to extinction or ecological exclusion of one or other of the competing species (Chapter 6). Only in very special circumstances (Section 6.6) can competing populations coexist. Under these conditions each population, as we have seen, exerts a depressive effect upon the growth of its competitor population; the intensity of suppression is a function of the relative densities of the competing populations and thus could be invoked as possibly providing some density-dependent regulation. However, populations rarely co-exist in continuing competition and interspecific competition would not normally be regarded as regularly involved in regulation.

Intraspecific competition, on the other hand, has long been recognised as one of the major driving forces in animal and plant population dynamics. We need to consider in this chapter the way intraspecific competition acts by affecting emigration, birth and death and the consequences for population stability, i.e. self regulation.

By definition, intraspecific competition shows at least an element of density dependence in the relationship between mortality (however expressed) and density. A generalised version of this relationship is shown in Figure 8.1a. Two points to notice in this graph are that there is a threshold density below which competition does not occur and that there is a level of density-independent 'background' mortality, from random causes. The hypothetical population graph in Figure 8.1b shows that in the fluctuations below the threshold density, intraspecific competition has no role.

The general relationship in Figure 8.1a can sometimes be detected in field data but it is helpful to consider the two extreme forms the relationship can take. These forms were first proposed by Nicholson (1954), who called them *scramble* and *contest* competition respectively. In the former case, all competing individuals are identical and they share the common resource equally. The consequence of this equal partitioning of the resource is that as soon as it is used up, all competitors die or emigrate. In contest competition, unequal sharing of the resource leads to an increasing proportion of individuals dying or emigrating as density increases; in 'perfect' contest competition, a fixed number of survivors remains whatever the density. It is helpful to express mortality from competition as a k-value so that it can easily be incorporated into population models to assess its role in stability. Table 8.1 includes data showing perfect contest and two other sets, one towards scramble and one towards contest; all three relationships are shown in Figure 8.2. Perfect contest plotted in this way gives a slope of $b = 1$ while $b > 1$ indicates a relationship nearer to scramble while $b < 1$ is closer to contest than scramble. In each case, a threshold of five organisms has been chosen, below which number competition does not occur.

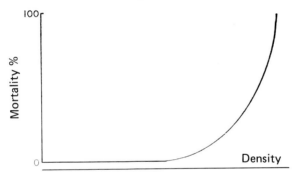

Figure 8.1a: The General Pattern of Mortality Resulting from Intraspecific Competition (cf. Figure 7.22).

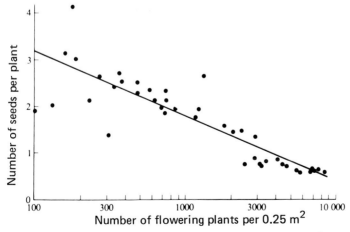

Figure 8.1b: Changes in Fecundity in Experimentally Manipulated Natural Populations of the Grass *Vulpia fasiculata*. Source: From Watkinson and Harper, 1978 and Silvertown, 1982.

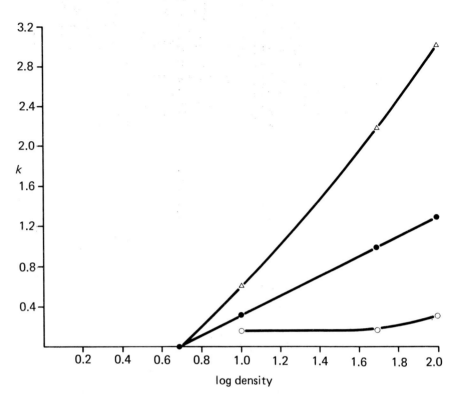

Figure 8.2: Relationship Between Mortality from Intraspecific Competition and log Population Density Derived from Table 8.1. ● = contest, ○ = closer to contest than scramble, △ = towards scramble.

Table 8.1: Relationships Between Mortality from Intraspecific Competition and log Density. C = contest, (C) = towards contest, (Sc) = towards scramble

Density	log density	Mortality			No. before mortality (N)/ no. after (S)			log N/S		
		C	(C)	(Sc)	C	(C)	(Sc)	C	(C)	(Sc)
5	0.7	0	0	0	5/5	5/5	5/5	0	0	0
10	1.0	5	3	7.5	10/5	10/7	10/2.5	0.3	0.15	0.6
50	1.7	45	17	49.7	50/5	50/33	50/0.33	1.0	0.18	2.2
100	2.0	95	49	99.9	100/5	100/51	100/0.1	1.3	0.29	3.0

We would expect real examples of competition to fall between these two extremes because of genotypic and phenotypic variation between organisms. Also, it would be surprising to find simple linear relationships like those shown in Figure 8.2. Some examples from the field are shown in Figure 8.3 while Figure 8.4 shows a case where, at least above intermediate densities, contest competition appears to be operating. There is obviously a time element in these

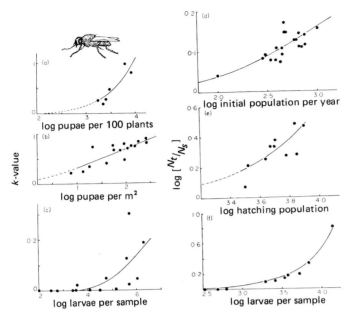

Figure 8.3: Density-dependent Relationships from Field Data and Their Description Based on Equation 8.2. (a) Cabbage root fly (*Erioischia brassicae*) pupal mortality. $a = 0.00011$, $b = 3.16$ (data from Benson, 1973). (b) Winter moth (*Operophtera brumata*) pupal mortality. $a = 0.6$, $b = 0.38$ (data from Varley and Gradwell, 1968). (c) Larch tortrix (*Zeiraphera diniana*) larval disease. $a = 0.000018$, $b = 0.11$ (data from Varley and Gradwell, 1970, after Auer, 1968). (d) Clutch size reduction in the great tit (*Parus major*) plotted against log maximum clutch size (log N_t). $a = 0.0064$, $b = 0.18$ (after Krebs, 1970). (e) Chick mortality in the partridge (*Perdix perdix*) plotted against log hatching population (log N_t). $a < 0.000005$, $b > 28.58$ (after Blank *et al.*, 1967). (f) Larval starvation in the Colorado potato beetle (*Leptinotarsa decemlineata*) plotted against log larval density (log N_t). $a = 0.00007$, $b = 30.95$ (data from Harcourt, 1971). Source: From Hassell, 1975.

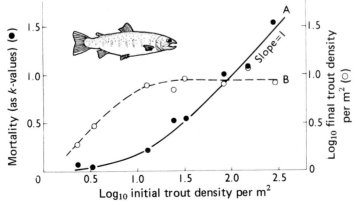

Figure 8.4: The Effects of Competition in Trout Fry. Curve A: Mortality as a function of the initial stocking density of trout. Note that the relationship at high trout densities has a slope of unity. Curve B: The surviving trout density at different initial stocking densities. Source: Data from Le Cren, 1973, from Hassell, 1976.

relationships and the mortalities shown appear only after a temporal threshold has been passed. Figure 8.5 shows this well, with no detectable effects after 21 days but a relationship close to contest competition after 63 days. This time component can be followed in more detail by following the fate of a single population starting at one density and by taking a series of 'harvests' to record numbers of survivors and their size. This approach was done for plants by Yoda *et al.* (1963) who showed that when the log of the mean plant weight (i.e. of the survivors) was plotted against the log of the density of survivors, the values for successive harvests produced a slope close to −1.5. This means that although the number of individuals in the population is decreasing, the weight of the total population increases. This increase more than compensates for the fall in numbers and is represented by the formula

$$w = cp^{-\frac{3}{2}} \qquad\qquad (8.1)$$

where w = mean wt. survivor, p = the density of surviving plants and c is a constant determining the position of the line, not its slope, and is influenced by the plant species' growth characteristics. Figure 8.6 shows a range of such relationships, all sharing the same slope.

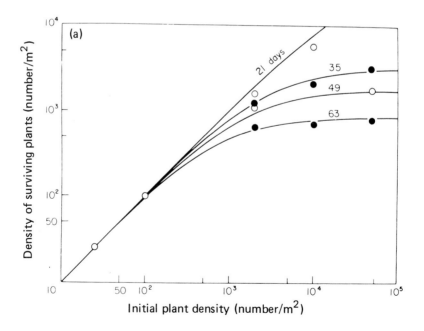

Figure 8.5: The Relationship Between the Density of Survivors and Initial Plant Density in Buckwheat (*Fagopyrum*) After Various Periods of Time. Source: From Yoda, 1963, after Harper, 1977.

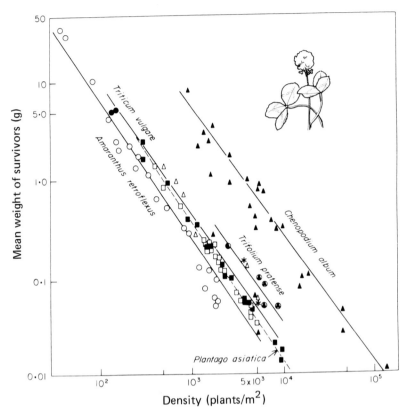

Figure 8.6: Changes in Plant Density and in Mean Plant Weight With the Passage of Time. Data for *Chenopodium*, *Amaranthus*, and *Plantago* from Yoda *et al*. (1963); data for *Trifolium* and *Triticum* from Harper and White (1971) after data of Black and of Puckeridge. Source: From Harper, 1977.

In terms of population stability, however, we need to return to the mortality/ density relationships and build population models from them on the assumption, for simplicity, that competition is the only mortality factor operating. We could use the simple linear model for k and log D in the previous chapter but Figure 8.3 shows that this is too simple a form for the relationship. The logistic population model in the last chapter was effectively a population model for contest competition in that the final numbers were fixed and determined by the environment's carrying capacity, not by the starting density as would be the case for a scramble model. However, its integral form shows that it is instantaneous and therefore as a competition model could not deal with time delays between competition and mortality. These are likely to be the norm, however, and Figure 8.5 shows it well; sometimes a generation may pass before the effects are evident, as shown by Dempster's (1971) data on competition, size and fecundity in the cinnabar moth (Table 8.2). Here, laboratory studies established a good relationship between the maximum number of eggs laid/female and pupal width and this line was used to provide the 'potential eggs' column in Table 8.2. The actual number of eggs laid in the following year is partly a reflection of pupal width but also includes components of emigration (1968) and immigration (1970), the former in particular a common result of intraspecific competition. A similar process in a plant population, where a subsequent generation's potential is affected, is shown in Figure 8.6.

Table 8.2: Larval Density, Pupal Width and Subsequent Adult Fecundity in the Cinnabar Moth, *Tyria jacobaeae*

Year	Density of young larvae	Mean pupal width (cm)	Potential no. eggs/	Actual no. eggs/	Proportion of potential/ aid
1966-67	1861	0.512	297	285	0.96
1967-68	16244	0.454	154	92	0.59
1968-69	14324	0.425	83	73	0.88
1969-70	62	0.511	295	561	1.90

Source: From Dempster, 1971.

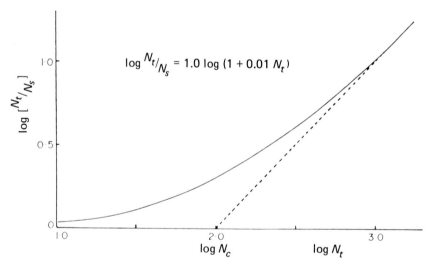

$$\log {}^{N_t}/_{N_s} = 1.0 \log (1 + 0.01\, N_t)$$

Figure 8.7: A Density-Dependent Relationship from Equation (8.2) where $a = 0.01$ and $b = 1.0$. Source: From Hassell, 1975.

The need for a model for intraspecific competition which includes such time delays makes difference equations particularly appropriate, although such delays can be incorporated in modifications of the logistic equation (p. 158). Some general single-species difference equation models were dealt with in Chapter 7 where we saw that the outcome for stability depended to a large degree on whether the natural periodicities or time delays in the regulatory mechanisms were short or long in relation to the system's response time. These models were general, however, and based on a few inflexible parameters, and did not include a specific function for mortality from competition derived from experimental data such as in Figure 8.3. The model of Hassell (1975) is based on such relationships which are represented by:

$$k = b. \log (1 + aN_t) \tag{8.2}$$

where $a = 1/$antilog of the intercept in Figure 8.7, b is the slope of the upper, linear part of the relationship and N_t is the density. These parameters enable us to stipulate two useful variables driving the relationship and the population model derived from it: a indicates the population density at which competition starts to become intense and b gives the intensity of the density-dependent mortality. We can now produce relationships which parallel those in Figure 8.2 in demonstrating the range from scramble to contest, but using this time a relationship closer to experimental results. A population model from equation (8.2) can be derived as follows:

$$\log \frac{N_t}{N_s} \text{ (i.e. } k) = b \log (1 + aN_t) \tag{8.2}$$

$$\frac{N_t}{N_s} = (1 + aN_t)^b \tag{8.3}$$

$$\frac{N_s}{N_t} = (1 + aN_t)^{-b} \tag{8.4}$$

$$N_s = N_t (1 + aN_t)^{-b} \tag{8.5}$$

where N_s = the number of survivors following the mortality's action; they produce the next generation when they reproduce at the selected rate F:

$$N_s = F.N_t (1 + aN_t)^{-b} \tag{8.6}$$

We can generate a range of population patterns from the model in equation (8.6), some producing perfect stability and others which reduce instability. Modification of the parameter a affects only the equilibrium level, not the way it is reached while changing b and F can, for instance, change exponential damping into chaos (Figure 8.8). Laboratory-generated population trends mirror very well the patterns from Hassell's (1975) model (Figure 8.9) but, as always, we should beware simple visual matching of population graphs. However the two strains of the beetle *Callosobruchus* in Figure 8.9 mimic very well the predicted effects of a changed reproductive rate. Hassell's model shows that high rates of increase and a tendency towards scramble lead to rapid population changes, marked oscillations and the possibility of strong under- and over-shooting

of the equilibrium, leading perhaps to extinction. This is a behaviour typical of
r-selected organisms while low rates of increase and/or weak competition lead to
gentle approaches to equilibrium, typical of *K*-selected organisms (see Section
10.3).

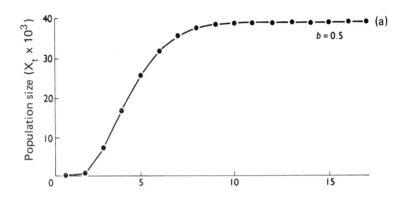

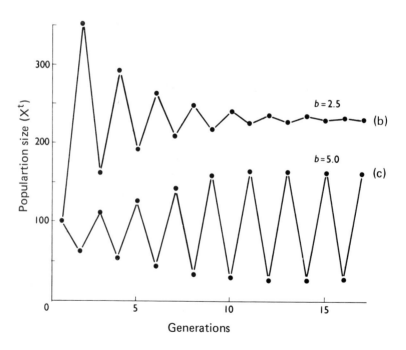

Figure 8.8: A Range of Population Behaviours Derived from Equation (8.2); *b* is one of the
equation's parameters. Source: From Hassell, 1976.

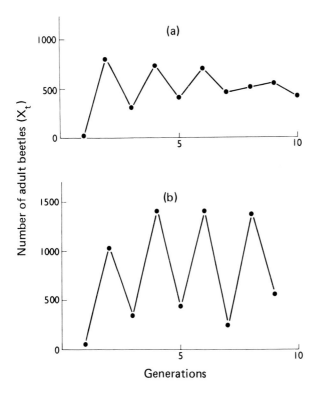

Figure 8.9: Examples of Single-Species Oscillations. (a) and (b) Two strains of the beetle, *Callosobruchus chinensis*, under identical culture. Source: From Fujii, 1968, after Hassell, 1976.

It would be surprising if field populations, which are essentially highly multivariate, showed clear reflections of the patterns in Figures 8.8 and 8.9. This does not invalidate the models, however, as in the field density-independent or other density-dependent events can obscure the pattern. Figure 8.10 shows this well for the sycamore aphid. A simple model based on competition predicted oscillatory damping but rain washed the newly hatched aphids from the tree's buds in the spring, disturbing the potential stability.

As was pointed out at the beginning of Chapter 7, much population theory and many models have derived from work on invertebrates. There are parallels in the vertebrates but data accrue slowly and the methodologies and analyses differ, so this group will be dealt with separately.

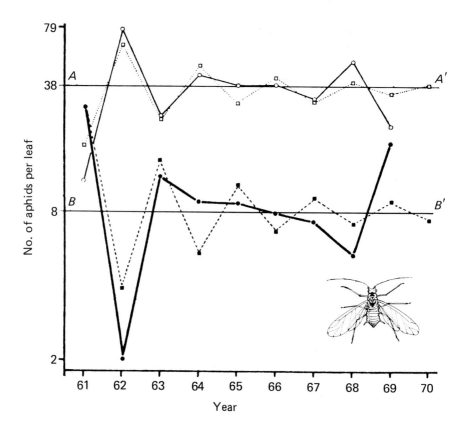

Figure 8.10: Peak Number of Sycamore Aphids per Leaf in the Spring (● — ● observed,
■ — — — — ■ expected) and Autumn (○ — ○ observed, □ — — — — □ expected) from 1961-69.
$A - A'$ and $B - B'$ = predicted equilibrium densities for autumn and spring, respectively.
Source: From Dixon, 1970a.

8.2 Regulation in Vertebrate Populations

It is now generally accepted that most vertebrate populations display some
degree of self-regulatory behaviour — that is that regulation is largely due to
intraspecific competition for a resource, but with an equilibrium established
well before the limiting resource starts to act as a finite shortage (cf. Section
7.5c, which gave examples for insects which contrasted with this view). In
practice, processes such as territoriality, aggressive behaviour etc. act to space
out individual population units (breeding pairs, individuals, sub-populations)
over available resources. Regulation can also act within the unit through changes
in patterns of birth, death and immigration and emigration. It is likely that the
latter two processes operate almost as safety valves and probably cannot

accommodate long-term changes. Fecundity can be increased in a range of subtle ways as competition pressure is ameliorated. Clutch and litter size may increase and the number of breeding attempts per year may also be increased. Most such examples come from work on birds but similar processes have been shown for North American white-tailed deer (*Odocoileus virginianus*) and for the grey squirrel (*Sciurus carolinensis*) (Chaetum and Severinghaus, 1950; Wood, 1977). Reduction of reproductive output can be achieved, by converse, by changes in size and number of individuals. Reduction in clutch or litter size may result from initial preparation of a smaller number of embryos or by later resorption of embryos; the latter occurs in the rabbit (*Oryctolagus cuniculus*) and in voles (*Microtus*; Pelikan, 1970). Most familiar examples come from food limitation as the ultimate shortage but space restriction can be equally important. When rats are crowded, birth rate decreases through direct physiological effects, hormonal effects on behaviour and through direct behavioural changes. These include, in females, decreases both in the number of young/litter and in the survivorship of the young; the female may come into oestrus less frequently and may need more copulations to effect fertilisation. Cannibalism among the new-born rats may also increase. In males, fertility and sexual drive may decline; the latter may be brought about by the pressure on time through increased fighting.

As in invertebrates, death rates may increase but some more subtle adjustments can be made to when and where deaths occur in the population; these more subtle effects are often called socially induced mortality and may range from the effects on the neonates, mentioned above, to stress shock. In the latter case, behavioural competition for limited resources causes an increase in adrenal activity (the fact that adrenaline is inhibitory to the sex hormones can account for the reduced sexual drive mentioned above). In extreme cases, adrenal hypertrophy can occur (Christian, 1950), resulting in fatty degeneration of the heart and in heart seizure.

The above range of subtle and graded behavioural and physiological mechanisms can act in a density-dependent way to bring about very stable vertebrate populations; the heron numbers in the UK (Figure 7.1) fluctuate by a factor of 2-3 over 40 years compared with the 200-fold changes in aphids shown over 10 years in Figure 7.46.

Some vertebrate populations do not show such stability, however; especially among small mammals, cyclical behaviour occurs. In extreme forms, this is seen not as regular oscillatory behaviour but as regular periodic cycles. These patterns in both vertebrates and invertebrates may well have intraspecific competition at their base so are considered here.

8.3 Population Cycles in Vertebrates

The established pattern of vertebrate cycles seems to be one of a 3-4 year periodicity for rodents and a 10-year periodicity for lagomorphs. Although workers in the past have occasionally suspected that such cycles could be generated by stochastic (random) processes (e.g. Cole, 1951) this is not now a popular view and, for the 10-year cycle at least, Moran (1953) showed that it could not be randomly derived. It should be remembered, however, that the famous cycles are not as clear as those computer-generated ones from simple models, such as in Figure 13.1. Equations like (7.5), with the incorporation of time-delays, can produce patterns very similar to wildlife cycles and early theories were compatible with this. This equation and related ones show clearly how a certain periodicity can be obtained which matches observed patterns simply by varying the lengths of the system's time delay in relation to the population's response time (Table 7.1). We need to consider whether this intra-specific approach explains enough and, in particular, the ecological and beha-vioural mechanisms behind it.

There have been propositions that density-independent events could drive the cycles but this cannot be the case as the 'crashes' occur when the population exceeds a certain level. Also, no density-independent parameter has as yet been found with a 3- or 10-year cycle. Further, less theoretical evidence for an intrinsic mechanism comes from Chitty's unpublished work on two populations of wood mice (*Apodemus*) only four miles apart and within the same young plantation; these nevertheless exhibited cycles clearly out of phase with one another.

Density-dependent theories based on extrinsic factors, notably predation, have had much publicity. The lynx and showshoe hare population oscillations are particularly well known (Elton and Nicholson, 1942). It now seems that the lag in the system which would occur before the predators' numbers responded to those of their prey would be too long for the observed 'responses' to occur so we must look to intrinsic factors. This type of explanation is given weight by the manipulative work of Krebs (unpublished) in which he translocated animals from a population nearing its peak to an area where populations were just beginning to increase. The translocated group continued to peak and then decline in phase with those left behind and did not parallel local population events.

Chitty (1960) was the first to suggest that if the traditional hypotheses involving predation (and also disease, climate and food depletion) were inade-quate, then the changes must come from within. He suggested that cycles of vole populations were due to changes in the selective advantages of various genotypes within the population as density changes. These changes, he suggested, were intimately connected with changes in the animals' behaviour, notably their aggression. The proposed system can be summarised as below.

High density leads to increased aggression which in turn results in selection for genotype x. Genotype x causes reduced viability and thus a decline in population density. Selection is now against genotype x and the population may increase in numbers once more. The most dubious aspect of this theory is perhaps the suggestion that a genotype can change sufficiently rapidly, but apparently this is possible. Krebs (1971) has established that a repeatable syn-drome of changes in birth, death and growth rates accompanies fluctuations in *Microtus* (short-tailed vole) populations in North America. The syndrome is complex and affects more than one age group. During it, male aggressive beha-viour changed throughout the population cycle and this level of aggression correlated, albeit weakly, with the rate of population increase. While it was not possible to study changes in the genetic basis of such aggressive behaviour, Krebs was able to demonstrate changes during the population cycle in a number of other genetic characteristics. Relative allelic frequencies were studied in two polymorphic systems. In both cases, Krebs showed clear changes in gene fre-quency associated with demographic change; clearly genetic change such as that required by Chitty's theory *can* occur within the time limits set.

In the above way, involving a genetic change and the changing action of selec-tion pressures, regulatory processes with fairly long periodicities take place, producing regular oscillations. However, such time delays must be common to many populations; yet these cycles are shown only by some species and popula-tions, and even within the same species some populations show stable point equilibria, others true cycles; clearly our explanations so far are insufficient in that they cannot account for the fact that cycles are observed only in certain circumstances.

Schultz (1964) proposed that in the well-known lemming cycles a parallel cycle in soil nutrients existed which explained the mammal cycle. He suggested that in such systems nutrients are extremely limited. When lemming numbers are low, nutrients are readily available to the vegetation which is thus productive and highly nutritious. The lemmings feed and reproduce well. With an increased population, an increasing proportion of the limited nutrients become incor-porated in the lemmings' tissue with a consequent declining proportion in the soil. With vegetation now less productive there is a decline in lemming reproduc-tive rates. Mortality is unaffected but mortality rates outweigh lowered recruit-ment leading to a population decline. Nutrients return to the soil and thus to the vegetation, returning the cycle to the beginning again. This theory gains support from Schultz's own studies which did show such a cycle of soil nutrients exactly in phase with the lemming cycle. An alternative theory was proposed by Freeland (1974). This theory was based on a parallel grazer-plant process but in this case dependent upon a preference by voles for non-toxic food. This preference could be less easily exercised as vole density increased, so increased quantities of toxin were ingested, leading to a reduction in growth and reproduction and an inhibi-tion of sexual maturity. Subsequent evidence which supports Freeland's pro-posal is that there are differences in the rates of population decline between sexes and age classes (Krebs et al., 1969), that young animals are shown to be more susceptible to toxins than were older animals and that males have increased liver microsomal enzyme activity (allowing faster degradation of toxins).

The theories of Freeland and Schultz both imply that the studied animal populations closely approach the carrying capacity of their environment, modify it and then suffer numerically until it recovers again. In this sense we are close to the Dempster and Pollard (1981) 'ceiling' model discussed in Section 7.5. Those authors considered a changing carrying capacity but here we have the added complication of a capacity changed by the organisms which use it. In other

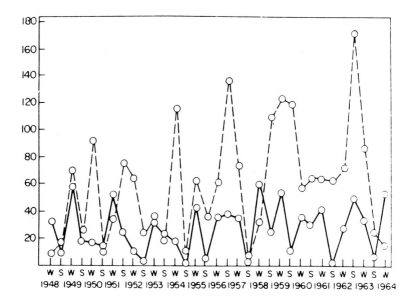

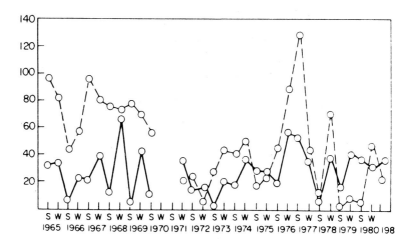

Figure 8.11: Indices of May-June Abundance for Bank Voles (*Clethrionomys*, - - - - -) and Wood Mice (*Apodemus*, ———), Showing Possible Cycles for the Latter Species. Source: From Southern and Lowe, 1982.

words, these populations approach and are in danger of exceeding their environmental capacity; it is this closeness of population requirement and environmental resource that, in these theories, leads to cyclic phenomena. This remains, however, a proximate answer only, and still does not really explain why certain populations should show cyclic behaviour (or at this level of analysis why certain populations should approach carrying-capacity to the level at which they themselves may alter that capacity) while others do not.

One apparent ecological common-denominator of the systems in which cycles have been described is that they are all relatively simple ecological communities. In more complex environments, intrinsic patterns such as we have seen could be obscured by other biotic and abiotic agencies, just as was demonstrated in the potentially stable sycamore-aphid system of Figure 8.10. In other words, it is quite possible that *all* populations show the potential for cyclic behaviour, but cycles are not expressed in more complex communities where the effects of interactions with other species may result in the development of different patterns of equilibrium.

We have already noted that, in theoretical analyses at least, stable limit cycles in single species populations will result automatically from time delays in self-regulatory responses to density if these delays are large enough. Time-lags of some degree are implicit in most theories of intrinsic density-dependent regulation, since these systems operate primarily through variations in birth rate, which, by definition, must take a generation before it is effective. As a result, intrinsic mechanisms of regulation of many vertebrate populations may be such as to promote cycles; see Figure 8.11.

The theories of Chitty and Krebs, while they may suggest a mechanism whereby the time-delays encountered in normal, non-cyclic populations may be exaggerated to a level where perhaps they are sufficient to cause cycling, cannot explain why this exaggeration should only occur in the particular circumstances in which we do observe cycles in the field. Similarly, the hypotheses of Schultz and Freeland supply only a partial answer and cannot explain why cycles are not encountered in all populations. Here, we are accepting that intrinsic mechanisms of population self-regulation may be such as to result in cycles in many if not all vertebrate populations, but that these may be masked by between-species population interactions in all but the simplest systems. In addition, although animal populations may reach equilibrium below carrying-capacity, this occurs normally when population numbers are affected also by predators or competitors; in simpler systems where populations are purely self-regulated and equilibria are established by intraspecific competition for limited resources, numbers usually stabilise at the carrying-capacity set by those resources; see Section 7.5c. Thus if there is any tendency for cyclic behaviour around this point, population changes may themselves influence carrying-capacity (as with Schultz's lemmings, and, once again, such an impact on carrying-capacity will be more likely in *simple* communities) exaggerating the cycles of the population with cycles in the resources themselves.

8.4 Population Cycles in Invertebrates

It is tempting to compare the cycles of vertebrate grazers in Figure 8.11 with those for insect grazers, some of which are shown in Figure 7.2, and to speculate that intraspecific mechanisms drive both. However, the great advantage of the insect data also leads to problems of interpretation — they span such a long period. Rigorous life-table data were in most cases not collected and virtually all we have are weather data together with minimal information concerning parasites. More detailed work on forest insects unfortunately usually combines more rigorous life-table data with a time-span too short to reveal obvious cyclical behaviour. The winter moth population and model in Figure 7.18 combine a long-term study with good data but the patterns revealed are hardly as clear as those for vertebrates in Figure 8.11. It seems from the rather 'coarse-grained' data from forest pests that time-delays again are operating, this time through parasitoids and/or changing plant condition. Whether the latter is changed by the insect grazers themselves is unknown; if it was, it would parallel the small mammal story in the previous section.

This brief section is limited by the type, not quantity of data collected. As we point out many times in this book, ecological analyses are limited in their scope by the methodologies chosen for initial data collection. Ecological data are usually hard-won and often not repeatable; if the questions and proposed analyses are clear at the outset, limited, unproductive and sometimes spurious interpretation is avoided; it is tragic that the potential for pest prediction embodied in the population changes in Figure 7.2, for instance, largely cannot be realised because of a lack of supporting data.

9 Predators, Parasitoids and Population Stability

9.1 Why Study Predators and Parasitoids?

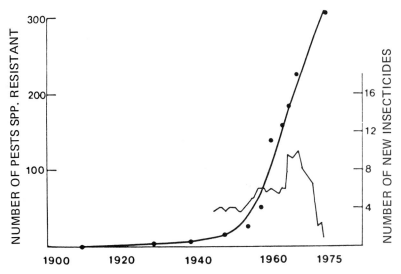

Figure 9.1: The Number of Insects Resistant to One or More Insecticides (Worldwide) and the Rate of Development of New Insecticides. Source: Data from Georghiou and Taylor, 1977, and Lewis, 1977; after Southwood, 1979.

One of the lessons from the analysis of populations in Chapter 7 was that natural enemies may sometimes cause density-dependent mortality of their prey and may be a major regulatory factor. This potential is one of the reasons why the role of such natural antagonists has been intensively investigated in the past 20 years or so. The other main reason is that there has been increasing interest in the biological control of pest and disease organisms, stimulated by the inter-related problems of increased pesticide costs, reduced rate of production of new agro-chemicals, increased pest and disease resistance to agro-chemicals, dangers of 'pest trading' and pest resurgence and concern about environmental disruption and pollution. The details of these applied aspects of ecology are beyond the scope of this book but are dealt with in Wratten and Watt (1984); however, Figures 9.1 and 9.2 and Table 9.1 illustrate some of the background to current

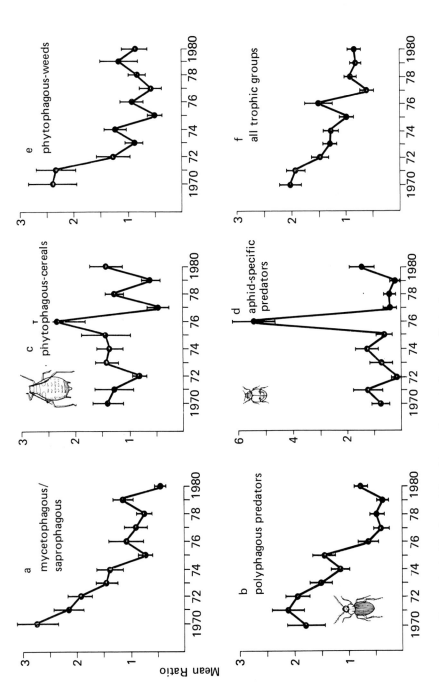

Figure 9.2: Changes in the Numbers of Cereal Invertebrates in Southern England over Eleven Years Shown According to Trophic Group. Source: From Vickerman *et al.*, 1983.

concern about the farm ecosystem. However, in some countries, such as the UK, biocontrol attempts in the field are a rarity and lag far behind the theory. Elsewhere, when ecologists have initiated biocontrol programmes, selection of the most appropriate enemy has often not been based on clear ecological criteria, follow-up of the consequences of release has been poor and success rate has not been good (Table 9.2). In the early days of biocontrol attempts in New Zealand, for instance, potential new biocontrol agents were released from the windows of moving trains and left to the mercy of the environment, with little scientific follow-up. There have also been cases where enormous numbers of adult ladybirds have been released into a crop, only to result in their virtual complete disappearance in a few days because their prey were at too low a density to induce them to remain in the crop. There is a clear need for a theoretical and experimental basis for biocontrol; the 4 per cent success rate in arable crops (Table 9.2) shows how far there is to go. This chapter will use theoretical, laboratory and field evidence to investigate the components of the predation process and the implications for the regulation of prey numbers, i.e. for the imposition of stability on the population. We will use the contrasting, laboratory-orientated analytic inductive approach and field-based, deductive methods.

Table 9.1: Total Numbers of Species in a Farmland Monitoring Scheme That Have a Positive (+ve) or Negative (−ve) Slope (b) when the Relationship Between log Numbers and Year is Examined by Regression Analysis

| | | Total no. species | | | |
| | | b | | b ($P < 0.05$) | |
Faunal group	Period	+ve	−ve	+ve	−ve
Cereal insects	1970-80	23	85	2	37

Source: From Vickerman et al., 1983.

Table 9.2: Cases of Biological Control of Insects by Imported Natural Enemies Grouped According to Habitat Characters (on r-K Continuum); Based on the Analysis of Table 12 of De Bach (1964).

| | All cases | | Cases graded 'Complete Control' | |
Habitat type	No.	% total	No.	% total
r-end				
Cereals and forage crops (wheat, lucerne, etc.)	6	4	0	0
Vegetables (tomatoes, asparagus, taro, etc.)	21	14	0	0
Sugar cane, cotton, pasture	19	12	2	8
Trees (citrus, coconut, oak, olive, grape, etc.)	107	70	23	92
K-end				

Source: From Southwood, 1977.

9.2 Analytical Models and the Components of Predation

The approach which has dominated this aspect of population ecology has used mathematical models based either on differential calculus or on difference equations. The former, like the simple, single-species model on page 130 (equation 6.2) are appropriate for organisms with overlapping generations and more or less continuous births and deaths. Organisms whose populations exhibit discrete generations and which show a succession of different developmental stages are best suited to difference equations. In these, population events progress in a 'stepwise' fashion, a step usually representing a generation. This approach will receive the main emphasis here, as it has been used in a large proportion of studies, especially of insects; we saw earlier that this group dominates population ecology theory.

Some of the work on the theory of how natural enemies respond to fluctuations in the density of their prey is actually more applicable to parasitoids than to predators. Parasitoid is the term given to a hymenopterous or dipterous insect which develops as a larva inside a host, eventually killing it; it is a 'predator' which usually kills only one 'prey' for complete development and, as we shall see later, this useful simplification helps make the models which are developed more tractable mathematically. For the sake of simplicity we will use the term 'predator' to encompass both, stating specifically when we mean one rather than the other. The history of models in this subject seems to have been dominated by differential equations for vertebrates and by difference equations for predators and parasitoids. True parasites have received much less attention but many of the population principles are similar. Examples of recent models in this area, using differential calculus, may be found in Anderson and May (1978) and May and Anderson (1978). The ecological characteristics of predators and parasitoids, alluded to above, which should influence a realistic model and which have led to the domination by parasitoid-orientated models are given below:

Predator	*Parasitoid*
1. Prey removed following attack and killed in a short time	Host remains *in situ* – not killed immediately

A ladybird larva may consume its aphid prey in seconds or minutes while a second-instar aphid host in which a parasitoid egg has been laid may move, moult three more times and even colonise new habitats before it dies. This has implications for re-discovery by the same or a different searching parasitoid and an ideal model should allow for these possibilities.

2. No further reproduction after successful attack	Host may reproduce after the parasitoid has laid its egg

If an insect is parasitised late in its larval life, significant reproduction may occur before the parasitoid larva within kills it. Some aphids, for example, can produce a large proportion of their potential under these circumstances. This makes modelling more difficult and is not allowed for in many models and restricts their applicability.

3. Larvae and adults may search   Only the adult females search for
   for prey   hosts. Larvae are usually internal,
                                    with reduced sense organs.

Considerations of the predator/parasitoid efficiency, crucial to most models, would need to take into account the way predator larval instars differ and also the way the adult's efficiency differs from this.

4. Many prey individuals are needed   Usually one new parasitoid arises from
   for predator development   each egg laid.

The way a predator larva's consumption changes in relation to prey density is of crucial importance in some types of model but this aspect is irrelevant for a parasitoid.

Holling (1964) suggested that a good ecological model should be realistic, precise and general. As we proceed through this chapter, these ideal aims should be borne in mind; however, the inclusion of all theoretical 'ingredients' in an analytical model fully satisfying Holling's three criteria would produce mathematical statements too complex to be of much use. Instead, attempts have been made to capture the 'essence' of parasite or predator behaviour in a limited number of mathematical functions which, when incorporated in an intergeneration model, show the contribution of the individual components to the stability of populations of predator and prey.

Most difference equation models (models which progress in a stepwise way, in this case, usually by generations) for parasitoid action are elaborations of a basic form:

$$N_{t+1} = N_t \, f(N_t, P_t) \tag{9.1}$$

$$P_{t+1} = cN_t \, 1 - f(N_t, P_t) \tag{9.2}$$

The upper model shows how prey numbers change from one generation (N_t) to the next (N_{t+1}) as a function (f) of their own and predators' numbers (P_t) and behaviour, driven by λ (lambda), the prey rate of increase per generation. λ may be constant or itself may change in relation to prey density through intraspecific events (see Chapter 8). In equation (9.2), c defines the mean number of predator progeny produced per prey attacked. The assumption in many models that this is unity brings them closer to parasitoid ecology, as stated above. In equation (9.1), $f(N_t, P_t)$ is the mortality function, so to obtain predator survival, $1-f(N_t, P_t)$ appears in equation (9.2). Models which are more appropriate to real predators can be represented by equation (9.3) below (Beddington, Free and Lawton, 1976):

$$P_{t+1} = P_t Q(N_t, P_t) \qquad\qquad (9.3)$$

in which the function Q defines the rate of increase per predator as a function of the density of the predators themselves and of their prey.

Given that most models start with the assumption that we are dealing with parasitoids, and allow themselves the biological luxury of the relationship: one parasitoid egg – one new adult female parasite, we need to start with an explanation of the factors influencing the number of hosts (prey) killed. This is important for predators too, but population models derived from these relationships are most appropriate to parasitoids. Our aim is to proceed from a 'dissection' of the components of an individual predator's response to prey density to a population response, looking at the consequences for stability for each 'ingredient' acting alone. We will start with the simple Nicholson-Bailey model, not only because of its historical interest but because many later models still depend on some of its basic assumptions. In other words, the next 20 pages are effectively a behavioural and ecological elaboration of the real components of the functions f and Q in equations (9.1), (9.2) and (9.3) above, and of their potential contribution to a predator's regulation of the numbers of its prey. This will lead us to the role of this 'piecemeal' approach to the prediction of an antagonist's potential role as a biocontrol agent. Then, just as we contrasted the analytical (life table) approach to population dynamics with that of simulation modelling, we will here examine the manipulative and field-deductive approaches to natural enemy action in an attempt to obtain a general statement on whether natural enemies are ever involved in population regulation of their prey. This is a slightly different emphasis from that in the population chapter (7), in which we asked which of all the biotic and abiotic possibilities (a) disturbed and (b) regulated animal and plant numbers but is similar to Chapter 8 in which we considered whether intra- and interspecific competition could promote stability.

9.2a A Simple Predator-Prey Model

Nicholson (1933) was aware that predators and prey appeared to exist in nature in a type of 'balance' in that extinction of one or the other is a rarity. His simple model was an attempt to mimic and explain such persistence; Lotka (1925) and Volterra (1926) had earlier produced the first mathematical version of predator-prey interactions, based on differential equations. With many constants and simple assumptions, their models produced persistence, through oscillations, of predator and prey populations. The ironical result of Nicholson's attempt, using in this case difference equations, was that, under most circumstances, increasingly unstable population fluctuations were produced, leading eventually to extinction of prey and host. In assessing the model for the reasons for this lack of stability, we need to decide whether the model is merely incomplete or that one or more of its assumptions are ecologically too simplistic, or wrong. If the latter is the case, we should be wary if later models have continued to depend on these assumptions. Nicholson's model requires the following to be the case:

(1) The predator and prey have synchronised generations. This means that they should be of the same length and start at the same time
(2) Predator and prey generations must each be discrete, i.e. no overlap between them. This was also a requirement for age-specific life tables (p. 166)
(3) One prey/host found leads to the production of one new natural enemy; nearer, but not very near, to a parasitoid than to a predator
(4) The predator searches randomly. This implies that there are no effects of prey density and spacing, the physical environment or of other searching predators
(5) The predator has an unlimited potential fecundity and cannot become satiated
(6) The predator has a characteristic, constant 'area of discovery' (*a*) i.e. a constant searching efficiency.

To see the consequences of such an impossible predator's action (bearing in mind that some later models still use an *almost* impossible predator) we need first to define 'area of discovery'. This has at least four definitions, all of which mean more or less the same thing:

(1) the total area searched in the predator's lifetime
or (2) the likelihood (probability) of finding a particular prey during the predator's lifetime
or (3) the proportion of the total area available which is searched
or (4) the average number of attacks/host.

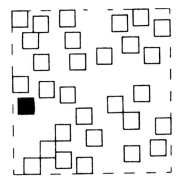

Figure 9.3: A Hypothetical Arena of 4m² With a Predator's Area of Discovery of 0.25m² (Black Square). White squares are prey.

If we envisage a mechanical predator which searches by placing a square search area down on an arena, like a rubber stamp, then Figure 9.3 helps us to visualise what area of discovery means. If the total area in Figure 9.3 is 4m² and the small black square is the predator's lifetime search area of 0.25m², then its area of discovery is 0.25/4 = 0.0625. The probability of a particular host (white squares) being captured, if the small square is placed randomly on the large one, is also 0.0625. This is also the average number of attacks/host if the host remains after attack. So the most useful definitions above (1, 2 and 3) all give $a = 0.0625$.

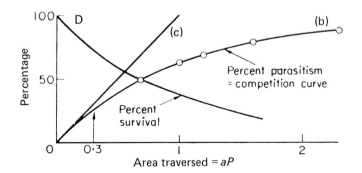

Figure 9.4: Nicholson's 'Competition Curve' (b). The straight line (c) would be obtained if the parasites searched systematically. Source: From Varley *et al.*, 1973.

Although Nicholson's predator has a constant search area, he did recognise that, through random movement, the rate at which new (i.e. previously unsearched) areas would be encountered would decline as the animal crossed its own tracks; this requires an almost impossible prey too, of course, one which never moves! Nicholson called the total area searched (some of it more than once) the area traversed; the area effectively searched by one or more predators will decline as the area traversed, or their number, increases, so the proportion of the prey population which is found will increase at a declining rate (Figure 9.4). Nicholson called this a 'competition curve' in that one predator searching randomly for static prey is effectively competing with itself by returning to previously searched areas. With these simple behavioural 'ingredients' we need to examine the consequences for predator-prey stability of predators searching in this way.

We begin with P parasites each with the same area of discovery, a; we predict the number of hosts parasitised; this number then becomes the population of searching parasites in the next generation. We use the Poisson distribution as a model to distribute attacks randomly between the prey units; the first (zero) term of the series is e^{-m}, where m is the mean, in this case the mean number of attacks per host. This term gives us the proportion of the population receiving no attacks. For *one* searching parasite, the mean required is a, defined above as the average number of attacks per host; with P parasites, the mean is aP. (The proportion of hosts attacked at least once is therefore $1-e^{-aP}$.)

Our population model for a prey species reproducing at a rate F per generation is obtained by substituting e^{-aP}, the proportion surviving, for $f(N_t, P_t)$ in our general equation (9.1) on p. 223. This gives:

$$N_{n+1} = F.N_n e^{-aP}$$

(9.4)

or $\log_e N_{n+1} = \log_e F + \log_e N_n - aP$ (9.5)

where N_{n+1} is the number of hosts in the next generation. For the parasite,

$$P_{n+1} = N_n - N_n e^{-aP}$$ (9.6)

i.e. the number of hosts minus the number of survivors.

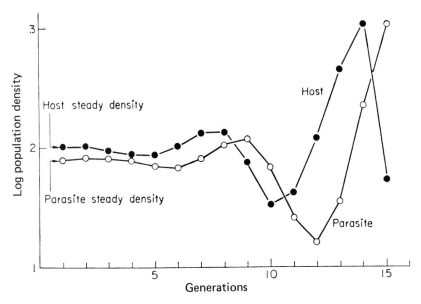

Figure 9.5: A Population Model Based on Nicholson's Theory (Equations 9.5 and 9.6). Source: From Varley *et al.*, 1973.

The unstable situation to which this model leads can be seen in Figure 9.5; this was the opposite of Nicholson's intentions but he suggested that, although local extinctions are likely to occur, immigration and emigration, occurring continually, would ensure that persistence of predators and prey over a more general area.

Stability through the Nicholsonian model could be brought about if the prey population exhibited some density-dependent regulation through intraspecific competition; we saw in the last chapter how this can stabilise single-species populations and it can also do so here; Beddington *et al.* (1975) showed how this can happen and we will return to their papers later when we discuss biocontrol successes and failures. It is not enough, however, to stabilise the Nicholson-Bailey model through prey population density dependence as we are still left with the glaring biological assumptions on which it depends; the parameter which has received a vast amount of experimental and theoretical exploration is represented by the simple letter a — the area of discovery. This is known to be an infinitely variable predator attribute and the way it varies may sometimes

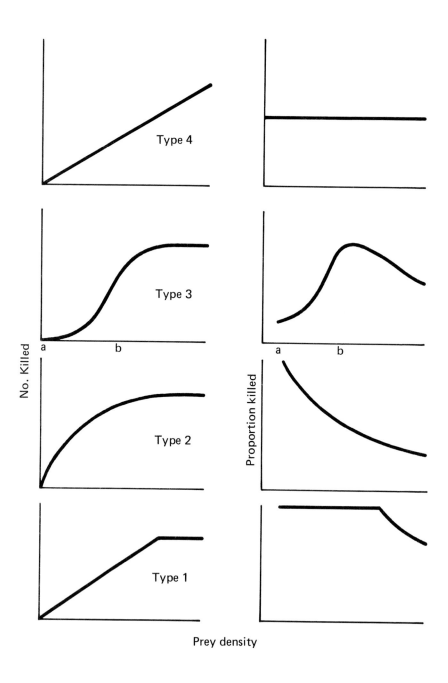

Figure 9.6: Four Types of Relationships Between Prey Density and the Attack Rate of a Single Predator (the Functional Response). Only type 3 yields density-dependent mortality, in the interval *a-b*; type 4 is that of a 'Nicholsonian' predator.

confer stability on a system. In examining the way a changes in real predator ecology, we should not forget that many of the models which examine its properties still leave unchanged many of the other Nicholson-Bailey assumptions.

9.2b The Relationship Between a Predator's Consumption and Prey Density – the Functional Response

Implicit in the Nicholson-Bailey model is the assumption that the predator exhibits a linear functional response (Figure 9.6a); this is impossible in practice. The four theoretical general forms the relationship can have are shown in Figure 9.6. The important aspect of these responses is whether or not they exhibit any density dependence as, considered in isolation, a functional response can only regulate a prey population if some density dependence is shown. Types 1 and 2 are partly or completely inverse density dependent so would tend to disturb rather than regulate a population, while type 3 shows some density dependence (between prey densities a and b). Type 1 is rare but type 2 was thought, until recently, to be the virtual norm for invertebrate predators while type 3 seemed to occur only among vertebrates. We shall see later that these distinctions are not so clear but we first need to consider why type 2 is common and what drives it. It is tempting to suggest that the declining curve in Figure 9.6b is simply a result of predator satiation – there must be a limit on consumption. In fact, Mills (1981) has modelled a ladybird/aphid functional response with satiation as an important component (see below). However, the asymptotes in Figure 9.6 can be reached experimentally when a predator is demonstrably still hungry; this led Holling (1959) to suggest that handling time could be the reason. Given that during the time it takes to quell, eat and perhaps digest the prey, searching ceases, it follows that as the number of prey captured in a fixed time interval increases, the proportion of time available for searching will decrease. This can lead to type 2 functional response and is the essence of Holling's model. We can express these ideas more formally: if Holling's explanation of the departure from linearity is correct, i.e. that an increasing proportion of searching time is occupied in dealing with prey as their density increases, then the simplest expression of the relationship is:

$$y = a'T_s x \qquad a'T_s = \text{area of discovery} \qquad (9.6)$$

where y = no. of prey eaten, x = density of prey, T_s = time available for searching, and a' is the instantaneous attack rate. To understand the parameter, a', consider two predators searching an area with prey density N and each with the same searching time T_s. If one predator eats 10 times as many prey as the other in the same time, this difference must be due to its higher attack rate. This could be, in a ladybird larva for instance, related to the distance at which it can perceive a prey, the distance between its forelimbs, its ability to quell a struggling prey, its speed of travel, etc. In the above notation, a' can be calculated as $a' = y/T_s x$. If T_t = total time for one experiment and b = time taken to deal with one prey item, then:

$$T_s = T_t - by \tag{9.7}$$

Substituting (9.7) in (9.6):

$$y = a'(T_t - by)x \tag{9.8}$$

which simplifies thus:

$$y = T_t a'x - bya'x$$

$$y + bya'x = T_t a'x$$

$$y(1 + a'bx) = T_t a'x$$

$$\therefore y = \frac{T_t a'x}{1 + a'bx} \tag{9.9}$$

Equation (9.9) is familiarly known as Holling's disc equation because Holling originally tested the idea using sandpaper discs at different densities on a card table and a blindfolded human 'predator' tapping with one finger tip. The way this model uses a changing value for a as T_s changes is a major difference from the constant a of Nicholson/Bailey's equations. However, it does assume constant values for a' and for handling time (b) and when data are fitted to Holling's equation, these parameters are rarely measured but rather are calculated empirically by fitting observed values for y to the model and calculating them by rearrangement. However, Holling's model does show that:

(1) the relationship between prey density and numbers captured by a predator need not be linear,
(2) a predator's area of discovery is not constant,
(3) a functional response of the above type (Type 2) cannot yield density-dependent mortality,
(4) handling time can be a major component in a predator's response to prey density, and while prey is being handled, searching ceases.

However, the model still assumes that the hypothetical parasite has no egg limitation, that the predator does not become satiated, that prey are randomly arranged and that searching is random.

One consequence of the dependence on random search with no prey replacement is that in practice, as soon as one prey item is removed, the density changes and the true probability that subsequent searching will contact a prey item has been reduced accordingly. There is also plenty of experimental evidence that a' and b are infinitely variable (e.g. Figure 9.7). This is not surprising as a' is likely to vary with the following, all of which will change diurnally or as the predator grows: (1) distance at which the prey can be perceived; (2) speed of movement;

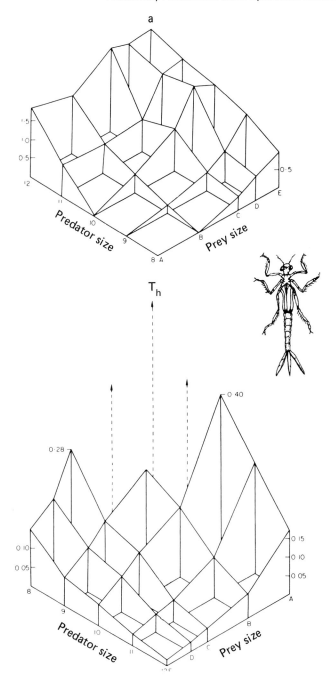

Figure 9.7: The Effect of Both Predator and Prey Size on a' and T_h Measued Under Standard Conditions for the Damselfly, *Ischnura elegans*, Feeding on *Daphnia magna*. Source: From Thompson, 1975.

(3) capture efficiency. Handling time (b) will also vary for many reasons related to predator development. In other words, if the total searching time (T_t) is large, the model overestimates the number captured. The disc equation can form the basis of better predator/parasite models if prey exploitation is taken into account by a randomly moving predator (Rogers, 1972). If this is done, however, one equation can no longer predict both predator *and* parasite 'consumption' because of the fact that parasites do not remove their host, so it can cause time-wasting in the future by the same or other searching parasites. Although the disc equation *and* Rogers' random predator and random parasite equation all generate a curve that fits the data, they give different estimates of the handling time b and attack coefficient, a', because they make different assumptions about the way predators and parasites search. The disc equation assumes *systematic* search (i.e. no re-searching of areas), Rogers' equations assume random search. Rogers' methods of calculating a' and b' are statistically dubious, too, and this is another reason why these calculated values may differ from observed ones and from those of Holling's equation. Experimental data can often still be fitted quite well by these models, however, so perhaps the potential problems with changing values of a' and b are not as important as they appeared. Such functional responses acting alone always result in instability of predator and prey populations, and the longer the duration of handling time in relation to T_t, the greater the instability (Hassell and May, 1973). This is not surprising in that larger values of b make the relationship more strongly inversely density dependent. In practice, the measured ratios are often very small, ranging in a survey by Hassell (1978) from < 0.001 to 0.1 for parasitoids and from < 0.001 to < 0.005 for predators. So perhaps the frequency of type 2 responses, despite insignificant handling times, suggests that satiation is a more frequent cause of this type of curve than previously thought. We are still dealing with an inverse density-dependent response, however, and even Mills' (1982) recognition and incorporation of the components of satiation still gives a type 2 equation with properties similar to those of Rogers' model.

Sigmoid functional responses (type 3) with their element of density dependence seem much more promising for the promotion of predator-prey stability, especially since they are now not the rarity among invertebrate predators which was once believed (Hassell, Lawton and Beddington, 1977). The rapidity of this changed awareness is emphasised by Figure 9.8, which appeared only a year earlier; the postulated sigmoidality in these responses is less than convincing. The reason for the occurrence of these sigmoid curves is not dissimilar for invertebrates and vertebrates. In both cases, one or more of the components of a' changes with increasing prey density; Figure 9.9 shows this. For vertebrate predators, too, a changed awareness in predators to the presence of their prey (sometimes called the development of a searching image) can lead to the seemingly important flexure of the curve (Figure 9.10). The reason why invertebrate sigmoid functional responses were shown to be fairly common much later than were those for vertebrates is related to the way functional response experiments are often conducted. It has been common to provide easily catchable,

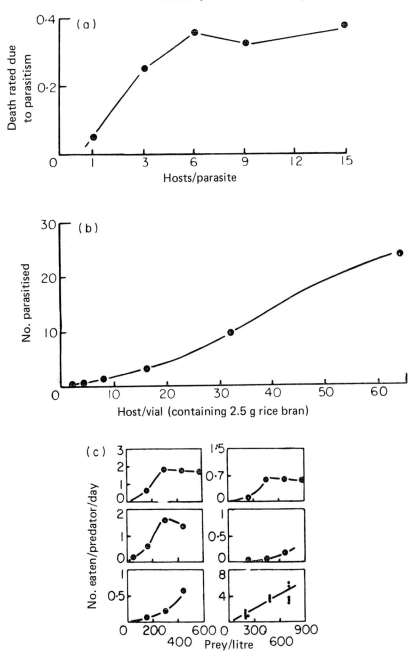

Figure 9.8: Possible Type 3 Functional Responses Suggested by Hassell *et al.* (1976).
(a) *Encarsia formosa* parasitising *Trialeurodes vaporariorum* (Burnett, 1964); (b) *Nemeritis*
(*Exidechthis*) *canescens* parasitising *Cadra* (Takahashi, 1968). (c) The predatory copepod
Cyclops bicuspidatus thomasi attacking a variety of different types of zooplankton (from
left to right and down the page: nauplii of their own species; copepodites of their own
species; *Diaptomus* nauplii; *Diaptomus* copepodites; Cladocera; rotifers) (McQueen, 1969).

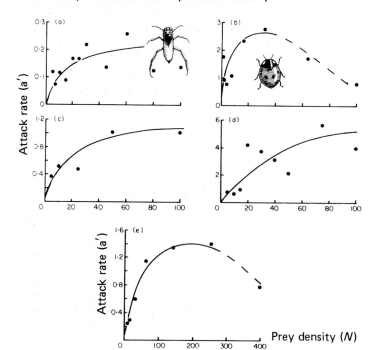

Figure 9.9: Relationship Between Attack Rate, a', and Prey Density (N), in Five Species with Sigmoid Functional Responses. The method used to calculate a is given in the text. (a) *Notonecta*, T_h = 0.091; (b) *Coccinella*, T_h = 0.025; (c) *Aphidius*, T_h = 0.018; (d) *Plea*, T_h = 0.029; (e) *Calliphora*, T_h = 0.0095. Attack rates are defined per experimental universe, with T = 1. Source: From Hassell *et al.*, 1977.

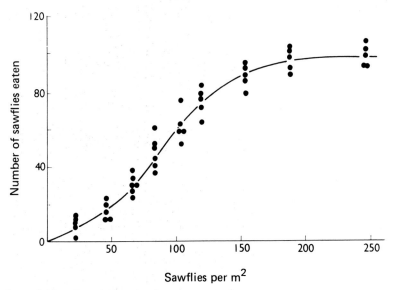

Figure 9.10: A Type 3 Functional Response of a Deermouse to its Sawfly Prey when an Excess of an Alternative Food (Dog Biscuits) is Also Provided: Source: After Holling, 1965.

often large, prey, whereas if they are more difficult for the predator to catch (e.g. smaller, so many are missed) a' may change at intermediate densities as the predator becomes more efficient through more frequent encounters. However, in the type of coupled, difference equation models with which we are dealing, it is *not* true that the density dependence in these responses leads to regulation. Hassell and Comins (1978) showed that, because the predator is prey-specific, there is a time-delayed, one-generation feedback between changes in predator numbers and the resulting effects on the prey. Models without time delays, such as the well-known Lotka-Volterra predator-prey model, *can* demonstrate stability resulting from sigmoid functional responses and once we move away from prey-specific predators with generations linked to those of their prey, then sigmoid functional responses again show regulation potential. A glance at Hassell *et al.*'s paper, however, reveals that many of the examples come from such systems as ladybird/aphid, waterbug/amphipod, parasitoid/aphid and even blowfly/sugar droplet! In these cases, it is difficult to see how the requirements of either the Lotka-Volterra or difference equation models could be satisfied, given the disparity of lifestyle and generation time of predator and prey and the fact that the responses were mainly derived from experiments in which uniformly sized prey items were offered to the predator. It therefore seems that the significance of even strikingly sigmoid functional responses in real population dynamics is still largely unknown. There is another problem in that some published putative sigmoid responses are not particularly convincing. Statistical tests based on Rogers' methods, mentioned earlier, are sometimes used to demonstrate that the data do not fit a type 2 and therefore must be a type 3. The statistics are doubtful and this 'negative proof' is not particularly logical either. Cook (1978) produced a graphical method in which $\log_e$ (number of survivors) is plotted against the number of prey eaten and a V-shaped curve signifies a sigmoid functional response. Problems arise when the shape of the graph does not exactly mimic the required pattern; how near does it have to be to 'prove' sigmoidality? Perhaps the best method is that of Griffiths (1983), in which polynomials are fitted to the curve in order to detect statistically the type-3 inflections; this is more rigorous than the other methods and, usually, no more than a third-order polynomial is needed for the test.

Sigmoid functional responses have been emphasised here because of their potential for regulation, albeit only in some types of predator-prey systems. A worrying question remains, however; do functional responses actually exist in nature? In the field, a predator moves between areas containing prey at different densities, feeding on prey of a wide size-range. A predator confined with a uniformly sized prey collection is meaningless under field conditions, so perhaps the uncharitable view of the functional response is that it tells us little more about the real world than some kind of maximum consumption rate. Whatever we believe, we are still dealing with a single individual's response so there are many other components of the predation process which may yield density dependence. What, for instance, are the consequences of more than one predator searching in the same area? Do they accumulate above a particular prey density

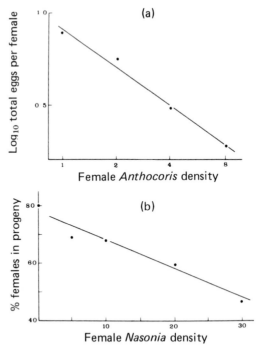

Figure 9.11: The Effect of Female Density on (a) the Fecundity of the Predatory Bug, *Anthocoris confusus* (Evans, 1973), and (b) the Sex Ratio of the Progeny of the Parasitoid, *Nasonia vitripennis* (Wylie, 1965). Source: From Hassell, 1978.

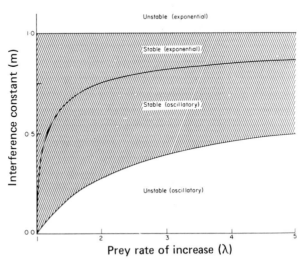

Figure 9.12: Stability Boundaries from Equation (9.11) in Terms of the Interference Constant *m* and the Prey Rate of Increase λ. The hatched area denotes the conditions for stability and is divided into two regions; one of exponential damping lying above that of oscillatory damping. The line between these regions indicates the conditions for the most rapid approach to the equilibrium. Source: From Hassell and May, 1973.

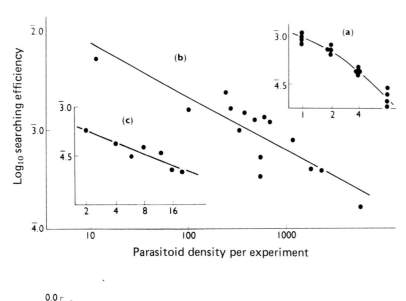

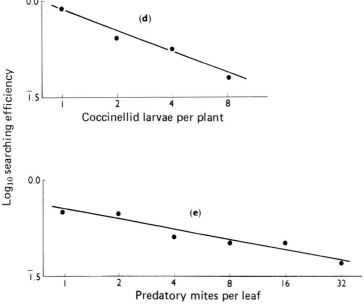

Figure 9.13: Interference Relationships Between Searching Efficiency (Expressed as the Area of Discovery) and Parasitoid (**a** to **c**) or Predator (**d** and **e**) Density per Experiment, on Logarithmic Scales. (**a**) *Pseudeucoila bochei* (data from Bakker *et al.*, 1967), (**b**) *Nemeritis canescens* (after Hassell and Huffaker, 1969), (**c**) *Cryptus inornatus* (data from Ullyett, 1949), (**d**) Coccinellid larvae (*Coccinella 7-punctata*) (S. Michelakis, unpublished), (**e**) Predatory mites (*Phytoseiulius persimilis*) (J. Fernando, unpublished). Source: From Hassell, 1976.

threshold or in response to prey patches, or lay eggs in a 'density-dependent' way? Having accumulated numerically or reproductively, do they compete and if so, is their efficiency reduced? We will consider these questions in the next section.

9.2c Numerical Responses to Prey Density, Predator Interference and Prey Aggregation

An early empirical model of the consequences for interference of a multi-predator system was based on the Nicholson-Bailey model and is that of Hassell and Varley (1969). This was based on the fact that, at least in laboratory systems, there is often a negative relationship between log searching efficiency and log predator density (Figure 9.11). The standard Nicholson-Bailey model does not allow for this, of course, in that a is fixed, so Hassell and Varley derived a new expression for a:

$$a = \log Q - m \log P \qquad (9.10)$$

where Q is the intercept on the y axis and m is the slope. If we then substitute this relationship, in the form $a = QP^{-m}$, in the Nicholson-Bailey population model, we obtain:

$$N_{t+1} = \lambda N_t e^{(-QP^{1-m})} \qquad (9.11)$$

$$P_{t+1} = N_t [1 - e^{(-QP^{1-m})}] \qquad (9.12)$$

This changes the Nicholson-Bailey model markedly in that certain values of λ and m can confer stability on the system; Figure 9.12 shows how they interact. Q, which is Nicholson's a for one predator (i.e. no interference) affects the equilibrium level but not stability. There are problems with the above model, however:

(1) the many experimentally derived values of m are largely from the laboratory and we should be wary of some of these manipulated densities; how common is it, for example, for 3500 parasitoids to aggregate in one square metre (Figure 9.13b)?
(2) the field-based examples, of which seven are listed by Hassell (1978), suffer from the fact that their calculation often breaks one of the 'rules' of regression in that predator density appears on both axes. This leaves us with only one valid field example, with a value for m of 0.39.
(3) the relationship between log a and log P may not be linear (Rogers and Hassell, 1974) but modified models can cope with this.

9.3 Predator Development and Accumulation

Apart from the above detailed problems, we are still dealing with randomly searching predators for randomly arranged prey, little different from Nicholson's model. We need to consider the evidence for predators' responses to prey density and distribution and the consequences for stability. For clumped prey, we should expect predator interference to be of importance again but relevant models will need to include this non-randomness, unlike Hassell and Varley's modification of Nicholson-Bailey. The three remaining ways predators can respond to their density and spatial arrangement of their prey are: (a) decreased development times; (b) increased survival rates; (c) accumulation through reproduction and movement.

9.3a Predator Development Time in Relation to Prey Density

There is ample evidence that, not surprisingly, predator rate of development increases with prey availability and the usual form of the relationship is similar to a type 2 functional response in that the rate increases at a decelerating rate. There must be a threshold prey density below which there is no survival so many of the examples in Figure 9.14 show an intercept on the x axis. There is no density dependence in these relationships acting alone, so we need to consider predator survival too.

9.3b Predator Survival Rate in Relation to Prey Density

Theoretically, predator survival has the following components (Beddington et al. (1976b) give the mathematical reasoning): relationships between survival rate and number of prey eaten/unit time; no. eaten/unit time and prey density and an overall relationship between survival rate and prey density (Figure 9.14). There is a hint of curvilinearity in some such relationships derived from real data although the data are of necessity from the laboratory and often involve a uniform-sized prey population; however, many again mimic in shape the familiar inverse density dependent type 2 functional response so under these circumstances a predator cannot regulate the numbers of its prey through its changing survival rate.

9.3c The Aggregative and Reproductive Numerical Responses to Prey Density

Unlike the ecologically primitive functional response, in which one predator individual searches at a fixed density of randomly arranged prey, we now permit the predators, in their numerical response, to move between areas. If these areas differ, as is likely, in their prey densities, an 'efficient' predator should spend more time, consume more prey and lay more eggs in the high-density areas. If the predators' responses, in any of the above ways, reveal density dependence,

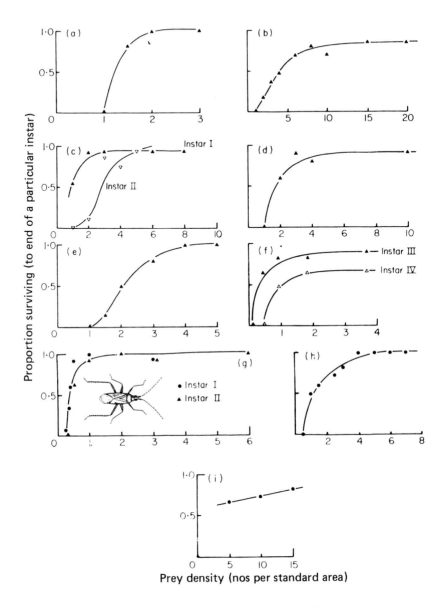

Figure 9.14: Relationships Between Mean Prey Density During Particular Instars and the Proportion of Individual Predators Surviving to the End of These Instars. (a)-(e) are all coccinellids. (a) First instar *Adalia bipunctata* (Dixon, 1970b); (b) all larval instars (c), and (d) individual instars of *A. decempunctata* (Dixon, 1959); (e) first instar, *A. bipunctata* (Wratten, 1973). (f) and (g) are Hemiptera. (f) *Notonecta undulata* (Toth and Chew, 1972); (g) *Blepharidopterus angulatus* (Glen, 1973); (h) a spider (*Linyphia triangularis*); survival through the first and second instar (Turnbull, 1962); (i) a mite (*Melichares dentriticus* (Berl)); survival through proto- and deutonymphs (Rivard, 1962). The x-axis in (f) is mg (dry weight) of prey presented per day per standard area. Source: From Beddington, Hassell and Lawton, 1976b.

the regulatory potential is realised. There is ample evidence that many predators show higher fecundity at higher prey densities, but few, if any examples of a density-dependent element in these relationships (Figure 9.15). It is in the realm of predators moving between prey density areas that the greatest potential lies. An idealised curve, propounded by Hassell and May (1974), shows a lower and an upper plateau where the predators do not distinguish between densities; in between is a range of densities, between which predators discriminate clearly and show a density-dependent response. As is often the case in the interface between theoretical and experimental ecology, finding field examples is not easy. A very clear field demonstration, however, comes from the work of Goss-Custard (1977) on the wading bird the redshank (*Tringa totanus*). Not only does Figure 9.16 show how the birds accumulate in response to prey density, but it shows the consequences in terms of prey consumption, achieved by careful observation of the birds' feeding in the field. Examples, mainly from the laboratory, certainly show a proportional increase in the predators' activities as the prey density increases and may be taken to indicate a part of the overall response curve. However, one of the more detailed field examples for an invertebrate does not show density dependence (Mills, 1982; Figure 9.17) so more evidence for invertebrates, and the field consequences, would be useful.

One consequence of predators' responses to prey heterogeneity has received increasing attention in recent years, that of its potential for inducing population stability. Another element of this behaviour which may contribute to its stability properties is that it is likely to lead to interference between the searching predators. This may occur in a more complicated way than in the simple Hassell and Varley (1969) model dealt with earlier. There is a suspicion, however, that regular, intensive interference may be largely a laboratory phenomenon, as hinted earlier. Models for its interaction with non-random search with respect to stability have been developed and a framework for how such predators should behave is given in Figure 9.18. Further details, and the current 'state of the art' can be found in Hassell (1978). Of greater importance is the numerical response itself to prey 'patches'; not only have there been analytical modelling attempts, as in other areas of predation ecology, but there is increasing evidence that this, of all the components of predation, may be one of the most powerful stabilisers with an important role to play in biological control. It is in this area particularly where the detail required of analytical models for them to mimic an increasingly complicated ecological system may be leading to their intractability; the attempts below, although of increasing mathematical complexity, remain simplistic ecologically.

One model which distributes the *proportions* of the predator and prey populations between areas is that of Hassell and May (1973). It is a modification of the Nicholson-Bailey model in which the fraction of the prey population in patch i is α_i and the predator proportion in the same patch is β_i. Stability can be generated by this if the prey are sufficiently unevenly distributed and predators aggregate to a sufficient extent in high-prey-density patches. The model is a little unwieldy, however, because it needs to be solved for each α, 1 β_i combination.

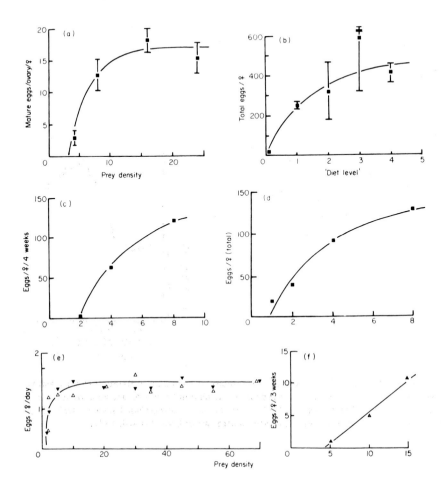

Figure 9.15: Fecundity in Various Predatory Arthropods as Functions of Prey Density or Food Availability. (a) The coccinellid *Adalia decempunctata* (Dixon, 1959). (b)-(d) Hemiptera. (b) *Podisus maculiventris* (Mukerji and Le Roux, 1969); (c) *Notonecta undulata* (Toth and Chew, 1972); (d) *Anthocoris confusus* (Evans, 1973). (e)-(i) Mites. (e) *Amblyseius largoensis* and *A. concordis* (Sandness and McMurtry, 1970); (f) *Melichares dentriticus* (Rivard, 1962). Source: From Beddington *et al.*, 1976.

An alternative model allocates prey to *one* high-density patch containing the fraction α of the total prey population and the remaining $n-1$ patches each have $(1-\alpha)/(n-1)$ prey. The allocation of predators/patch was given by: $\beta_i = c\alpha_i{}^\mu$. μ defines the predator's aggregation and ranges from 0 (= random search) to ∞ (all predators in the one high-density patch). Stability was again the consequence and was influenced by:

(1) μ: higher levels give greater stability

(2) λ: prey rate of increase. As this increases, stability declines

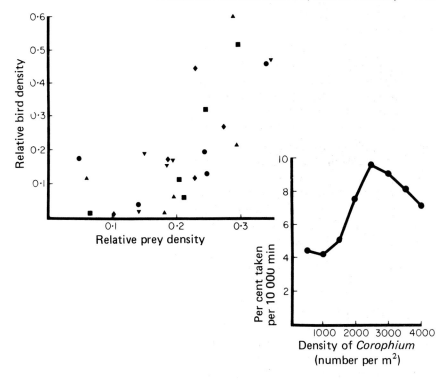

Figure 9.16: (a) The Relative Density of Redshank in a Transect in Relation to the Relative Density of Their Crustacean Prey (*Corophium*). Each point refers to one zone/period and symbols refer to different periods. (b) The Proportion of *Corophium* Taken Per Unit Time by Redshank in Relation to Prey Density. Source: After Goss-Custard, 1977.

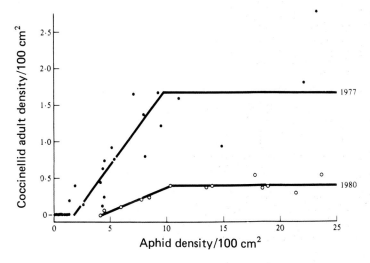

Figure 9.17: The Aggregative Numerical Response for a Ladybird Feeding on Lime Aphids. Source: From Mills, 1982.

(3) α: the proportion of prey in the high density patch. Increasing α promotes stability but it interacts with μ and λ.
(4) $(n-1)$: the number of patches of prey at low density; i.e. the greater the number of spatial sub-units, the greater the stability.

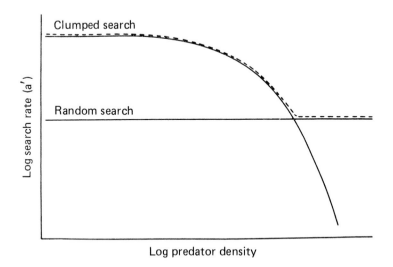

Figure 9.18: A Schematic Picture to Illustrate Both the Increased Searching Efficiency at Low and Intermediate Predator Densities and the Apparent Interference Relationship that Arise from Predators Always Aggregating in Patches of High Initial Prey Density Rather than Searching at Random. The broken line illustrates a more prudent predator strategy in which aggregation gives way to random search at high predator densities. Source: From Hassell, 1978.

The above models still deal with *proportions* of predators and prey rather than numbers. Models based on frequency distributions would be better at allocating *numbers* of individuals differentially between areas: the negative binomial frequency distribution has been used to this end by Hassell and May (1974) and Murdoch and Oaten (1975) among others. These models differ in their assumptions, despite being based on the same frequency distribution, but these and others lead to some general conclusions about stability in relation to aggregation. The main one is that prey in low density areas are effectively in partial refuges, in that predation rate is lower there than in high density patches. In prey patch systems where a fixed *proportion* of prey occur in particular refuges, the proportion protected in the low-density refuges is critical; if too small or too large, stability breaks down. Fixed-number systems have more potential (and are probably ecologically more realistic) because at low densities a greater proportion of the total population will be protected. The famous experiment by Huffaker (1958) in which he showed empirically the role of spatial refuges for invertebrates usefully vindicates these recent models (Figure 9.19).

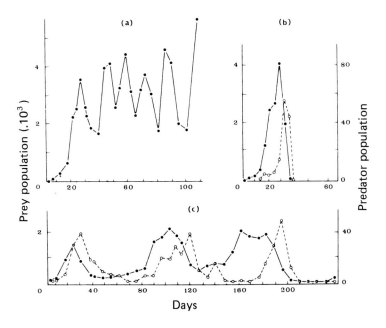

Figure 9.19: Predator-Prey Interactions Between the Mite, *Eotetranychus sexmaculatus* (●) and its Predator, *Typhlodromus occidentalis* (○). (a) Population fluctuations of *Eotetranychus* without its predator. (b) A single oscillation of predator and prey in a simple system. (c) Sustained oscillations in a more complex system. Source: After Huffaker, 1958 and figure from Hassell, 1976.

9.4 A Theoretical Basis for Biological Control

This chapter began with the suggestion that a desire for better biocontrol is the main influence on the development of predator-prey theory. Historically, how-ever, there is little evidence that biocontrol attempts have had a clear theoretical foundation rather than just common-sense principles concerned with the enemy and pest's phenology, diet etc. In fact, of the 107 attempts classified as at least partially successful (Southwood, 1977) only 16 per cent were completely successful, so there is room for improvement. A more informative analysis of the same data shows that for pests of forest trees, 21 per cent were completely controlled compared with only 4 per cent for attempts to control pests of arable crops (Table 9.2). Many of the models discussed in this chapter emphasise inter-generation stability for coupled predator/prey systems and their use, if any, in predicting events in the field may be limited to more permanent systems such as forests. So are there any theoretical lessons to be learnt which can provide a basis for future biocontrol attempts? Beddington *et al.* (1978) sur-veyed a range of population models and biocontrol success and tried to match the extent of pest suppression achieved by the latter to model parameters. Their measure of pest suppression was q, which was calculated as:

$$q = \frac{N^*}{K}$$

where N^* = the new equilibrium level after the enemy's action and K the maximum prey population level reached before the enemy's action. The first interesting result of their survey was that, despite the large number of biocontrol attempts mentioned above, it was difficult to obtain values of N^* and K for most of them, suggesting that accurate before-and-after assessments of biocontrol attempts are rare. Secondly, although the laboratory examples involved

Table 9.3: Parameters from Predator/Prey Models and their Contributions to Stability i.e. Potential Value in Biological Control. Applies to: (a) *monophagous* predators/parasitoids in a more or less (b) *stable* crop e.g. a forest.

	Parameter	Optimum for biological control	Effects and limitations of parameter
$a^/$	attack rate	High	Reduce average population level; no effect on stability.
a	area of discovery		
Q	etc.		
*$a^/$		*or* $a^/$ changes with prey density	Sigmoid functional response — gives stability only in models with no time-delays e.g. in differential equations but *not* in models *with* time-delays e.g. difference equations.
T_t	total searching and feeding time	High	Reduce average pop. level; no effect on stability.
T_h	handling time, as proportion of T_t	Low	Slight increase in stability; slight increase in average population level.
*m	mutual interference constant (Hassell and Varley, 1969)	0-1	Optimum value depends on prey reproductive rate (F or λ) — increase in stability but large values of q only. Field relevance?
*μ	predator aggregation index (Hassell and May, 1973)	High	0 = random search, ∞ = all predators in one patch. Increase in stability but influenced by α (proportion of prey in the one large patch) and λ (prey rate of increase).
*k	predator aggregation index (k = exponent of negative binomial distribution) (Hassell and May, 1974)	All values of $k<1$ (range is $0 \rightarrow \infty$)	Stability but needs some prey intraspecific density-dependence if $k \ll 1$.

* = parameters giving stability.

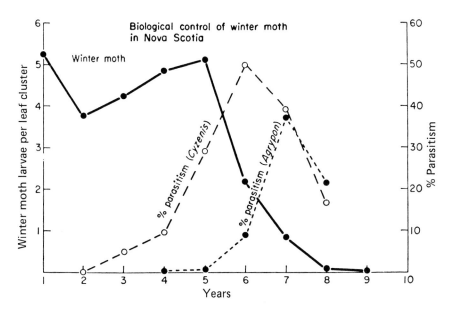

Figure 9.20: After the Establishment in Nova Scotia of the Introduced Parasites *Cyzenis* and *Agrypon* the Winter Moth Population Declined as the Percentage of Parasitism Increased (Embree, 1966). The graphs are average figures for seven localities. The time scale for each place was counted from the time *Cyzenis* was first observed there. For most places year 1 was 1954. Source: After Varley *et al.*, 1973.

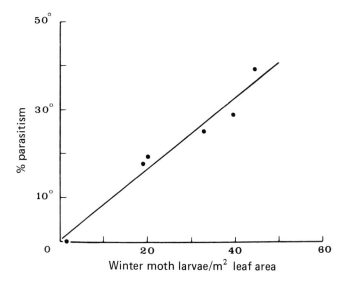

Figure 9.21: The Relationship Between the Percentage Parasitism (Angular Transformation) of Winter Moth Larvae by *Cyzenis albicans* and the Winter Moth Larval Density on Different Trees. Source: After Hassell, 1968.

significant declines in prey numbers, q in these cases was one or more orders of magnitude higher (i.e. less suppression) than in the field examples. What properties, therefore, of field systems are absent from laboratory ones and, as shown by models, can confer these levels of stability on predator-prey interactions? Table 9.3 summarises the role of some of the model parameters in generating stability; this type of survey led Beddington *et al.* to conclude that only spatial heterogeneity and predators' responses to it could provide the q-values observed; other supporting evidence is given in the paper. It is certainly true that in three of the four laboratory-generated q-values, the hosts were not significantly uneven; however, field evidence would be more convincing. One intensively studied and well-publicised example, that of the winter moth (*Operophtera brumata*) in the UK and Canada, does demonstrate heterogeneity but also offers a cautionary tale concerning how optimistic we can afford to be in proposing scientific, non-empirical evaluation of natural biocontrol agents. A fly and a hymenopterous parasite of this moth were introduced into Nova Scotia as potential control agents well before detailed life tables (see Chapter 8) for the moth had been completed in the UK. The natural enemies combined to reduce the moth population to very low levels (Figure 9.20) and models derived from life tables were vindicated in that they mimicked these events fairly well (Figure 7.18). The fly parasite (*Cyzenis*) also shows a strong aggregative response to different densities of its host (Figure 9.21), providing circumstantial evidence supporting the suggestion that responses to spatial heterogeneity by enemies may be a 'common denominator' of field biocontrol successes. However, the 'cautionary tale' stems from the fact that *Cyzenis* in the UK showed no sign of acting in a density-dependent way (Figure 9.22) whereas in Canada it clearly regulated the numbers of its host. The key to this was that in England, *Cyzenis* was killed when the moth pupae were eaten on the woodland floor by vertebrate and invertebrate predators and this predation was density dependent. In Canada, there was thought to be no such pupal predation (Figure 9.23). This difference led Hassell (1978), a proponent of using theory in biocontrol research, to make the rather poignant statement that '. . . with such subtleties sometimes determining the outcome of biological control programs, it will be difficult to escape completely from the present *ad hoc* basis of natural enemy introductions', which is where we came in, except that recent work emphasises that we cannot ignore a major class of predator with which, by definition, the above difference-equation approach cannot cope. These are the polyphagous predators such as spiders, staphylinid and carabid beetles, harvestmen, earwigs etc. which, especially in arable crops, may play an important part in reducing pest numbers. The analytical approach has begun to deal with this group but there is a range of survey and experimental data which give more direct information about this group's potential.

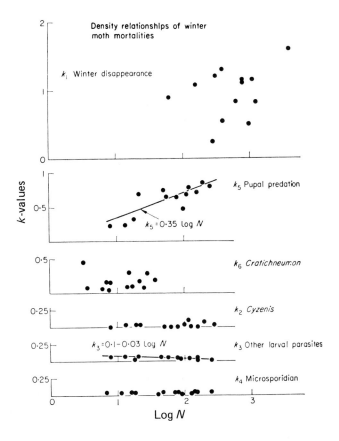

Figure 9.22: The *k*-values for the Different Winter Moth Mortalities Plotted Against the Population Densities on Which They Acted. k_1 and k_6 are density dependent and vary quite a lot; k_2 and k_4 are density independent but are relatively constant; k_3 is weakly inversely density dependent, and k_5 is quite strongly density dependent. Source: From Varley *et al.*, 1973.

9.5 Polyphagous Predators and Analytical Models

A polyphagous predator will have the ability to move between areas of differential profitability where not only the density of prey will differ but its species too. This is similar to the aggregative responses we have already discussed but the lack of coupling and therefore of time-delays between the predator and one prey species means that any sigmoidality in the response curve may contribute to regulation. The difference between a functional and numerical response becomes less clear here if we assume that sigmoid responses result from a change of place as well as species eaten. The sigmoidality comes about through the proportion of a prey type taken changing from less than to greater than expected as the

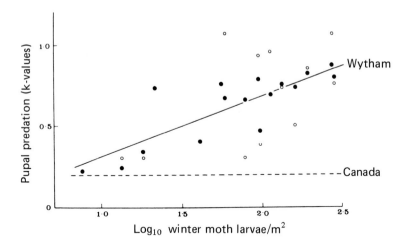

Figure 9.23: The Relationship Between the Pupal Mortality of Both the Winter Moth and *Cyzenis* Ascribed to Predation (Expressed as k-values) and the Density of Winter Moth Larvae Each Year. (●) Winter moth mortality in Wytham (Varley and Gradwell, 1968); (○) *Cyzenis* mortality in Wytham. The broken line is the assumed winter moth and *Cyzenis* pupal mortality in Nova Scotia (Embree, 1966). The solid line is the regression for the Wytham winter moth data ($y = 0.37x$; $p < 0.01$). Source: From Hassell, 1977.

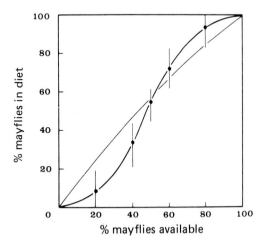

Figure 9.24: The Percentage of Mayfly Larvae (*Cloën dipterum*) in the Diet of an Adult *Notonecta glauca*, Shown as a Function of the Availability of Mayflies in Relation to the Alternative Prey, *Asellus aquaticus*. Means and ranges from five replicates are shown, with the switching curve fitted by eye and the expected proportion of mayflies in the diet calculated using Murdoch's (1969) index: $c = 1.28$. Source: After Lawton, Beddington and Bonser, 1974; from Hassell, 1978.

proportion of that prey available increases (Figure 9.24). The behavioural basis of such changes of diet may involve 'preference' and/or the development of a 'search image' or may simply be a 'passive' function of prey density, availability and the predator's capture efficiency. Chapter 10 elaborates this. A stability model, modified from the disc equation of Holling in that a' changes with prey density (Hassell and Comins, 1978), can give large regions of stability.

The above theoretical evidence that polyphagous predators can exert a large effect on the numbers of their prey is still, however, a long way from showing that they can do so in a biocontrol context. It is here that we must turn to the less precise, more empirical but *field* evidence from survey and manipulative studies. These represent a completely different approach to the investigation of the role of predators, involving less theory, completely different field and laboratory techniques and more deduction than induction.

9.6 Field Studies of the Role of Polyphagous Predators

The field evaluation of polyphagous predators may be divided into the following broad approaches: (a) 'hints and correlations' (b) density and diet information and laboratory-field extrapolations, (c) manipulation of population levels and (d) simulation. The last approach is increasingly common in predation studies but as the example in the last chapter included a predation analysis further case studies will not be catalogued here.

9.6a Hints and Correlations

The accumulation of field evidence that polyphagous predators may be important in reducing the numbers of pest insects in agricultural land is a relatively recent event; the work referred to below was mostly carried out in the last few years. In a major study of the cereal ecosystem in Southern England, Potts and Vickerman (1974) obtained large amounts of data from suction sample and pitfall trap surveys of arable land. One way of analysing such data broadly is to investigate the correlation between the numbers of selected pests (in the case below, aphids on cereals) with the faunal diversity (Williams' α; see Section 12.2) of the rest of the catch. For different crops and at different times, this correlation was negative and significant (Figure 9.25). The complexity of the relationship, if any, between diversity and stability will be discussed in Chapter 12; in Potts and Vickerman's work, a further, simple analysis related the proportion of the individuals captured which were predatory to the diversity index; there was a positive correlation (Figure 9.26) which may help explain the relationship in Figure 9.25. These relationships were established at times when many of the 'classical' aphid-specific predators were at low population levels, so that polyphagous ones were implicated in the aphid number/diversity relationship.

Other circumstantial evidence from the same ecosystem took a slightly different form; Chambers *et al.* (1982) noted that early-summer aphid populations were significantly higher in the centre of early-sown fields than at the edge, but this difference did not occur in the late-sown field. Pitfall-trap catches of poly-phagous predators were higher at the field edges than in the centre in early-sown fields, but not in the late-sown. This implicated these organisms in reducing the prey numbers. In contrast, the numbers of aphid-specific predators such as ladybirds in the early-sown fields correlated *positively* with the numbers of their prey, although prey rate of increase was negatively related. This difference in correlations between prey and predator numbers, depending on natural enemy type is shown by the data of Potts and Vickerman and by Edwards *et al.* (1979) (Figures 9.27 and 9.28). It suggests that prey-specific enemies exhibit a numerical response to the varying densities of their prey late in the latter's population development and, having arrived, begin to affect the prey population growth rate. Earlier in the season, spatial heterogeneity in the density of polyphagous enemies, not initially caused by variation in aphid density, means that 'back-ground' predation densities can be high. An aggregative numerical response (Section 9.2c) by these predators is not excluded, but they are such a diverse group with variable diet spectra that we would not necessarily expect them to correlate positively with the density of one prey species.

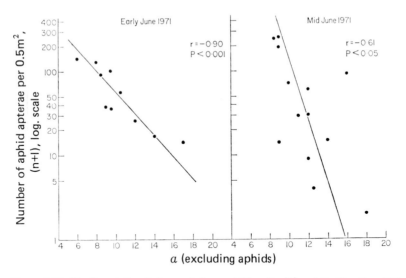

Figure 9.25: The Correlation Between Arthropod Diversity (α) and the Density of Wingless Aphids in Spring Barley, 1971. Source: From Potts and Vickerman, 1974.

9.6b Density and Diet Information

The above correlations tell us nothing about which polyphagous species may be implicated in the reduction of prey numbers, neither do they give the frequency with which each species consumes a particular prey. They also provide no data on consumption rates and their temporal pattern, nor give any clues to possible density relationships. The work of Sunderland and Vickerman (1980) is closer to

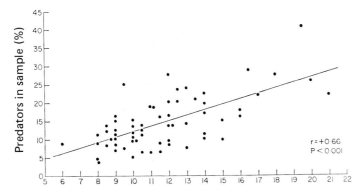

Figure 9.26: The Correlation Between Invertebrate Faunal Diversity (α) and the Percentage of Individuals Which were Classified as Predatory in Cereals. Source: From Potts and Vickerman, 1974.

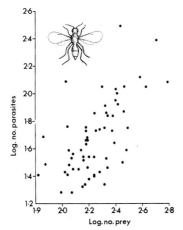

Figure 9.27(a): The Correlation Between the Density of Hymenoptera-'Parasitica' and the Density of Other Insects (Excluding Collembola) in Cereals, Southern England, 1971. Source: From Potts and Vickerman, 1974.

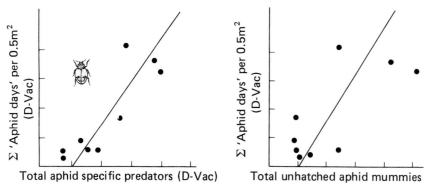

Figure 9.27(b): Regression Between Aphid Specific Predators' and Cereal Aphids' Numbers. Source: From Edwards *et al.*, 1979.

answering some of these questions. These workers collected a wide range of polyphagous predators in cereals over a six-year period by suction sampling, pitfall trapping and careful surface searching in quadrats from May to September each year. For predators which chew their prey (many beetles) rather than feeding on body fluids (e.g. spiders), gut dissection was used to reveal the presence of prey exoskeleton, claws etc. in the predators' diet. In this study, c. 12,000 individuals were dissected for evidence of aphid remains and a 'league table' was produced based on the proportion of individuals in each predator species which contained aphid remains; it was impossible to quantify the number of prey eaten by each predator using this method. This league table took no account of the predators' abundance so a 'predation index' was obtained by multiplying predator density (obtained from the quadrat surface searching) by the proportion containing aphid remains (Table 9.4). This was the first major attempt to design an empirical field-based criterion to rank predators in an arable crop. A robust ranking is necessary because, if manipulation of predator levels as a component of integrated control is ever to be fully realised, then some knowledge of which organisms have most potential will be necessary. Also, as new agrochemicals of varying toxicity to natural enemies enter the market, the ecological consequences of a decline in the numbers of particular 'non-target' organisms need to be known. Having constructed the crude ranking in Table 9.4, Sunderland and Vickerman were able to use the same predator dissection data in a search for any density relationships. They plotted the percentage of individuals containing prey remains against prey density classes to produce a wide range of relationships (Figure 9.29). These field-generated relationships are difficult to interpret using the framework of Sections 9.2b and 9.2c but must include elements of both aggregative numerical and functional responses; some show intriguing increases in 'consumption rates', reminding us of similar-shaped curves seen in our discussion of functional and numerical responses.

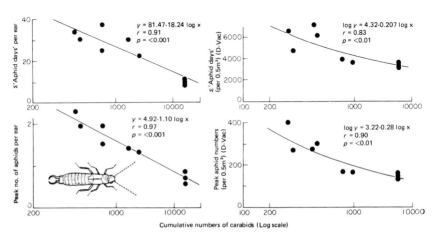

Figure 9.28: Regression Between Polyphagous Predators' and Cereal Aphids' Numbers. Source: From Edwards *et al.*, 1979.

Table 9.4: Proportion of Predators Containing Aphid Remains During the Cereal Aphid Increase and Decrease Phases and Indices of Early Predation. (Data for aphid densities > 1000 m^{-2} are not included.)

	Proportion containing aphid remains at aphid densities of:			Number examined	Mean density of predators (for aphid increase 1-1000 m^{-2})	Predation* index
	1-1000 m^{-2} (increase phase)	1000-1 m^{-2} (decrease phase)	Limit 1000 m^{-2} (increase plus decrease phases)			
Demetrias atricapillus	0.253	0.136	0.230	113	1.23	0.311
Agonum dorsale	0.236	0.336	0.257	653	1.28	0.302
Forficula auricularia	0.278	0.165	0.220	236	0.61	0.170
Tachyporus chrysomelinus	0.051	0.054	0.052	346	2.39	0.122
Tachyporus hypnorum	0.024	0.034	0.260	778	4.50	0.108
Bembidion lampros	0.082	0.167	0.093	989	1.23	0.101
Amara familiaris	0.033	0.019	0.027	255	1.47	0.049
Amara aenea	0.034	(0.00)	0.034	176	1.42	0.048
Nebria brevicollis	0.086	(0.00)	0.085	531	0.48	0.041
Notiophilus biguttatus	0.040	0.013	0.031	228	0.67	0.027
Asaphidion flavipes	0.048	0.000	0.044	114	0.31	0.015
Amara plebeja	0.016	0.000	0.014	147	0.88	0.014
Harpalus rufipes	0.053	0.056	0.054	147	0.14	0.007
Pterostichus melanarius	0.161	0.073	0.101	346	0.03	0.005
Loricera pilicornis	0.007	0.019	0.011	442	0.27	0.002
Calathus fuscipes	(0.000)	0.098	0.085	47	0.02	0.000

*Values in column one multiplied by values in column five. ( ) denotes sample size <10.

Source: From Sunderland and Vickerman, 1980.

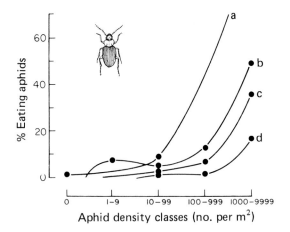

Figure 9.29: Relationships Between the Percentage of Individual Predator Species Containing Aphid Remains and Cereal Aphid Density. (a) *Calathus fuscipes*; (b) *Nebria brevicollis*; (c) *Harpalus rufipes*; (d) *Loricera pilicornis*. Source: After Sunderland and Vickerman, 1980.

The above crude ranking of predators, based as it is on only two criteria, is but one possible way of evaluating and ordering the very large number of potential biocontrol agents in arable land. A big problem with Sunderland and Vickerman's method is that it cannot deal with fluid feeders like many staphylinid beetles and spiders and gives no clue to consumption rates. Other complementary methods, of varying degrees of empiricism, could include the assessment of the following:

(1) *The extent to which the species' populations are disrupted by agronomic practices*, such as agrochemical use, ploughing, burning, weed control etc. The carabid *Agonum dorsale* in Table 9.4, for instance, enters the field in May in the UK and returns to the field boundaries before harvest, so seems well adapted to avoid the traumatic events of a cereal harvest.

(2) *Season of predatory activity.* A high predation rate when the prey species is at its peak or later may be of no consequence for useful control; the predator's phenology is of obvious importance.

(3) *Response to prey spatial heterogeneity.* We saw in Section 9.4 that this may be crucial in a predator's potential to regulate prey numbers. Field and laboratory experiments are being carried out now in the UK to compare predators on this basis.

(4) *Spatial distribution in the field.* Because of their polyphagy, many species probably find a wide range of prey in the field boundary so their distribution may be markedly influenced by this. Earwigs (Dermaptera), for instance, hardly penetrate agricultural grassland from the boundary and in barley, a less dense crop than grass, they enter late and hardly any individuals enter further than

25 m from the boundary. The earwig's position as third in Sunderland and Vickerman's ranking indicates how easily a species' ranking may change with more information.

(5) *Where do they feed: plant or soil surface?* The dissection technique cannot separate scavenging of dead prey from active predation. Some beetles are known to be poor plant climbers yet contain many prey remains. Recent evidence has shown that a significant proportion of aphids on wheat can be found alive on the soil surface (Griffiths, 1983). Why they are there and their ultimate fate is unknown, but their presence implicates even non-climbing predators.

(6) *Assessment of fluid feeders.* Animals like spiders, harvestmen (*Opiliones*) and some beetles are fluid feeders so are not included in Sunderland and Vickerman's ranking. Techniques like electrophoresis or immunological analyses will enable these groups to join the 'league table' and their presence may alter existing rankings substantially.

The next methodological stage, however, is to demonstrate other than by a correlative method that these predators can reduce prey numbers. Manipulation of their numbers has been used to good effect in this way.

9.6c Manipulation of Population Levels

The potential advantage of manipulating the population level of polyphagous predators is that, with careful methodology, any subsequent change in prey populations can be related to the changed predator population. There are disadvantages and limitations however, which will be discussed later. In agricultural systems, sometimes the significance of the changed prey population levels can be assessed agronomically by recording changes in plant damage levels, too. In the cereal system in which Potts and Vickerman and Chambers *et al.* worked, Edwards *et al.* (1979) set up two types of barrier plots in each of the months March, April and May in the UK. Each plot had a polythene wall 60 cm high around its perimeter and some had pitfall traps plus an insecticide treatment while others had traps only. In the latter, only beetles of the family Carabidae were permanently removed while in the insecticide plot, the numbers of all categories of polyphagous predators were reduced by the chemical and traps combined. The consequences for the studied aphid populations are shown in Figure 9.30; it seems that predator removal had an effect and that carabids alone may have been contributing to pest suppression. For a virus-transmitting aphid on sugar beet in New Zealand, Wratten and Pearson (1982) used a similar technique and investigated the consequences of predator population reduction for the number of aphids/plant, percentage infested plants, virus incidence and root sugar yield. There was up to a 40-fold difference in aphid numbers/plant with only a 50 per cent reduction in predator numbers (Figures 9.31 and 9.32); the percentage infested plants never exceeded 30 per cent in the controls but approached 100 per cent in the predator-reduction plots; incidence of virus too was affected; only sugar yields remained unchanged in what was a year of generally low aphid numbers. New Zealand agricultural land has a very low carabid population (only three caught in 300 trap/weeks) yet the effects of other predators such as spiders and staphylinids were great, emphasising again that it is not just the chewing predators which can be important. Populations of pest species as diverse as butterfly larvae on Brussels sprouts (Dempster, 1975) and eggs of the cabbage root fly around cauliflowers (Wright *et al.*, 1960) have been shown to be influenced by the experimental reduction in polyphagous predator numbers (Figure 9.33). These results, however, share with those above the following problems associated with manipulative work:

(a) crude barrier or insecticide-reduction techniques do not identify which predators are mainly responsible for pest population reduction;
(b) experiments which boost enemy numbers to ascertain the effect on pest populations may create artificially high predator populations which may deplete the numbers of alternative prey, causing them to turn to the less-preferred study organism;
(c) if insecticides are used, there may be effects on the numbers of the studied prey too, but at least this would make the observed differences attributable to predators minimal;
(d) the dynamics of the predator-prey interaction remain unknown; the relative roles of numerical and functional responses could not be separated;
(e) higher prey levels achieved through predator population reduction could 'attract' other, flying predators, again leading to a minimal, rather than incorrect statement about the role of the manipulated predators.

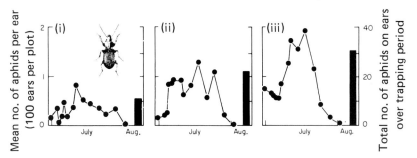

Figure 9.30: Effect of Changed Arthropod Populations on Numbers of Aphids on Ears. (i) Control, (ii) Carabids trapped out, (iii) Polyphagous predators trapped and killed. Source: From Edwards *et al.*, 1979.

We are now in a position to contrast briefly the range of approaches to the study of natural enemy action and attempt a synthesis of the current knowledge of predation biology. It may be that attempts to use and test theories from analytical models, with their concepts of inter-generation stability, in predator-prey systems in short-duration habitats, are doomed to failure. In these systems, such as arable agriculture, perhaps a lowering of the prey's equilibrium level, with no reduction in the amplitude of its fluctuation (i.e. no true regulation) is the most likely result of natural enemy action. This 'biological insecticide' effect is equally as satisfactory in an applied sense as regulation in that predators could

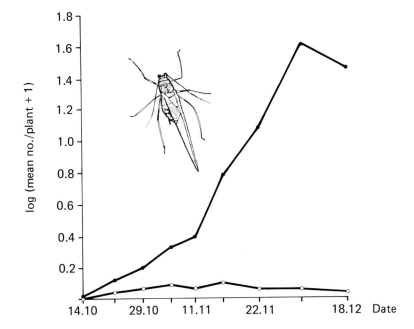

Figure 9.31: Consequences of Experimental Changes in Predators' Numbers for the Numbers of the Aphid *Myzus persicae* on Sugar Beet. ● = predator numbers reduced; x = control. Source: From Wratten, unpublished.

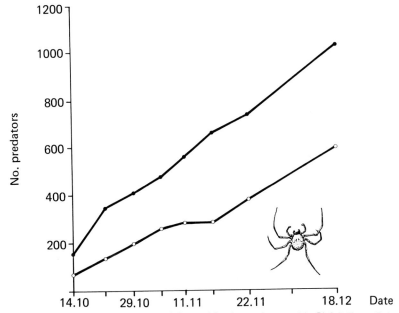

Figure 9.32: Cumulative Numbers of Ground Predators Captured in Pitfall Traps Following Experimental Manipulation of Their Numbers in Sugar Beet. Source: From Wratten, unpublished.

maintain prey (pest) numbers below a damage threshold while regulation *sensu stricto* is irrelevant if the system is destined to be reduced to the burnt, ploughed, chemically treated habitat which can be the state of an agricultural field after harvest. We saw in the lime aphid simulation model in Chapter 7 that these 'background' levels of natural enemy action can be 'escaped' from under some circumstances. This is demonstrated in Southwood and Comins' (1976) synoptic population model in which the likelihood that enemies will suppress the numbers of their prey is shown to depend on the position of the prey on the r-K continuum (see Chapters 7 and 10). The 'natural enemy ravine' (Figure 9.34) can be seen to be most pronounced in the life histories of K-selected organisms. Extreme K-selected species, however, would be less influenced because of their large size and high competitive ability while extreme r-strategists develop so rapidly that enemies probably cannot colonise quickly enough. We must remember the distinction made above between the 'background', diet-switching predators (prey numbers negatively correlated with theirs) and the prey-specifics (positive correlation between prey and predator numbers) when we consider attempts at synthesis, such as Figure 9.34, in predation ecology.

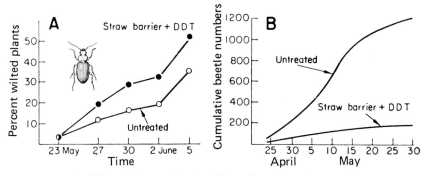

Figure 9.33: The Effect of Surrounding Cauliflower Plants by a Straw Barrier Soaked in DDT. **A** In the treated plot a larger percentage of plants wilt from the attack of larvae of the cabbage root-fly, *Erioischia*. **B** Fewer of the ground beetle *Bembidion*, which eat cabbage root-fly eggs, are found in pitfall traps within the treated plot. Source: After Wright *et al.*, 1960; From Varley *et al.*, 1973.

Despite the range of approaches to predation ecology, in each of which arises the danger that methodological limitations may affect the ecological conclusions, population ecology has advanced rapidly since the 1950s, when the semantics of density dependence and related concepts seem to have dominated the subject. The fact that, in the UK, the awareness that polyphagous predators may be important has arisen in only about 5 years shows how fast the subject is moving. It makes a pleasant change for experimental and observational data to be in the advance in some areas, awaiting theoretical support; 20 years ago, predation ecologists seem to have been constrained by incomplete or incorrect theory. We are now in an exciting field with many new methodologies and insights still to make major contributions, with the integration of theory, experimentation and observation the goal.

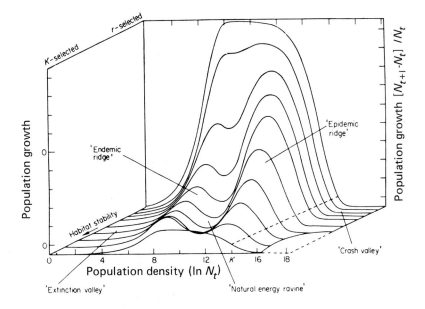

Figure 9.34: Southwood and Comins' (1976) Synopsis of Organisms' Population Dynamics in Relation to Their Position on the *r-K* Continuum.

9.7 The Effects of Predation on Prey Productivity and Community Structure

Our considerations of predation have so far concerned themselves largely with the potential regulatory effects of predators upon the populations of their prey. Yet these interactions of predator and prey populations clearly concern the predators as well as the prey – analyses of the mechanics of predation, for example, can be interpreted just as well from a consideration of their implications to the predator, as from examination of a predator's potential regulatory influence on its prey. In this section we will reconsider the predation and predator-prey relationships from the viewpoint of the predator. Perhaps the easiest way to approach this is to consider how best to become an efficient predator.

For a predator to be truly efficient it must adopt a strategy which offers it maximum return. Clearly it must first 'select' a predation strategy which avoids, as far as possible, competition with other predators – but once having selected its niche in this way, it must develop a method of exploitation of that niche which returns a maximum yield from its prey for minimum outlay of effort. In this, it is not sufficient for the predator merely to become successful at catching and killing its prey. It must also know when to stop: for the predator to ensure a sustained yield from its prey community, predator and prey populations must remain in equilibrium and, if possible, the action of the predator should be such as to maximise the productivity of the prey community. How might these ends be achieved?

9.7a Stability of Predator-prey Relationships

In our analysis of predator and prey population patterns earlier in this chapter (stable oscillations, oscillatory damping, increasingly unstable oscillations etc.) we came to the conclusions that simple predator-prey interactions were rarely stable in themselves, but that these potentially unstable systems could be stabilised by an increase in biotic or physical complexity (e.g. spatial hetero-geneity).

The consequences of increasing the system's physical complexity in this way were clearly seen in the Huffaker orange mite experiments (Figure 9.19). We also saw (in Section 9.5) how a simple two-prey species system can generate stability, the predator turning its attention to different prey species as the relative abundance of each fluctuates. Neither will necessarily be driven to extinction because, for the predator, it will be uneconomical to hunt down the last survivors of a rare prey when an alternative, more abundant prey is readily available. The increased stability is effected whether the predator operates strictly on considerations of relative abundance or, with more complex behaviour, shows the classical 'switching' of Murdoch (1969). While a one predator-two prey community is better than a single species system, if one of the prey popula-tions becomes scarce or even extinct for other, unconnected reasons, we revert to the dangerous one-to-one interaction. Clearly then, the more prey species the better; equally a multispecies prey community can afford to support more than one species of predator: thus stable, multi-predator-multi-prey communities may develop.

9.7b The Productivity of Prey Populations

An efficient predator should not merely act to ensure that its prey does not die out; it will also act to maximise the yield of its prey community. It may be shown that predation can actually enhance the productivity of the exploited populations and of the prey community overall. The most efficient predation strategy would be the one that maximises these effects.

Within individual prey populations, predation acts to reduce numbers. In so doing it reduces competition within the population, and, if the population is self-regulating, effectively can release the density-dependent brake on population increase. Reproduction increases, the population age-structure shifts towards the younger animals, which have a faster growth rate and higher food conversion efficiency. Productivity of the population rises. This increased productivity of a population under predation will only occur if (i) the prey species is normally regulated by intrinsic, density-dependent mechanisms so that it can respond to a decrease in numbers by an increase in reproduction, (ii) the population is at the time limited by environmental resources and is thus already self-regulating.

Clearly a population below its environmental ceiling is already at its maximum rate of increase and a further decrease in numbers cannot boost production further. (It is suspected that this latter state of affairs may obtain at present amongst many marine fish stocks.) In the same way the increase in productivity can only continue up to that level of exploitation at which the population reaches its maximum rate of reproduction. However, for populations which are self-regulatory, exploitation up to that maximum level should increase productivity. Such effects are now well-established in many laboratory populations (e.g. Nicholson, 1933; Watt, 1955; Slobodkin and Richman, 1956; Silliman and Gutsell, 1958); the level of exploitation which promotes maximum productivity of the prey is easily calculated from data on life expectancy and fecundity, as the maximum intrinsic rate of natural increase of that population (Caughley and Birch, 1971).

Number of immature individuals in three weeks

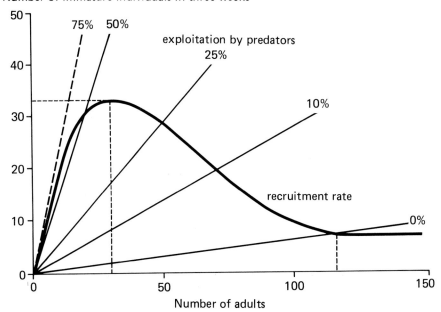

Figure 9.35: Recruitment Curve and Hypothetical Exploitation Rates for Aquarium Populations of Guppies. In the absence of predation (0 per cent exploitation curve) the natural mortality of adult guppies would stabilise the population at about 120 individuals. The maximum exploitation rate possible is just under 50 per cent per three weeks, at which point an adult population of about 30 and a yield of 33 would be maintained. A 75 per cent exploitation rate is more than the population can bear. Source: From Gulland, 1962, after Ricklefs, 1971.

However, while a predator can remove animals at a rate up to this maximum, it is not necessarily the most *efficient* level of predation. Productivity of the prey may be at a maximum, but actual yield to the predator may not: predation at the level of the maximum rate of prey increase usually involves a reduction of yield to the predator below that achieved by a somewhat lower rate of exploitation. As Slobodkin (1968) points out: 'The process of predation increases the rate of "manufacture" per unit of prey population but decreases the size of the prey population. An extremely small prey population cannot produce the same total quantity of yield to the predator as a somewhat larger one, even if the larger one is producing at a slower rate per animal.' Such considerations lead to a redefinition of optimal rates of exploitation, not as those which maximise productivity of the prey, but those which produce maximum sustained yield to the exploiter.

Table 9.5: Optimum Rates of Exploitation for Several Laboratory Populations of Small Organisms.

Species	Exploitation rate (% of adults/day)	Source of data
Blowfly	99	Nicholson (1955)
Daphnia	23	Slobodkin and Richman (1956)
Algae	13	Ketchum *et al.* (1949)
Flour beetle	3	Watt (1955)
Guppies	2	Silliman and Gutsell (1957, 1958)

Source: From Watt, 1962, after Ricklefs, 1971.

In an analysis of the work of Silliman and Gutsell (1958) on guppies (*Lebistes reticulatus*) Gulland established that the maximum sustainable yield to a predator from these particular populations was reached at an exploitation rate of approximately 40 per cent per three weeks (three weeks being the average turnover time of each generation of guppies under the conditions of the experiment), with a yield of 33 fish in the three-week period (Gulland, 1962) (Figure 9.35). Optimal rates of exploitation for other populations may be calculated and may be expected to depend, *inter alia*, on prey biomass and generation time as well as fecundity, natural mortalities and the strength of density-dependent processes acting on the population. Watt (1962) summarises the optimal levels of exploitation, as percentage of adults per day, for various laboratory organisms (Table 9.5).

If these figures are recalculated from the original sources and reduced, where sufficient information is available (Ketchum *et al.*, 1949; Nicholson, 1955; Watt, 1955; Silliman and Gutsell, 1958), to the more *relative* scales of percentage exploitation of the total population per generation turnover time, the wide spread of Table 9.5 diminishes considerably as optimal rates of exploitation cluster around 40 per cent (blowflies, 60 per cent; algae, 45 per cent; *Tribolium*, 41 per cent; guppies, as before, approximately 40 per cent). While this may be a function of the particular examples selected, theoretical considerations involving the exploitation of model populations of a wide range of fecundities and natural mortalities (modelled by Leslie matrix algebra: p. 163) support the idea of a more limited range of optima than might be anticipated; exploitation rates for maximum yield, calculated again per total population per generation time, fall between 25 and 45 per cent (Putman, unpublished). It is possible therefore that an optimal exploitation rate may be, to some extent, a relatively independent figure − only weakly dependent on population parameters: such a conclusion, if further substantiated, must have considerable implications.

There is one further complication. If exploitation is selective, removing preferentially those individuals of least use to the prey population (post-reproductive adults, surplus young, surplus males in harem species), optimal levels of exploitation and yields may well be higher than those cases in which the predator acts in a random fashion. Further, by removing from the population those animals which are the ones most likely to die anyway, the predator is altering the natural *pattern* of mortality as little as possible and is thus less likely to upset the overall equilibrium of that population (Slobodkin, 1968), and risk its extinction. Beddington (1974), again for populations modelled with matrix algebra, found that as a general rule, for maximum yield, at most two age classes of a population should be cropped: one suffering a partial exploitation, the other complete removal.

To what extent do natural predators act in such a way as to maximise the stability and yield of their prey populations in this manner? From our arguments so far, *any* rate of exploitation below a certain maximum will enhance the productivity and yield of the prey population to an extent, but how close does natural predation come to the theoretical optimum? Clearly over-exploitation, which diminishes yield to the predator and may lead ultimately to the extinction of the prey, must not occur and, since the predator, of necessity, will be shaped by evolution to be as efficient a killer as possible, such over-exploitation is presumably prevented by physical and biotic complexities of the system. But whether natural predators operate at or below the optimum level of exploitation is more difficult to establish. Very few estimates of predation level have been made in this way. Mech (1966), for example, calculated that wolves kill about 25 per cent of the moose population on Isle Royale each year; for a population of white-tailed deer in Algonquin Park, Canada, Pimlott (1967) estimated that wolves were taking approximately 37 per cent per year. However it is not clear how such levels relate to the theoretical optima for those particular populations − although they are curiously close to our theoretical 'constant' of 40 per cent − and in many situations exploitation must remain well below the 'ideal'.

What little evidence we have on prey selection by predators *does* suggest that they select specific classes of individual and groups of individuals from amongst their prey populations, and that those prey selected are indeed those of least use to the prey population. Schaller's (1972) figures for prey selection by lions in the Serengeti show that from a total of 697 kills, of a variety of species, only 177 (25.4 per cent) were reproductive adults, the remaining 520 (74.6 per cent) being post-reproductive adults or juveniles. Sinclair (in Schaller, 1972) provides information on the sex ratios of adult wildebeest (*Connochaetes taurinus*) killed by lions, showing that of 183 kills, 59 (32.2 per cent) were females, 124 (67.8 per cent) males – a heavy preponderance of males in an essentially polygynous species where many males are indeed surplus. The selectivity suggested by these results is borne out by evidence from other predator-prey systems: amongst wolves (Mech, 1972) and in the dog-whelk, *Thais lapillus* (Connell, 1961) etc. At least in this regard, the predator appears to be operating in the way which best increases the productivity and fitness of its prey.

Predator-prey interactions rarely occur in isolation. We have already invoked considerations of multispecies systems to prevent over-exploitation of prey populations. A predator may be selective in regard to the age class and sex of prey taken – but can this effect be maintained in a multi-predator system, even if each predator in isolation would attempt to select in this way, since in such a multispecies system the predators themselves must alter their individual strategies to avoid competition with each other? In the Serengeti ecosystem from which we have taken our examples so far, five main predators hunt essentially the same ten species of prey. Yet separation of hunting strategies appears to be with regard to more behavioural elements: selection of habitat, time of day, different killing technique (Table 9.6) rather than in selection of prey individuals (Kruuk and Turner, 1967) and the overall effect still appears to be one concentrating on surplus prey individuals (Schaller, 1972) although the evidence is now less clear-cut.

Such an outcome results not from altruism, nor forward-planning on the part of the predator, but from the application of an operational rule of maximum return for minimum effort. It is almost as easy for a whelk to bore through the test of a large barnacle as a small barnacle; greater returns may be obtained by attacking the larger individual. But barnacles are space-limited and the removal of larger individuals releases the density-dependent brake on population growth further than would the removal of small individuals; thus the prey population as well as the predator 'benefits' (Connell, 1961). In the same way, on the Serengeti plains of East Africa, the old and the young amongst the herds of plains antelope are the easiest individuals for the lions to kill. Surplus males of harem species will be the weaker males, since they are the ones that have been beaten in struggles for female groups. In addition these animals are usually pushed out onto poorer ground by competition from the reproductive groups and thus their condition tends to deteriorate further. They, too, are *easier* victims for the predators.

While the predator may not be calculatingly prudent about its exploitation of prey populations, such considerations may well lead one to speculate as to whether the prey species may 'manipulate' the predator's selection to its own advantage – evolving so that it is the animals they have least use for that become more attractive to predators, or more readily available to them. It has been shown, for example, that the postreproductive life of some aposematic species of saturniid moth (*Dirphia* spp.) is extended beyond that of related species which do not rely on this method of protection, thus increasing the proportion of non-useful individuals within the population and decreasing the depredations

Table 9.6: Different Hunting Strategies of Large East African Predators

Predator	Habitat	Time of Activity	Prey species	Hunting behaviour	Usual max. prey size
Lion	All habitat types but particularly open woodland	Nocturnal	Wildebeest Zebra Thompson's gazelle Buffalo	Solitary or in prides	Stalk and short chase	900 kg
Leopard	Dense vegetation	Nocturnal	Impala Thompson's gazelle Grant's gazelle	Solitary	Stalk	60 kg
Cheetah	Plains, open woodland	Diurnal	Thompson's gazelle	Solitary	Run down prey very fast over short distances	60 kg
Hyaena	Plains	Nocturnal	Wildebeest Zebra Thompson's gazelle	Solitary or in small groups	Run down prey	300 kg
Hunting dog	Plains	Dawn and dusk	Wildebeest Zebra Thompson's gazelle other gazelle	In packs	Run down prey fast over very long distances	250 kg

Source: After Krunk and Turner 1967.

of predators on reproductive animals whilst learning to avoid the species (Blest, 1963).

9.7c The Productivity of the Community

Since the prey community is composed of a number of prey species populations, a predator acting to increase the productivity and yield of individual prey populations increases the productivity of the whole community by the same token. In addition, however, predation may have a direct effect on the community itself, by increasing its diversity. In a study of an intertidal system in California, Paine (1966) defined a community of fifteen species, dominated by a major carnivore, the starfish (*Pisaster ochraceus*) (Figure 9.36). The effect of this predator was such as to keep the numbers of all its prey well below the level at which fundamental resources (food, space etc.) would become limiting on their populations. With superabundant resources, potentially exclusive species could coexist without competing (Chapter 6). When Paine removed the predator from the system, the various prey populations increased in numbers and began to compete; the new community which ultimately developed contained only eight species (only four of the original eleven species preyed upon by *Pisaster*) and in terms of numbers was further heavily dominated by two main species: a mussel, *Mytilus californianus* and a goose barnacle *Mitella polymerus* (Figure 9.36).

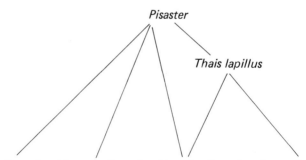

Chitons: 2 species Limpets: 2 species Mussels: 2 species Acorn Barnacles: 3 species
(plus 4 further species not directly consumed by *Pisaster*)

Figure 9.36a: A Schematic Diagram of the Relationships Between the 15 Species of a Rocky Intertidal Zone Community of Californian Coasts. After Paine 1966, 1969.

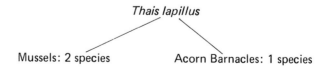

Figure 9.36b: The Same Community after the Removal of *Pisaster*

Predation thus acts not only to enhance the productivity of individual prey populations, but to create and maintain diversity within a community; we have already suggested above that increased diversity amongst a prey community may promote stability of predator-prey interactions and hence that of the predator's own position.

10 Evolution and Adaptation

10.1 Evolution and Ecology

It seems important in any introduction to ecology to stress the relationship between ecological study and evolutionary theory. Ecology is the study of the interaction of organisms with their environment. That interaction, and the adaptiveness of the organisms to their biotic and abiotic surroundings, are, respectively, the driving force for and the result of evolutionary change. The ecological pressures acting upon an organism *are* the selection pressures which direct evolutionary change and adaptation. For this reason it is essential always to maintain an evolutionary perspective in ecological analysis; it would almost be fair to say it would be dangerous to try and treat ecology in isolation: for ecology is really the study of selection pressures, or the study of adaptation. The effects of evolutionary pressure on organisms may be reflected in their own adaptation. Equally changes within individual organisms may have wider implication for the system of which each organism is part and evolutionary change may be reflected in a shift of emphasis or structure of the entire system as a whole (Chapter 11).

10.2 Adaptation

If we believe in evolution at all, we believe that all organisms are constantly under pressure (through competition against other species, or variants within their own species, through interactions with other organisms in predation or parasitism) to be maximally efficient in occupying their particular niche – or at least better adapted to it than any current competitor. The action of evolution upon the ecological system is not always apparent: for the major role of selection is a stabilising one, 'weeding' out variants which (once the organism *is* well-suited to its niche) can only be inferior by comparison. Selection will only promote actual *change* if environmental conditions (biotic or abiotic) are altered in some way. But within the ecological system, the effects of such pressure may be observed in terms of effective adaptation.

Every student of biology, let alone ecology, must have at his fingertips literally dozens of examples of remarkable adaptations of morphology, physiology, biochemistry: all equally the result of the natural selection pressures arising from its environment. Yet such adaptation is so well established, so axiomatic, that it would be pointless to labour that here. Instead we will restrict our considerations to more ecological adaptations, where both stimulus and response fall within the ecological sphere and where perhaps the adaptation may be far more subtle and difficult to identify.

Thus, for any organism there are often a number of possible ecological or behavioural solutions to a given ecological problem — equally practicable strategies but of different ecological efficiency under different circumstances. Selection will favour the adoption by the species of the better alternative for its particular conditions. Perhaps the most fundamental adaptation to be developed is in terms of reproductive strategy. Natural selection acts upon the ability of the individual to reproduce; those organisms which by chance mutation or chance genetic combination, are better fitted for survival will leave more offspring to succeeding generations, at the expense of their less fit peers; as generation follows generation, each dominated by the offspring of the fittest, genes controlling advantageous characters spread through the population.

If selection acts upon the reproductive ability of the individual, evolution must, in all species, be reflected in adaptations towards more efficient reproduction. In practice, there are two alternative strategies in this regard. A species may 'opt' for relatively low reproductive recruitment, in the production of relatively few offspring, but ensure high survival of these offspring by investing heavily in parental care. Or alternatively, the species may be adapted for maximum recruitment. Since the number of offspring, and the degree of parental care which may be offered to each are inversely related (the higher the number of offspring, the lower the possible investment in care, both in absolute terms and per offspring) survival of the young of these more prolific species may be greatly reduced. Since low survival can itself only be compensated for by large initial numbers, evolutionary processes will act to accentuate the strategy — minimising parental care, maximising recruitment. Thus the two strategies must polarise at extremes: many offspring, little care; few offspring with considerable parental care. Either method is valid, and different species 'adopt' either alternative depending on their circumstances (but see Law, 1979). An understanding of the alternatives in this most fundamental decision — what Southwood (1976, 1981) has coined a 'bionomic strategy' — and of the influence and implication of such primary decisions in later evolutionary choices, offers considerable insight into the development and integration of an individual's ecology.

10.3 Bionomic Strategies

These alternative strategies of rapid growth, massive reproductive output at the expense of short lifespan, or longer life span with slow growth and very low recruitment, while potentially equally efficient, are in practice favoured by rather different circumstances.

Thus organisms colonising or characteristic of temporally or spatially unstable environments — environments subject to sudden and drastic change — are under heavy selection pressure for a short life-cycle, and maximum reproductive output; the first, in order that resources may be exploited while still available, the second so that *some* offspring will survive to locate and exploit another area of similar environment. Such organisms are commonly referred to as being '*r*-selected': selection for reproductive potential being the most powerful of all selection pressures acting upon them. By contrast, in colonisation of stable and persistent environments, there is little premium in rapid reproduction. Rather, longevity and the ability to spread reproduction over a longer period will be favoured; selection will act towards balance with the stable environment and its carrying capacity ('*K*'-selection). Clearly, all organisms are subject to the forces of both '*r*' and '*K*' selection: no organism is ruled purely by one or the other. But one pressure must prove the more important, and once again, once a slight inequality has crept in, commitment to one or other strategy will enhance differences; once again organisms tend to gravitate towards the ends of the continuum.

Predominantly *r*-selected organisms show a number of specific (and interdependent) characteristics — some as a direct result of adaptations to unstable systems, others as consequences of these primary adaptations. Thus, *r*-selected organisms are characteristically small, short-lived creatures, with rapid maturity and abbreviated reproductive cycles. They tend to show low juvenile mortality, high adult mortality, and engage in what is eloquently described as 'scramble' rather than 'contest' competition (Nicholson, 1954). Organisms which are predominantly under *K*-selection pressures are larger, longer lived. Reproduction is extremely low — just sufficient to replace low adult mortality: populations are characteristically *extremely* stable. Adult competition is intense, and of 'contest' type (Chapter 8, p. 203).

Note: the stability of *K*-selected populations, and relative instability of rapid-growth *r*-strategists can be explained fairly easily. Given the differences in population parameters outlined above, the population growth curves of the extreme *r*- and *K*-strategists will be of very different forms (Figure 10.1). The *K*-strategist will tend to have a stable equilibrium point (S in Figure 10.1) to which the system returns after moderate disturbances; but if the population declines below some lower threshold (x) it cannot recover, and it decreases ineluctably to extinction. Conversely, the *r*-strategist's population grows rapidly at low densities, has an equilibrium point (S in Figure 10.1) about which it is liable to oscillate, and crashes down from high densities. Natural enemies are unlikely to be important for these extreme strategies, for the *r*-strategist because enemies will be unlikely to colonise in sufficient numbers sufficiently quickly; for the *K*-strategist because of its large size and high competitive ability. For those organisms which are in an intermediate position natural enemies will have a significant role, and may provide at least one further equilibrium point part-way up the population growth curve (Southwood, 1976).

Predominantly *r*-selected organisms with the characteristics just described of course make ideal colonisers in succession. Early seral stages are usually pretty hostile environments in which to live, and are by their very nature impermanent. By contrast, the more stable environments offered by climax communities are ideally suited to K-strategists (r-selected organisms could wreak absolute havoc in such systems by overshooting the carrying-capacity of the environment and damaging essential resources). It is no coincidence that the characteristics of r and k-strategists correspond closely to those characteristics of organisms of early and late successional stages listed in Table 4.2.

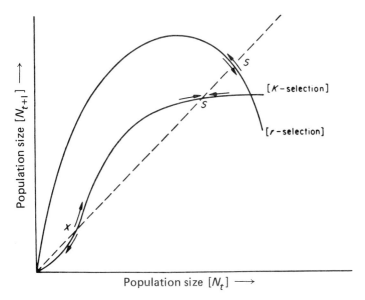

Figure 10.1: The Population Growth Curves of r- and K-strategists, Showing Stable Equilibrium Points (S) and an Unstable Point, the Extinction Point (x). Source: From Southwood *et al*., 1974.

Nor is this correlation with ecological succession the only evidence of the relationship of 'r' and 'K' strategies to environmental stability. A number of examples are available from the literature in support of this. Perhaps the most elegant is that derived from the work of Abrahamson and Gadgil (1973) on golden rod (*Solidago* sp.). Golden rod occurs in a number of different species in the eastern USA in habitats ranging from open dry, disturbed sites to established woodland areas. The amount of energy allocated to reproduction, as represented by biomass allocation to flowers, can be shown to be related accurately to the stability of the environment (Figure 10.2). Nor is this effect evident merely between different species. Convincingly, the same increase in reproductive effort in less stable environments is shown even within a single species – *S. speciosa* occurring in a range of sites. A similar increase in allocation of biomass to reproduction in disturbed sites is shown for *Taraxacum officinale* by Solbrig and Simpson (1974).

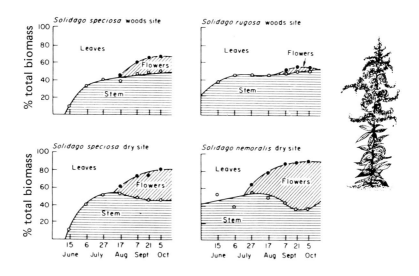

Figure 10.2: The Percentage of Biomass Allocated at Various Seasons, to Stems and Flowers in Four Golden Rod Populations. Source: From Abrahamson and Gadgil, 1973; after Southwood, 1976.

It is apparent that whole taxonomic groups of animals and plants tend to be largely '*r*' or '*K*' strategists. Thus, amongst animals, insects are usually *r*-selected, while mammals and birds tend to be *K*-selected. Amongst plants, almost all annual species and taxa are predominantly *r*-selected; forest trees like oaks and beech are of course clear *K*-strategists. This is not, however, a fixed rule and within each category there are gradations and exceptions. While birds are generally more *K*-influenced than insects, they themselves may be more or less *K*-dominated. Southwood (1976) quotes the wandering albatross (*Diomedea exulans*) as an extreme *K*-strategist. This species breeds only in alternate years and has a clutch size of one. Its period of immaturity (of 9 to 11 years) is the longest for any bird (Elliott, 1971). Population numbers are low but remarkably stable; one breeding population, on Gough Island, has remained constant at approximately 4,000 individuals since 1889. By contrast, the blue tit (*Parus caeruleus*) referred to in Chapter 8, is much more an *r*-strategist. It is an opportunist species with rapid reproduction, heavy mortality and widely fluctuating populations. Amongst insects there are again more and less *r*-selected species. Animal carrion must represent *par excellence* an environment which is both temporally and spatially unstable. Insects colonising such resources may show extreme *r*-selection. Thus Calliphorid blowflies show as adaptations to this impermanent habitat (Putman, 1977): high fecundity, rapid development of juveniles to adult (shortened generation time), low juvenile mortality. Intraspecific competition leads to a reduction in pupal size rather than leading to mortality, at low levels of competition. These reduced pupae are perfectly viable and give rise to small but perfectly reproductively active adults (Nicholson, 1950; Ullyett, 1950).) Further 'r' characteristics are early age of first reproduction, small body size and short life-span (Putman, 1977; Beaver, 1977). By contrast, various tropical butterflies are noted for their longevity and the stability of their populations. These butterflies are often large and territorial. Some *Morpho* species take 10 months to reach maturity. Fecundity is low, and adult survival is relatively good (Young and Muyshondt, 1972).

10.4 Implications of *r*- and *K*-selection

As noted earlier, once selection pressure has established any organism as more of a *K*-strategist or an *r*-strategist this 'decision' rapidly becomes accentuated and the organism continues to develop towards one extreme or the other, rarely remaining in an intermediate position (Figure 10.3). *r*-selection confers small size and short life-span as the cost of rapid reproduction; but of course if you are saddled with a short life span and early death, then it becomes still more critical to have massive fecundity while you are alive: the strategy spirals to extreme. (Likewise, any organism even marginally *K*-selected will also gravitate further towards the *K*-extreme. In order for the population to stay in balance with the carrying capacity of the environment it is necessary for individuals to reproduce slowly, over a prolonged period. But longevity in itself costs energy, so that less may be set aside for reproduction; with still lower reproductive rate, the organism has to live even longer and so extreme strategy develops.)

But the adoption of *r*- or *K*-strategy has repercussions in addition on all other aspects of the organism's ecology (Law, 1979). As an example of such influence, in a very fundamental way, we may note a close relationship between 'choice' of *r*- or *K*-strategy and use of assimilated energy.

Assimilated energy for all life processes can only be used for production or respiration, but the relative amount used for each differs in different organisms (Section 3.3). McNeill and Lawton (1970) (after Engelmann, 1966) calculated a ratio of P/R for the use of energy by different organisms for Respiration and Production over the course of a year. Although, of course the spread of P/R ratios was over a wide range, they found that there was a great tendency for organisms to cluster at one extreme or the other. In Figure 10.4 which shows a plot of log R upon log P for their data, this 'gravitation' towards extremes of high or low P/R is represented by the two regression lines drawn. Data on P and R values of different organisms continue to accumulate. With more figures at his disposal than had McNeill and Lawton, Humphreys (1979) has presented a more sophisticated analysis of P/R ratios. The same conclusion emerges: that organisms fall into a limited number of groups, each characterised by relatively few regression relationships of P upon R. It is almost as if of the whole continuous range theoretically possible in values of P/R, only certain positions are functionally tenable. More importantly, there are certain general characteristics of the various groups created. Organisms with high P/R ratios tend to be short-

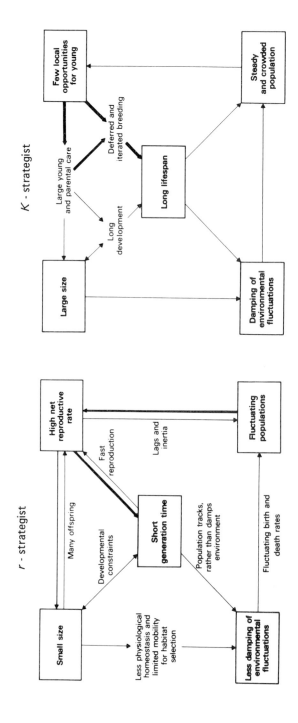

Figure 10.3: Positive Feedback Reinforces Both (Left) r- and K-selection (Right). These diagrams summarise the relationships and interactions between different aspects of each life-strategy, showing how the effects self-reinforce. Source: From Horn, 1978.

lived and, if animals, ectotherms; endothermy, a respiratorily expensive strategy results in a relatively high R, and thus, with less energy left available for production, a lower P/R ratio.

Of course, either strategy can be perfectly efficient. Organisms with low P/R ratios can put little energy into growth and reproduction, but because they tend to be relatively long-lived (one of the reasons R, over the course of a year tends to be fairly high) they can accommodate slow growth, and have the potential to reproduce in more than one season. Thus the longevity, while to some extent accentuating the size of R, compensates by allowing P to accumulate over considerable periods. Organisms with high P/R ratios, while short-lived, may grow rapidly and express a very high reproductive rate. Either strategy is viable — and once committed to one or another, each can be improved upon by exaggeration; which is perhaps why organisms do gravitate towards the extremes of the P/R continuum. In practice this alternative of life-strategy (to be short-lived, with rapid growth and massive reproductive potential, or longer-lived, with slow growth and reproduction) of course relates to, indeed is a direct consequence of, the more fundamental dichotomy already discussed of '*r-*' or '*K-*'strategy. Endothermy/ectothermy, high or low P/R ratio are merely practical consequences of this more basic division of strategy which has, as we shall see, many other implications in addition.

Once selection pressure has established any organism as a *K*-strategist or *r*-strategist, this decision has considerable repercussions for other aspects of its ecology. We have discussed its effects on the use of assimilated energy. Earlier we have established that *r*-selection as side effects confers small-size, short life span and so on. Differences in degree and type of intraspecific competition (contest, or scramble) between *r*- and *K*-selected organisms may, in addition, have profound consequences for population dynamics and stability (Chapter 8). More specifically in this context, we can observe a relationship between *r* and *K*-strategy and form of population regulation.

In our discussions of population regulation we have stressed that for dynamic balance of animal populations some form of density-dependent regulatory mechanism must be in force, whether intrinsic or extrinsic to the population to be regulated. We noted however that certain authors (e.g. Andrewartha and Birch, 1954; Andrewartha, 1959, 1970) claim that no such density-dependent mechanism is required. The controversy between proponents of density-dependent and density-independent theories of regulation was a long and a bitter one. Yet — need either mechanism exclude the possibility of the other? Could each not operate under different circumstances? The persistence of populations through the action only of density-independent events, if it occurs, results in irregular fluctuations in numbers of great amplitude, completely un-'controlled'. If a population crash is too great, the whole population may become extinct (yet local extinctions and recolonisations are something we have already accepted, in Chapter 2). Density-dependent regulation, because of the way it compensates with respect to population size, results in a relatively stable population with comparatively small fluctuations in numbers: a population, in short,

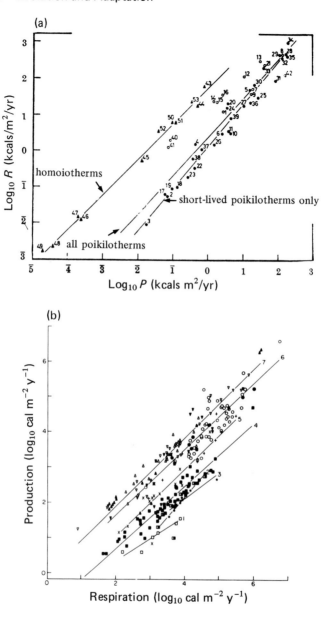

Figure 10.4: The Relationship Between Annual Production (P) and Annual Respiration (R) in Animal Populations. (a) (After McNeill and Lawton, 1970) ▲, Endotherms; ○, long-lived ectotherms; ●, short-lived ectotherms. (b) From Humphreys, 1979. The lines are numbered 1 = insectivores, 2 = small mammal communities, 3 = birds, 4 = other mammals, 5 = fish and social insects, 6 = non-insect invertebrates and 7 = non-social insects. The symbols denote: ▢ insectivores, ▣ small mammal communities, ■ other mammals, * birds, + fish, x social insects, ○ molluscs, ● Crustacea, ▼ other non-insect invertebrates, △ Orthoptera, ▲ Hemiptera, ▽ other non-social insects.

Table 10.1: The Relationship Between 'r' and 'K'-selection and Density-dependent, Density-independent Behaviour of Populations. Many examples offered in the literature for regulatory processes are confused, or indeterminate. Only those definitely attributable as density-dependent, or independent are included below.

Density-dependent regulation. [Environmental limiting factor does not act directly but is mediated w.r.t. density]			Population control is density-independent. [Environmental limiting factor acts directly and is density-independent]		
Example	Authority	'r' or 'K' selected	Example	Authority	'r' or 'K' selected
Reindeer	Scheffer, 1951	K	*Drosophila*	Chiang and Hodson, 1950 Sang, 1950	r
White-tailed deer	Severinghaus, 1951 Chaetum and Severinghaus, 1950	K	*Urophora jaceana*	Varley, 1947	r
			Metriocnemus	Lloyd, 1937, 1943	r
			Thrips imaginis	Davidson and Andrewartha, 1948	r
Elk	Cowan, 1950 Davis, 1951	K			
Rat	Christian, 1950, 1959	K			
Laboratory rodents	Christian and Davis, 1964	K			
Red Grouse *Tribolium*	Watson, 1967 Park, 1948	K (open to interpretation)			
Blowflies (natural popns)	Nicholson, 1933 Putman, 1977	r, but over short time scale	Blowflies (artificial popns)	Nicholson, 1933 Ullyett, 1950	r
Daphnia (Lab. popns)	Frank, Boll and Kelly, 1957	?	*Daphnia* (Lab. popns)	Pratt, 1943	?
Balanus	Hatton, 1938 Deevey, 1947	K			
Great tits (*Parus major*)	Kluyver, 1951 Krebs, 1970	?	Great tits (*Parus major*)	Perrins, 1971	r

in balance with the carrying-capacity of its environment.

Any K-selected organism therefore, because it must, before all else, maintain a balance with its environment, must surely show density-dependent regulation. But r-strategists are by their nature, selected for maximum reproduction. They have little need for regulatory mechanisms since they are in any case short-lived, and because their environment is so unstable anyway, it is of no real consequence if they do damage it. Further, the environment is by its very nature changeable: thus it is virtually impossible for a population to be regulated to what becomes a variable and rapidly decreasing carrying-capacity. If one reviews all examples of animals known to show density-dependent regulation, it is clear that all, without exception (whether mammals, birds or insects) are essentially K-strategists. If by contrast, one examines all those cases cited by those who would claim that certain populations may persist solely due to the effect of density-independent processes, such species are equally clearly highly r-selected organisms (Table 10.1). (This is not intended to suggest that there are such things as populations which do not show any form of density-dependent behaviour; merely that if this does occur, we would expect to encounter it only amongst r-selected organisms that do not require it.)

Even within the clearly density-dependent group Stubbs (1977) has shown that there are significant differences in the strength of the response between more r-biased and K-biased organisms, with K-strategists showing a steady if not particularly powerful mortality, and r-selected organisms showing either very weak or very strong density-dependent mortalities, or as frequently both, at different stages in the life cycle (Section 7.5).

This influence of one strategy-choice upon another occurs over and over again. Within their bionomic 'choice' of r- or K-strategy organisms are faced with other choices: animals may have alternatives available in hunting-method, of social organisation, of reproductive strategy, whether or not to hibernate: for every possible problem there are regularly alternative and equally practicable solutions. In practice the particular solution adopted may be, as we have suggested, dictated by choice in a previous 'decision'. At every stage in 'designing' its particular niche any organism is faced by such alternatives, so that the final life-strategy adopted by any individual animal is a function of a whole host of interacting pressures. And it's all something of a vicious circle. Adoption by any organism of a certain bionomic or ecological strategy will necessarily have profound influence on other aspects of its ecology/biology. We have already shown that 'choice' of r- or K-strategy affects use of assimilated energy, influences size, life-span, endo-/ectothermy; it also sets restrictions on decisions adopted in other subsequent 'choices' of strategy.

10.5 Adaptiveness of Foraging Strategy

One of the most important considerations to any organism is efficiency in obtaining its food. Yet there is a variety of ways in which, for example, an animal may choose to forage. We have already noted in Chapter 9 that a predator, by its actions, may increase the stability of its position by increasing the diversity of its prey community, and may further increase the yield of those individual prey populations to itself, by stimulating their separate productivities. In the present context we must consider within that context, how the method of operation of a particular predator may be redefined to maximise its own *intake* from such a prey community. It is clear that a predatory species may evolve to increase the potential food intake of its individual members — in avoidance of competition or in selecting prey which offer maximum return for minimum effort. But there is literally more than one way of killing a cat. What are the merits and demerits of the variety of available hunting strategies which a predatory species may adopt?

In a theoretical study of the optimal size for various types of predator (defined as that size at which the predator takes the least amount of time to satisfy its energy requirements), Schoener (1969) examined the cost-effectiveness of various different predation strategies, and their implications for the predator. The potential costs of any foraging strategy (not just predation) have been summarised by Lawton (1973) as those of maintaining a food supply, those of locating a food supply, energy costs of gathering or catching the food, energy costs of processing that food and of eating or swallowing, and finally, energy costs of transporting and storing food (Table 10.2). Clearly not all strategies involve all costs and all involve differing proportions of the various cost-elements. With respect to costs such as these, Schoener defined three possible types of predator. Type I predators are predators which passively locate their prey by scrutinising an area surrounding a vantage point and then pursue that prey irrespective of its size or distance. The predator is considered to expend no time or energy specifically in prey search since, while watching for items to enter its possible field of prey capture, it is engaged in other activities (grooming, maintaining territory etc. etc.). Type I predators expend energy only in pursuit — and correspond to what are now commonly known as 'sit and wait' predators. Type II predators utilise time and energy in searching for food that is expended for that purpose alone. (This implies discrete periods of time during which the predator does nothing but forage and eat.) Such predators expend no energy in pursuit; once a food item is located it may be eaten directly. All energy is expended in search. Type IIa and IIb predators are predators which expend energy and time in both search and pursuit. Type IIa predators are a special case of Schoener's model where the relative distance between predator and prey for pursuit is constant; type IIb is the more general expression of the class where pursuit distance may vary. All predators are assumed to spend time and energy in handling and swallowing prey. Note: Schoener's Predatory types I and II are distinct from, and do not in any way correspond to Holling's (1961) types I and II predatory response.

Schoener's models offer some fascinating predictions: a possible explanation, for example, for sexual dimorphism, suggesting that such dimorphism might be expected in Type I and Type IIa, IIb predators, but not in predators of Type II.

Table 10.2: Summary of the Main Processes on Which Animals and Man Expend Energy in 'Food-gathering'

	Activity	Examples for animals other than man	Examples for man
i	Energy cost of maintaining food supply	Small or non-existent: some exceptions e.g. ants which maintain aphid populations, and possibly part of the cost of territorial defence in some birds (Stiles, 1971) could be attributed to this category	Most (but not all) of the activities embraced by the word 'farming' e.g. crop protection, animal husbandry, etc.
ii	Energy cost of locating food supply	Probably of major importance in animals which have to hunt or forage for suitable food, and hence move about actively (Schmidt-Nielsen, 1972)	Non-existent in modern farming, but important in primitive man, as in hunting animals
iii	Energy cost of gathering or catching food	Chasing, catching, overcoming and killing prey (Salt, 1967); energy used during grazing in herbivores (Golley and Buechner, 1968)	Harvesting in modern farming: very important
iv	Energy cost of processing food	Probably very small for most animals; opening nuts in rodents (Rosenzweig and Sterner, 1970), cracking snails (thrush), or 'chewing cud' (ruminants)	Very important, particularly in modern man, where food may be extensively processed before eating
v	Energy cost of eating or swallowing food	Probably very small for most animals; may be important in species which swallow relatively large prey (Schoener, 1969); may be difficult to distinguish from iv	Small or negligible
vi	Energy cost of transporting and storing food	Probably small for most animals; transport back to nest (bees, birds) or winter food-stores	Very important, particularly in modern man, where food may be transported long distances and stored for long periods

Source: From Lawton, 1973.

Such a conclusion derives from the observation that there are in fact two possible optimal sizes for predators (Types I, IIa and IIb) which contain an element of pursuit in their strategy rather than pure search, and that, in avoidance of intraspecific competition, dimorphism might be expected to evolve to exploit this 'double-niche'. More important within the present context are his conclusions about the relative cost-effectiveness of the different strategies and the restrictions that adoption of one or other strategy may impose upon the predator. Energetic considerations are such that, for example, type I predators must be non-selective — eating any type of food they may catch. Type II predators will tend to eat a particular item of food if the energy gained per unit time is greater than the energy gained per unit time by ignoring it and searching for and eating a more favourable item (cf. considerations of 'optimal foraging' discussed in Section 10.6). The reintroduction of an element of pursuit in type IIb predators — and the fact that energy is lost both in search and pursuit — forces the return of generalisation. The predator is unselective with respect to prey size and distance and will therefore tend to eat a particular item if the energy expended in its capture and ingestion is less than the energy *return* from the prey.

Schoener's study, however, does not distinguish between any of the different strategies in overall efficiency. Lawton (1973) examines the ratio of energy return to energy expended in actual examples of different organisms adopting a variety of foraging strategies. Each strategy has a different breakdown of costs within the various categories of Table 10.2; each has different levels of return. His results (Table 10.3) are rather surprising in that — despite the inclusion here of all three types of Schoener's 'predator' — the overall ratio of energy gained to energy expended is very much the same for all organisms and all strategies. Similar results are discussed in a more recent study by Huey and Pianka (1981) of foraging patterns amongst desert lizards. Such lizards are typically either widely foraging or sit-and-wait predators; and clearly the type of prey which can be taken by the lizards is related to their foraging mode. Huey and Pianka note that daily energy expenditures of widely foraging lizards appear to be 1.3-1.5 times greater than those of sit-and-wait lizards in the same habitats, but gross food gains are also about 1.3-2.1 times greater.

Clearly energetically expensive techniques of food-gathering can only be applied to food sources which guarantee a high rate of energy return, while equally, only animals capable of meeting the expense may exploit such high energy sources. Such considerations may partly explain the apparent convergence in the ratio of energy gained to energy expended in all strategies examined. Overall it is clear that in the right context, any of a variety of foraging/predation strategies may prove highly efficient. The most efficient strategy which a species may adopt in exploitation of a particular resource will however depend on and be dictated by the characteristics of the resource itself. (Widely foraging lizards in the Kalahari desert, the Western Australian desert and the North American desert, generally are found to be feeding on prey which are sedentary, unpredictably distributed and clumped (e.g. termites) or that are large and inaccessible (inactive scorpions). In contrast, sit-and-wait lizards eat more prey that are active (Huey and Pianka, 1981).)

Table 10.3: Summary of the Energy Costs of 'Food-gathering' in Various Animals

Animal	Author	Type of organism	Type of food and method of feeding	Absolute rate of energy expenditure in feeding (cal min^{-1})	Energy gained / Energy expended
Eugenes fulgenes	Wolf and Hainsworth (1971)	Tropical hummingbird	Rapid, hovering flight at experimental vials of sugar in the laboratory	32.9	7–70
Amazilia tzacatl *Phaethornis superciliousus* *Thalurania furcata*	Wolf, Hainsworth and Stiles (1972)	Three species of tropical hummingbird	Rapid, hovering flight at natural flowers in the field to collect nectar	16.1–21.5	3.8–22.2
Spiza americana	Schartz and Zimmerman (1971)	Dickcissel; a north American finch	Flitting through low vegetation in search of insects and seeds	15.6	12.8
Bombus vagans	Heinrich (1972a)	Bumblebee	Flight between *Epilobium* (fire weed) flowers for nectar; in cold weather, energy is also expended in maintaining a high thoracic temperature	0.32–0.46	4.4–20.2
Pyrrhosoma nymphula	Lawton (1971a)	Freshwater damselfly (Odonata) larva	Lying in wait for prey (other small aquatic insect larvae and crustacea) to swim past	5×10^{-5} 5×10^{-6}	1.1–3.6
Micropterus salmoides	Glass (1971)	Largemouth black bass; freshwater fish	Swimming after small prey – fish (guppies) in the laboratory	2.2–3.0	3.8–10.3

Source: From Lawton, 1973.

10.6 Optimal Foraging

Such considerations may be extended to any foraging behaviour and are not restricted to predatory species alone. In addition, we may note that while selection over many generations will result in the development by a species of a foraging strategy which is best for that species under average circumstances (i.e. those obtaining over an evolutionary time scale) refinements *within* that species strategy in the shorter-term may be effected by the individual during its own lifetime by behavioural change.

10.6a Search Images, and Switching :Other Behavioural Mechanisms for Rapid Adaptation of Foraging Strategy

Many polyphagous predators seem to show frequency-dependent preference for different prey types: that is, rather than feeding upon them in direct relative proportion to their occurrence within the environment, they take a disproportionately high number of abundant prey, and correspondingly disproportionately low numbers of the rare prey (Figure 9.25). This classic 'switching' behaviour (Murdoch, 1969; Murdoch and Oaten, 1975) has profound implications for the effects of predators in stabilising or regulating populations of their prey (Chapter 9), but may also be a mechanism by which the predator may maximise its own foraging efficiency.

One learning mechanism which can produce switching is the development of a 'searching image'. Faced with an abundance of one prey type the predator becomes significantly better at recognising that particular type of prey even against a cryptic background, and will thus be able to continue to recognise them more easily than alternative prey even when they become relatively less abundant (Dawkins, 1971).

Such short-term behavioural adaptations may enable individual predators to increase their own instantaneous foraging ability within the overall species strategy established by evolution. The adaptiveness of the individual in this way, the flexibility of foraging strategy adopted in a given set of circumstances, and the way in which the animals respond to optimise their foraging success has become the core of a whole branch of behavioural science.

10.6b Optimal Prey Selection

Under the influence of intraspecific competition with its fellows, every animal is expected to optimise its potential forage intake with regard to some critical factor: usually regarded to be energy. An enormous body of literature has been devoted to examination of how animals arrive at an optimal feeding strategy; a range of simple models or possible sets of 'decision rules' (Krebs, 1978) have been developed and tested in experimental studies. Any prey item eaten by an animal has, as we have seen, a cost in terms of the time and energy taken to capture and eat it, and a benefit in terms of its actual food value. The net profit of any item, divided by the handling time required to subdue and eat it is considered the profitability of the food, and it is presumed that predators are able to distinguish between items of differing profitability and to select the more profitable types. It is clear that most predators do feed selectively and are preferring the more profitable food types (Figure 10.5). But, as Krebs (1978) points out, while it is easy to see that an optimal predator should prefer the most profitable prey, it is less obvious to what extent it should also include less profitable items in its diet. Assume, with Krebs (1978), that a predator spends its foraging time either searching for or handling prey (cf. Schoener, 1969, above). Intuitively one can see a trade-off: if a predator selects only the best items it has a high rate of food intake per unit handling time – but it also has to spend a relatively long time searching for each item. A predator which included in its diet relatively less profitable but more available prey might increase overall its food intake per unit time. If predators are to optimise their own individual intake, the diet should consist of a mixture of the most profitable, and less profitable items in direct proportion to the simple balance of expected gain over expenditure for each item. Three general predictions emerge, if predators are, as the model assumes, attempting to maximise energy intake per unit time. (a) Predators should prefer more profitable prey; (b) they should feed more selectively when profitable prey are abundant; (c) they should include relatively less profitable items in the diet when the most profitable food is relatively scarce, but should ignore unprofitable food, however common, when profitable prey are abundant.

These predictions have been tested by Krebs and coworkers (1977) who used caged great tits (*Parus major*) as predators, with large (profitable) and small (unprofitable) pieces of meal-worm as prey. They estimated the profitability of the two prey types as (weight/handling time) and controlled the encounter rate with each type very precisely by presenting them to the predator on a moving belt. The results conform well to predictions: when large and small prey were present at a low density the birds were unselective as predicted, but when the density of large prey was increased to a level at which the birds could do better by ignoring small prey, they became highly selective. Finally, keeping the density of large prey constant, Krebs *et al.* increased the density of small prey so that they were twice as common as large ones. The birds, as predicted, remained highly selective and essentially ignored the small but abundant prey.

Similar results were obtained by Werner and Hall (1974) for bluegill sunfish (*Lepomis macrochirus*) hunting for three size classes of *Daphnia* in a large aquarium, while in test of the hypothesis in the field, Goss-Custard (1977) studied the selection by redshank (*Tringa totanus*) of different sized polychaete worms (*Nereis*, and *Nephthys*) on mudflats. As with the previously discussed laboratory studies, Goss-Custard estimated the profitability and availability of different size classes of prey. Then, by comparing the rate of feeding by the redshank on large and small worms at various different study sites, he was able to show that the largest, most profitable, prey were eaten in direct proportion to their own density, while the smallest worms were not taken in relation to their own density, but at a rate inversely proportional to the density of large worms. In other words, as the density of large worms increased, the redshank became more selective, and further, they tended to ignore small prey regardless of their own density as long as large worms were common.

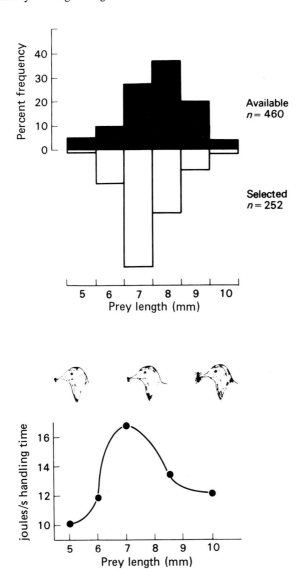

Figure 10.5: Selection of Flies by Pied Wagtails. The histogram shows that the wagtails select flies in a distribution different to their actual availability. Consideration of the profitability of the different prey sizes (as energy return per second of handling time) explains why. Source: After Davies, 1977.

10.6c Optimal Patch Selection

Actively searching predators usually hunt for food which is clumped or patchy in distribution. Just as with choice between profitable and unprofitable prey items, one can see that an optimal predator should forage preferentially in 'rich' patches, and include less profitable patch types only when the availability of good patches is low (Royama, 1970). (Once again this assumes that the feeder aims to maximise food intake per unit time.) An optimal predator should in fact stay in each food patch until its rate of intake drops to a level equal to the average of intake for the habitat as a whole; the predator should not stay in a patch when it could do better by travelling to another one. As before the model makes a number of predictions: (a) each patch once occupied should be reduced in food content to a constant level which (b) should be equal to the average food content of the habitat as a whole. Cowie (1977), in one of the most rigorous tests of the optimal patch model, studied captive great tits foraging in a large indoor aviary for small pieces of mealworm hidden in sawdust-filled plastic cups on the branches of five artificial trees. Cowie tested six birds individually, and arranged the experiments to last for a time short enough to exclude any effect of revisiting patches. Each bird was tested in two 'habitats', with a short and long 'travel time' between patches, the prediction being that birds would adjust the time spent per patch in relation to the travel time as predicted by the optimal foraging model. The predator should spend longer in each patch when the travel time is long. (The 'travel time' was manipulated by making it easy or hard for the birds to start foraging in a new patch, by placing a loose or tight-fitting cardboard lid on each plastic cup.) Cowie measured, for each bird, the travel time in the hard and easy environment as well as the curve of cumulative food intake within a patch, which was the same in both environments, as patches always contained the same number of prey. From these, he could predict the optimum relationship between travel time and time in patch. The observed relationship is quite close to the predicted, but birds in fact tend to spend too long in each patch (Cowie, 1977).

Real animals may in any case be expected to deviate to some extent from optimality in this regard as indeed in optimal prey selection since the animal is not omniscient: in order to acquire the information on which to base its foraging strategy, the predator must first 'sample' the environment — and continue to sample at intervals to ensure that conditions have not changed in the meantime. As a result, foraging strategy observed — which must include occasional sampling — must deviate to some degree from the theoretical optimum.

A study which has shown empirically that predators sample and can use the information when the environment changes, is that of Smith and Sweatman (1974) who trained captive great tits to search for hidden mealworms in six patches containing different prey densities within a large aviary. The tits soon learned to concentrate their foraging effort in the most profitable patch, but they also spent more than the optimal amount of time in the various less profitable patches (Krebs and Cowie, 1976). Smith and Sweatman showed that when the best patch was suddenly reduced in quality, the tits switched to foraging primarily in the second best patch, which seems to show that the birds had sampled each place and stored up information about its relative profitability.

10.6d Return Times

Predators exploiting renewable supplies have the opportunity to return to a previously depleted area to feed again; the problem, however, is to time the return visit optimally. If the predator returns too soon, replenishment will not be complete; if it returns later than necessary, it risks travelling further or spending more time on its foraging bout than if it had returned earlier. Either way its foraging efficiency for the bout is decreased.

As Charnov *et al.* (1976) point out, the utility of returning to a renewed resource is limited by the amount of competition a predator faces. To achieve an optimal return time, the predator must have exclusive use of (and hence reliable information about) the resource. For this reason, we are most likely to find 'return' strategies among territorial and socially aggregating species.

Territorial nectar-feeding birds are good examples of 'return' strategists. Territory size in golden winged sunbirds (*Nectarinia reichenowi*) and Hawaiian honeycreepers (*Loxops virens*) is geared to the rate of nectar renewal within depleted flowers (Gill and Wolf, 1975; Kamil, 1979). Territory owners regulate their patterns of revisits to enhance their net rate of energy intake over a foraging bout. Perhaps the best illustration of optimisation of return time comes from the work of Davies (1976). Pied wagtails (*Motacilla alba yarelli*) feeding on insect debris washed up along a river bank were shown to patrol the banks in a methodical way, moving systematically up one shore and then returning along the other – a behaviour quite distinct from that of other individuals feeding on insects in pasture. Davies showed that the pattern of search: regular patrolling of the river banks, served to maximise the birds' return time to previously depleted areas; he also showed that with such a predictably renewable resource, the birds were far more alert in evicting intruders to their territories than were pasture feeders – again as we would predict (Davies and Houston, 1981).

10.6e Nutrient Value of the Diet

Most models of optimal foraging have tended to assume that what is maximised by any foraging organism is net rate of energy intake per unit time. Yet this is not necessarily the feature of the diet which is most critical to the feeder (page 80) and models based on energy return alone may be misleading.

Sometimes specific nutrients or other qualities of the prey may be more important. A possible example of selection for a specific nutrient, is found in the redshank studied by Goss-Custard (1977b). When both nereid worms and the crustacean *Corophium volutator* are available in the mudflat, birds select *Corophium*, even though they are energetically less profitable than any size class of worm. One possibility is that *Corophium* is rich in some nutrient, like calcium, which is required by the birds.

Selection for specific nutrients is well known in breeding animals. During periods of egg-laying, spotted flycatchers take a number of calcium-rich prey (e.g., woodlice (Isopoda), millipedes (Diplopoda) and snail shells (Gastropoda)) which do not occur in the diet at other times (Davies, 1977). Grass-feeding bugs have been shown to switch to a low-energy, high nitrogen diet when preparing eggs (Gibson, 1977; page 80 here). Recognition of the importance of such nutrients, beyond mere energy value of the food, has begun to be accommodated in the optimal foraging literature in recent years (e.g. White, 1978; Stenseth and Hansson, 1979; Pulliam, 1975b; Westoby, 1974; Greenstone, 1979).

This section on optimal foraging is based in large part on a review by Krebs (1978). The reader wishing to pursue this subject further is recommended to read this excellent review in full.

Considerations of optimality need not, however, be restricted to foraging strategy. The same adaptiveness and flexibility, the same idea that there may be more than one, equally efficient, solution to any given problem may be observed in other contexts.

10.7 Reproductive Strategy

The evolutionary adoption of a specific reproductive strategy may equally well be shown to be strictly influenced by environmental 'circumstances'.

Thus adoption of a particular reproductive strategy out of a host of equally plausible alternatives − monogamy, polygyny, promiscuity, even polyandry − is dictated by ecological circumstances.

We have already noted a distinction between so-called 'r'- and 'K'-selected organisms: those more or less ordered by the differing selection pressures for maximum recruitment or for population stability, whose resultant reproductive strategies polarise towards totally different responses. But these same alternatives of strategy are available, in reduced form, *within* a species too, in terms of the strategy adopted by the different sexes. Natural selection acts differently on male and female. Within that class of organisms that opt for some degree of parental investment there is competition between the sexes to minimise their own contribution; the resolution of this sexual conflict is reflected in the different reproductive strategies of monogamy, polygyny, polyandry or promiscuity.

Once the evolutionary decision to invest in parental care, to whatever small degree, has been taken, the *species* is committed − for as soon as you start to make an investment of this sort, as soon as you have devoted time or resources to that cause, then you stand to lose more by abandoning the offspring in which you have invested than by continuing even if in continuation you have to invest still more (the so-called 'Concorde Fallacy' of Dawkins, 1976). For in desertion, after all, you lose all the investment you have committed so far − and still have to repeat that initial investment in any later breeding attempt. But, if the levels of investment committed by the two separate sexes are different, however subtly, then the sexes will compete to offer least. Such analysis of the degree of parental investment commited by each parent to a breeding attempt in an evaluation of the origin of different mating systems was first proposed by Trivers (1972). Other, more descriptive or qualitative explanations had earlier been presented (by, for example, Orians, 1969) but Trivers' imaginative analysis provides a more generally applicable hypothesis and offers an evolutionary mechanism.

Trivers argues that any animal will opt for the mating system which maximises the number of its offspring which will survive while minimising any investment on its own part which will hamper its fathering or mothering of other additional offspring. Trivers defines such investment as 'any investment by the parent in an individual offspring that increases the offspring's chance of surviving (and hence reproductive success) at the cost of the parents ability to invest in other offspring in the future'. (Dawkins and Carlisle (1976) stress that it is this reduction of investment in the future, rather than the degree of investment in the past which is important.) Investment includes not only the metabolic costs of producing gametes, but later activity such as foraging for or protection of young, milk-production in mammals etc. etc.

At any stage each individual is under heavy selection to adopt that strategy which maximises its own production of young. Thus at any point in time the individual whose cumulative investment is exceeded by his partner's is theoretically tempted to desert, especially if the disparity is large. (More strictly: the partner who has invested most, who would thus have to invest most in recommencing a new breeding attempt, but has in fact less left for such future investment, is *less* likely to desert.) This temptation occurs because the deserter loses less than his partner if no offspring are raised and the partner would therefore be more strongly selected to stay with the young. Any success of the partner will, of course, benefit the deserter but meanwhile he has another chance of success in another breeding attempt. For example, desertion by the lesser investor right after copulation will cost him very little, even if no offspring are raised, while the chances of his partner raising some young alone may be great enough to make the desertion worthwhile (Trivers, 1972).

Clearly there will always be selection pressures upon the partner who has contributed least hitherto, to desert. For in thus reducing its investment to that family, it may be able to increase its own eventual production of viable offspring by going off to spawn elsewhere. Since the greater investor in the original alliance has more to lose by deserting as well, it will continue to rear the offspring of that first mating even if unaided. And, even if the survival of one-parent-reared offspring may be less than the survival chances of a two-parent family, as long as some *may* survive then it will pay the lesser investor to desert and try his luck elsewhere — for it will increase overall the number of offspring he leaves to posterity.

We use the word 'his' advisedly — for amongst most vertebrates (and it is the higher vertebrates particularly which tend to go in for parental care) it is the male who is the lesser investor. His initial contribution to a breeding attempt — a mere package of sperms — is insignificant beside the investment of a female in preparing and laying an egg, or in protecting and nourishing an embryo throughout pregnancy. The score can be balanced: after all, since survival of a one-parent family is never as high as that of young reared by both partners, selection should act on the female (who is also concerned with maximising *her* surviving offspring) to try to retain the male and increase his investment. There are a number of ways in which increased investment by the male can be so elicited: if he must, for example, establish a territory before he may win a female. And once he is committed, then his interests too are better served by continuing that commitment rather than by deserting. Amongst most birds, the investment of each partner is usually roughly equal. On the whole, most birds tend to be monogamous (although in some cases, females have cashed in on the male's increased commitment and have turned the tables, deserting themselves to become truly polyandrous). Thus the greater initial investment of the female may be compensated for by some increased commitment by the male − but only up to a point. Amongst mammals, the initial discrepancy is too great. Not only must the female carry all responsibility for the offspring throughout gestation; even after birth, mammalian physiology dictates that it is the female who shall continue to feed the young through their infancy. Male birds may be able to share equally in the effort of foraging for and feeding their chicks; male mammals cannot produce milk. It is amongst mammals, of all organisms, that differences in investment between the sexes reach extremes.

From our arguments so far, we should thus predict that amongst the mammals, natural selection should 'encourage' promiscuity in the male. Clearly, selection will also act upon the female to some extent in an effort to retain the male and increase his investment; but the distance is too great, and even by way of such compromise we may expect in the final outcome promiscuity, perhaps tempered to polygyny or polygamy. In practice such mating systems predominate within the mammalian world. Amongst birds, monogamy becomes the norm, although polyandry *may* result in certain circumstances, or in yet others polygyny may occur. (If the advantages to the female of mating with an already-mated male (perhaps because he defends a very high quality territory) outweigh the disadvantages of the fact that his postnatal care must be shared with others, females may well opt for polygynous matings.) In general, once more, the adoption of any one of a multitude of strategies is highly adaptive to maximise reproductive success of each individual, whatever sex, under differing ecological circumstances.

(The arguments here have of necessity been abbreviated and oversimplified. For a fuller exposition the reader is referred to Orians, 1969; Trivers, 1972; Dawkins and Carlisle, 1976; Maynard Smith, 1977 and Ridley, 1978.)

10.8 Adaptiveness of Social Group

Reproductive style is adopted within the framework of an established social organisation. This too may be shown to be highly adaptive.

In 1966 Crook and Gartlan published what has now become a classic paper relating group size and social organisation of various species of primate to habitat. In a review of published data they categorised species according to five crude descriptors of habitat (Forest, Forest Edge, Tree Savannah, Grassland, Arid zones); within each category they found striking similarities in social organisation. Thus forest dwellers at the one extreme were found to be primarily solitary animals with small ranges. Monkeys of forest edge moved in small family groups, while in open, grassland or arid zones, the species tended to occur in large troops, covering a large range.

Such a suggestion aroused considerable excitement and criticism. Clutton Brock (1974) notes that closer analysis reveals considerable variation in social organisation within each 'ecological' category: indeed differences between groups are probably less impressive than differences *within* groups. Further, Clutton Brock points out that different species may well react to similar environmental pressures in different ways. Thus a common selection pressure will not necessarily draw the same response from different species: the particular response of any species to any selection pressure will be influenced by its own phyletic history. In brief, many interspecific differences in social organisation may represent different methods of overcoming the same problem — rather than reflecting different ecological pressures. Conversely, an apparently similar adaptation in distantly related species may be in response to quite different selection pressures. Clutton Brock concluded that the relationship between social group size and ecology needed closer evaluation. Choice of more precise, more functionally related ecological parameters to categorise species might overcome within-group heterogeneity (there are after all a number of different ways of living in a forest), while studies of closely related species of the same phylogenetic inheritance would offer more convincing grounds for concluding that differences in social behaviour were indeed attributable to differences in ecology.

Clutton Brock's own study of red colobus and black and white colobus monkeys (*Colobus badius* and *Colobus polykomos*) in Tanzania, offered the first incontrovertible evidence for such relationship (Clutton Brock, 1974), and the flexibility of group size has since been confirmed in a number of other studies (e.g. Jarman, 1974, working with the assemblage of African antelope and gazelle, Figure 10.6; Putman, 1981 for British deer). The close relationship between group size and ecological circumstance is shown even more clearly by the fact that even within the same species, different populations may show variable group size under different circumstances. Jarman recognised distinctly different social structures for African buffalo (*Synceros caffer*) of forest and savannah habitats; similar differences have been noted for other species (e.g. Sika deer in Britain: Horwood and Masters, 1970; Mann, 1983). Even the same individuals within a population may show different social behaviour in different contexts (Horn, 1968; Caraco and Wolf, 1975). The social organisation is clearly in some way adaptive.

There has however been considerable speculation as to what is the adaptive significance of such changes in group size, and to what ecological pressure they

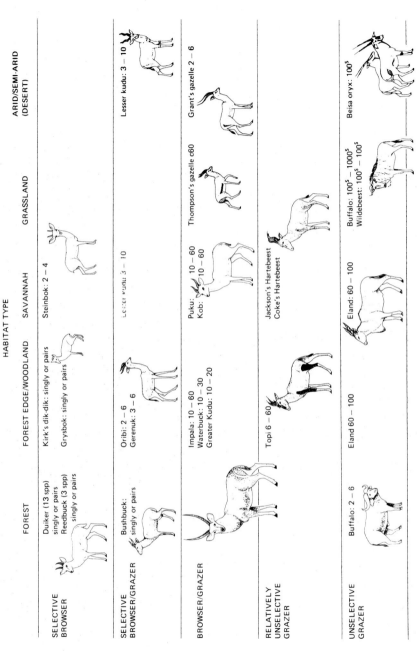

HABITAT TYPE

	FOREST	FOREST EDGE/WOODLAND	SAVANNAH	GRASSLAND	ARID/SEMI-ARID (DESERT)
SELECTIVE BROWSER	Duiker (13 spp) singly or pairs Reedbuck (3 spp) singly or pairs	Kirk's dik-dik: singly or pairs Grysbok: singly or pairs	Steinbok: 2 – 4		Lesser kudu: 3 – 10
SELECTIVE BROWSER/GRAZER	Bushbuck: singly or pairs	Oribi: 2 – 6 Gerenuk: 3 – 6	Lesser kudu: 3 – 10		Grant's gazelle 2 – 6
BROWSER/GRAZER		Impala: 10 – 60 Waterbuck: 10 – 30 Greater Kudu: 10 – 20	Puku: 10 – 60 Kob: 10 – 60	Thompson's gazelle c60	
RELATIVELY UNSELECTIVE GRAZER		Topi 6 – 60	Jackson's Hartebeest Coke's Hartebeest		
UNSELECTIVE GRAZER	Buffalo: 2 – 6	Eland 60 – 100	Eland: 60 – 100	Buffalo: 100^s – 1000^s Wildebeest: 100^s – 100^s	Beisa oryx: 100^s

Figure 10.6: Relationship Between Social Organisation and Ecology. Group size amongst East African antelope is related to feeding style and habitat. Source: Data from Jarman, 1974.

are related. Sociality in a general way can confer advantages in a number of ways. Clutton Brock and Harvey (1978) summarise these as:

(1) Increased detection, avoidance, or defence against predators. Formation of a large group offers several advantages in this regard: more eyes and ears for detection and advance warning of predators in the area; true 'safety in numbers' – in that the greater the number of prey available the lower the probability for each that he will be selected; a confusion effect – it is difficult for a predator to select any individual in a milling throng where his attention is easily distracted; and finally the potential within a group for group defence or even positive attack. (There are many examples in the literature for each of these potential benefits (e.g. Patterson, 1965; Carl, 1971; Vine, 1971; Hamilton, 1971).)

(2) Increased efficiency of food finding, food handling, or exploitation (Ward, 1965; Krebs *et al.*, 1972; Krebs, 1974; Schaller, 1972; Cody, 1971).

(3) Reproductive advantages including regular access to the opposite sex (Emlen, 1973; Brown, 1975) and the presence of peers during the period of social learning (Hinde, 1974).

Both *food distribution* and *predation* have been cited as possible factors involved in the more precise determination of the specific social organisation to be adopted in any given context: in determination of actual optimum group *size* as opposed to a more general 'whether or not to be social at all'. Crook and Gartlan, Clutton Brock and Jarman suggest that group size may be determined in large part by food availability and distribution. Such a conclusion also receives support from the studies of Horn (1968) and Caraco and Wolf (1975). Horn (1968) working with Brewers blackbirds (*Euphagus cyanocephalus*) showed that the entire social structure of these birds could be modified purely in relation to the distribution of prey within the environment. Thus in areas where food was uniformly distributed the blackbirds set up and defended exclusive territories; where food was abundant but patchily distributed, the birds foraged in large groups (Figure 10.7). Caraco and Wolf (1975) have also shown clear changes in the foraging group sizes of lions depending on the particular prey species being hunted, such that group size is adjusted to offer maximum return to each individual lion from the prey pursued. However the pressures of efficient foraging are not the only ones which might explain such a relationship: Jarman considers that predation may constitute a more critical determining factor, suggesting that although the relationship between group size and food distribution is a valid one, social organisation may be more markedly influenced by predation pressure. Jarman himself concludes (1974): 'When prey are fast enough to avoid predators once detected, or strong enough to defend themselves so that a group will not be subjected to inescapable, repeated attacks, they will benefit by herd formation. The advantages of herd formation, inducing animals to form groups, and the dispersion and availability of food items, limiting the maximum size of these groups, will between them account for the group size typical for each species.'

On reflection however, this suggestion that predation may be the dominating pressure is perhaps unlikely at least amongst animals (though it will of course be a major influence on distribution of plant species). Any animal must eat to survive — irrespective of the risks of predation. In addition, since the avoidance of predators may be achieved through increase *or* decrease of group size, and is thus essentially a more flexible response than that to food dispersion, it seems likely that restrictions of feeding ecology on social organisation might take highest priority, with predation pressure further adjusting social structure within the limits set by foraging requirements. In many instances however it is impossible to tell which pressures are involved and if both, which indeed has the dominant influence. In summary it is clear however, that group size, and social organisation are highly adaptive and closely related to specific ecological requirements.

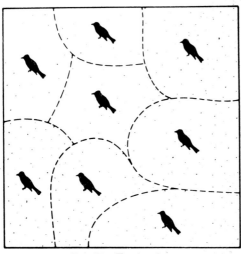

Feeding Territories

Feeding Flocks

Figure 10.7: Horn's Analysis of Group Foraging in Brewers Blackbirds. If food is more or less evenly distributed through the environment and can be easily defended, it is energetically most efficient to occupy exclusive territories. But if food occurs in unpredictable patches, it becomes more efficient to forage as a group.
Source: From Horn, 1968.

10.9 Optimality and Evolutionarily Stable Strategies

Other strategies of ecology or behaviour can be analysed in the same way – in terms of their adaptiveness: foraging strategies for example have been examined in Section 10.6 while the more general (but equally adaptive) concept of how animals allocate their time between different activities has been discussed by, for example MacFarland (1977), McCleery (1978). In all cases analysis is in terms of *optimality* of the strategy or pattern of response: balancing return over investment. But in all cases the choice of strategy may have been predetermined by choice in a more fundamental decision.

Section 10.5 suggests that work by Schoener (1969) has shown that a predator's size may be closely linked to its choice of hunting strategy. Schoener presents considerations of optimum size for a number of predator types. But change in size, or adoption of a certain size has its own implications. For example, adoption of a certain size restricts an organism's 'choice' of positions on the 'r'/'K' continuum. Increase in size, and concomitant increase in metabolic cost lead to a decrease in the proportion of energy which can be set aside for reproduction. As a result, generation time increases and rate of reproduction decreases. (Figures 10.8 and 10.9 demonstrate the clear relationships between generation time, and r_m with body size for a range of animals species – after Southwood, 1976.) Yet size is separately dictated by considerations of r- and K-selection from other environmental cues. Colonisation of a particular sort of environment forces adoption of r- or K-biased lifestyles. Choice of r-strategy requires small size; yet size, and thus *degree* of commitment to r may also be influenced by other considerations: such as choice of a certain foraging strategy, a particular prey-size forced upon the organisms by competition, problems of heat loss and so on. It is, as we note, something of a vicious circle, or at least a complex interaction of selection pressures. At the end, an organism presents an apparent 'compromise' in every strategy-choice, taken separately. No strategy in isolation will be optimised, but overall, the total life-strategy must be efficient; selection would not permit otherwise.

Such apparent compromise in any one strategy viewed in isolation is of course illusory for while a 'choice' *may* appear suboptimal when circumstances relating only and immediately to that strategy are considered, it is of course unrealistic to consider any one element of an organism's ecology out of the context of the rest of the selection pressures acting upon it. Further, the strategy ultimately adopted is also not developed by the individual organism in isolation. Optimal ecological strategies for the individual *in vacuo* are further modified in practice by what other organisms around it (of the same or different species) may be doing. The organism's *realised niche* (page 106) is derived from the fundamental niche under the effects of intraspecific competition, interspecific competition and predation.

Students of evolutionary theory have considered this same problem of what is optimal in isolation and what is best in the real world. They define an Evolutionary Stable Strategy (Maynard Smith, 1976b, 1977) as that strategy

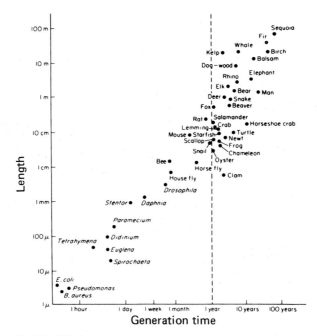

Figure 10.8: The Relationship Between Length and Generation Time for a Variety of Organisms on a log-log Scale. Source: From Bonner, 1965; after Southwood, 1976/1981.

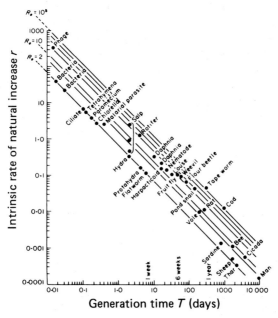

Figure 10.9: The Relationship Between Intrinsic Rate of Natural Increase and Generation Time for a Variety of Organisms. Diagonal lines represent values of r_0 from 2 to 10^5. Source: From Heron, 1972; after Southwood, 1976/1981.

which, if adopted by the majority of individuals within a population cannot be bettered by any alternative ('mutant') strategy which may arise within that population (even if the mutant strategy might appear to benefit the individual were it in isolation). In brief, an Evolutionarily Stable Strategy (ESS), is that strategy which pays best under all circumstances.

In such analysis the niche-strategy adopted by the individual may be regarded as an ESS in each dimension. We noted above that organisms tend to cluster in a limited number of ecological positions on any strategy continuum (P/R, or r- and K-strategies) while the whole continuum is potentially available. Such observation suggests that certain positions are most stable – and thus organisms will tend to gravitate towards these limited stable positions. The concept of a limited number of stable strategies ties in well with this concept of the ESS.

10.10 The Evolution of Stable Strategies

This concept of stable strategies is a sound one: indeed once it has been explained, appears a blinding glimpse of the obvious (as indeed it is); it has however been of tremendous benefit in analysis of the evolutionary development of any particular characteristic. In all cases, the strategy adopted by an organism need not in practice represent the optimum for the individual in isolation, may not be as would be expected for maximum adaptiveness of the individual under natural selection. Because the individual does not exist in isolation but surrounded by others it must 'adopt' that strategy which optimises its fitness in the light of what all around it are doing. The strategy is stable because in a curious way it is optimal for the individual in its complete environmental context; in addition it has stability in its own right in that it cannot be overthrown by a mutant strategy. Intuitively this too is sound – it is based on the same sort of logical argument that considers the position of the 'cheat' in an altruistic society – and the influence of the possibility of cheating on the evolution of altruistic behaviours (Trivers, 1971).

The concept – first put into the literature by Maynard Smith (1976b) – was developed in the first instance to deal with situations of aggression and conflict, but may be extended to analysis of almost any behaviour in terms of evaluating it as an 'unbeatable strategy'. The real strength of the concept is in enabling us to understand how something that appears sub-optimal may in practice be the optimal strategy. At the same time it enables us to examine with a rigorous mathematical approach the evolution of ecological strategies.

The theory of ESS first rose from a philosophical examination of game-theory – basically an analysis of any contest in terms of likely losses and likely pay-offs. Suppose animal adopts strategy I and opponent adopts J. Then pay-off to I will be $EJ(I)$ (or expected gain to I); this pay-off is the change in I's fitness as the result of the contest. From this we may derive an ESS more rigorously. Suppose a population consists of individuals adopting strategy I or J with frequencies p and q (with $p + q = 1$). Thus in any contest, any individual (whether

I or J strategist) stands p chance of being pitted against an I and q chance of coming up against a J. Thus:

$$\text{Fitness of } I = p\, E_I(I) + q\, E_J(I) \tag{10.1}$$

$$J = p\, E_I(J) + q\, E_J(J) \tag{10.2}$$

An ESS will be that strategy which *when common* maximises fitness. Thus

I is an ESS if

$$E_I(I) > E_K(J) \tag{10.3}$$

or $E_I(I) = E_I(J)$ and

$$E_J(I) > E_J(J)$$

(for where $q \to 0$ this gives Fitness I maximum)

One obvious situation in which we may examine this idea is in relation to our discussions above of reproductive strategy in terms of polygyny, promiscuity or monogamy. Selection favours in every case desertion of the young by whichever parent has invested less in the reproductive attempt as we noted on pages 290 and 291. Yet it also acts on the other partner to prevent such desertion — because if assisted in the breeding attempt, she herself has then to invest less. The result is usually some form of compromise between the conflicting selection pressures, an evolutionarily stable strategy.

In Section 10.7 we derived the 'evolution' of monogamy, polygamy, even polyandry, on the basis of relative selective advantage to each partner and a 'compromise' between the two when necessary. Such argument can be treated more rigorously in terms of ESS theory (as Maynard Smith, 1977). What we are looking for is a pair of strategies I_m and I_f for male and female respectively, which are evolutionarily stable in the sense that: If most males adopt I_m it would not pay a female to do other than I_f and if most females adopt I_f, then it would not pay a male to do other than I_m.

We have two basic strategies for either sex — desert, or guard. Maynard Smith considers the pay-off matrix for all combinations (male and female guard; male guards female deserts; female guards male deserts; both desert) and for three simple models (Maynard Smith, 1977). Model 1 assumes discrete breeding seasons. The success of a pair varies only with post-copulatory investment. This corresponds to the situation typical of birds, in which the number of young successfully raised depends on the capacity of the parents to guard and feed the young, but is not limited by the number of eggs a female can lay: it would not help her to lay any more eggs. Model 2 also assumes discrete breeding seasons. The success of a pair varies both with post-copulatory investment, and also with the pre-copulatory investment made by the female; a female who invests heavily in eggs cannot guard as well as one who lays fewer eggs. Model 3 assumes continuous breeding. The pay-off matrix for model 2 is shown by way of example in Figure 10.10. P_0, P_1, P_2 are the probabilities of survival of offspring which are guarded by no parent, by one and two parents respectively. A male who deserts has a chance p of mating again. A female who deserts lays W eggs while one who guards lays w eggs; $W \geqslant w$. The pay-off in Figure 10.10 shows that there are four possible ESSs.

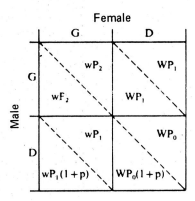

Figure 10.10: The 'Pay-off' Matrix for Different Mating Strategies Under ESS Theory. Figures relate to pay-offs under assumptions of Maynard Smith's Model 2 (see text). Source: From Maynard Smith, 1977.

ESS 1. D♀ and D♂

requires $WP_0 > wP_1$, or the female will guard, and $P_0 (1 + p) > P_1$, or the male will guard.

ESS 2. D♀ and G♂ (what Maynard Smith refers to as the 'stickleback' ESS)

requires $WP_1 > wP_2$, or the female will guard, and $P_0 > P_0 (1 + p)$; or the male will desert.

ESS 3. G♀ and D♂, the 'duck' ESS.

requires $wP_1 > WP_0$, or the female will desert, and $P_1 (1 + p) > P_2$, or the male will guard.

ESS 4. G♀ and G♂

requires $wP_2 > WP_1$, or the female will desert, and $P_2 > P_1 (1 + p)$, or the male will desert.

For given values of W, w, P_0, P_1, P_2 and p, ESS 1 and ESS 4 can be alternative possibilities, as can ESS 2 and ESS 3.

The 'stickleback' ESS is favoured if

(i) $W > w$; that is, if the female does not guard, she can lay more eggs, and

(ii) $P_1 \gg P_0$; P_2 not much greater than P_1; that is, one parent guarding is better than none, but two parents are not much better than one.

However, the 'duck' ESS is often an alternative possibility, especially if one parent is an adequate guard and if a male who deserts has a good chance of mating again.

It is a feature of all three models that two alternative ESSs are often possible; the 'duck' strategy in which the male deserts and only the female cares for the young, and the 'stickleback' strategy in which only the male cares for the young. These two possibilities are likely if one parent is almost as effective as two in caring for the young, and if the prospects of a deserting parent mating again are good.

In those species in which the ESS is for one parent to desert, it will tend to be the male which cares for the young if the female has invested so much in eggs that she cannot effectively do so, or if there is an excess of males; it will tend to be the female which cares for the young if the timing of mating is such that a deserting male has a better chance of re-mating than a deserting female, or if there is an excess of females.

In contrast, if two parents can raise twice as many offspring as one, or if the chance that a deserting parent will re-mate is small, then monogamy with both parents caring for the young is the likely ESS.

These conclusions overlap with those of Trivers (1972), based on investment assessment and it is interesting that we obtain essentially the same results by treating the considerations mathematically and searching for an ESS. However they also lead to one surprising conclusion, not clear from Trivers' more qualitative, discussive analysis, namely that the sex which contributes most to producing the young may contribute least to caring for them (Maynard Smith, 1977).

Although we have, as example, applied the concept of ESS to our previous considerations of reproductive strategy, the approach is clearly one which is of enormous value in a great many contexts in explaining the origin and evolution of any life-strategy. Further the concept of stable strategies which may withstand invasion by alternatives, helps explain the curiously restricted range of solutions which seem to be available to evolutionary problems.

11   Coevolution

The influence of evolutionary change in ecology is not just upon the adaptiveness of individuals to their particular niches, but can affect the dynamics, structure and function of whole systems.

Selection acts upon the fitness of the individual, or population – not upon an ecosystem. Although an individual's 'inclusive fitness' may be extended to include related individuals sharing common genetic material (Hamilton, 1963; Maynard Smith, 1964) and the potential for selection at the level of entire populations has been much debated (e.g. Wynne Edwards, 1962, 1963; Wilson, 1975; Maynard Smith, 1964, 1976; Dawkins, 1976) no one would try and justify the evolution of whole systems. Despite that, whole systems do appear to be adaptive, or at least, the species within them co-adapted to the extent that whole ecosystems may indeed appear to show evolutionary change (May, 1978).

Such *coevolution* of whole systems derives from the fact that the evolution of the individual is under selection pressure from its *complete* environment: that environment includes not only abiotic factors, but other organisms. Evolutionary change in one species changes the selection pressures acting on organisms associated with it, causing them, too, to change; these latter changes in turn, then result in further change in the coevolving species. Thus, in many instances, the separate evolutions of two or more species serve to influence each other, and form an interactive coadapted system – as the multispecies predator-prey community of page 266.

Perhaps this predator-prey interaction is the clearest example of such coevolution. Predation is, after all, an extremely potent selection force upon both predator and prey: the predator must be successful in order to survive, while prey survival relies upon their escape from predation. Under pressure from its predators the prey adjusts and evolves to reduce that predation by improved crypsis, heightened senses giving earlier warning for escape, or simply by becoming fleeter of foot. So, gazelles become faster to escape the cheetah; but that increase in the speed of its prey acts in turn as a powerful selection pressure upon the cheetah to run still faster itself, to stalk more quietly. Each 'advance' made by predator or prey acts as a selection pressure to evoke change in the other: each becomes involved in what has been compared to an 'arms race'.

11.1 Insect-plant Interactions

Entirely analogous to this predator-prey arms race is the relationship between phytophagous insects, and the plants on which they feed. Phytophagous insects can inflict severe damage to their food plants – in extremes leading to complete defoliation. As such they constitute perhaps the single greatest selection pressure upon the plants upon which they feed. In response, the plants themselves develop defences. r-selected species: small annuals, early colonisers in succession, rely on small size, scattered distribution and short life-span to escape 'predation'. Longer-lived K-selected species, more exposed to such insect attack, must develop alternative defence. Many develop physical defences: thick, toughened cuticles resistant to piercing mouthparts, sticky hairs, thorns and prickles. Others rely upon chemical defences.

All plants contain chemicals which seem to have no obvious role in primary metabolic pathways such as respiration and photosynthesis (30,000 such compounds have been identified so far: Harborne, 1977). These chemicals are sometimes called secondary metabolites but this is an unsuitable phrase because many have been shown to be involved in the synthesis of compounds which the plant does need, such as enzymes, pigments etc. Whatever the origin and present metabolic role of these chemicals, there is no doubt that many have acquired a further role, that of defence against grazing (see Edwards and Wratten, 1982). All plants are toxic to something, even if a wide range of insects are able to feed on them; for instance, the secondary chemicals conferring on cabbage its characteristic odour are toxic to insects not adapted to this plant group. The level of commitment of a plant to defence, however, does seem to vary and ecologists have devoted much energy in recent years in seeking the patterns and strategies of this variation. Cates and Orians (1975) for instance, related this commitment to the plant's position on the r-K continuum (Chapter 10). They used slugs as a grazing bioassay of plants' palatability and compared plant species from different positions in the successional series. They showed that annual, early-successional species were grazed more heavily than mid-successionals, which in turn were favoured more than late successionals, mainly trees. This may be interpreted as the development of defence by organisms according to their 'need'; an early-successional species, in which the r-strategy characteristics such as short generation time, high fecundity etc. are emphasised, devotes large amounts of metabolic energy to growth and reproduction and little to defence because it will be dead before large numbers of grazers can find and exploit it. A late-successional tree, on the other hand, with perhaps a 50-year generation time and occupying large areas, is ecologically 'conspicuous' and is very likely to be discovered and exploited by a range of potential grazers.

As soon as a plant acquires chemical defence, through a mutation in its biochemistry, a selection pressure has appeared which will deter current and potential grazers. If some of these adapt to the defence and overcome it thus defoliating it and reducing reproductive fitness they then impose a new selection pressure on the plant itself and it is driven to change again.

One of the most elaborate examples known of the consequences of coevolution of plants and plant-feeding insects concerns the butterfly genus *Heliconius* and the flowering plant family Passifloraceae (Gilbert, 1975). Heliconiid butterflies are common and conspicuous insects of lowland tropical forest in the New World. The adult butterflies, which may live for up to six months, feed on the pollen and nectar of various plants, but particularly on vines of the genus *Anguria* (Cucurbitaceae).

The larvae of *Heliconiid* butterflies feed exclusively on species of *Passiflora* and closely related genera. Each of the 45 or so species of *Heliconius* is a specialist on only a few passifloracean species and within one locality each *Heliconius* species usually feeds on a different *Passiflora* host. This precise partitioning of the host species among the various insects strongly suggests the outcome of chemical coevolution between the butterflies and the Passifloraceae. However, it has been shown in the laboratory that several *Heliconius* species can be reared successfully on *Passiflora* species other than their usual larval host, implying that non-chemical factors may also be important in determining the species chosen for oviposition. Since vision and visual memory are important in the behaviour of the adult butterfly, it is likely that the egg-laying females recognise the host visually; Gilbert describes how female insects often spend a considerable time 'inspecting' a host plant before they lay their eggs.

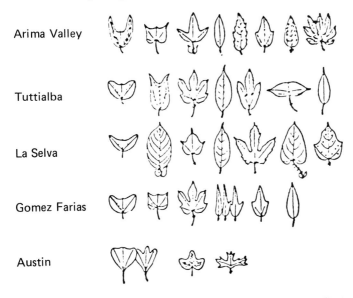

Arima Valley

Tuttialba

La Selva

Gomez Farias

Austin

Figure 11.1: Leaf Shape Variation Among Sympatric Species of *Passiflora*. The localities are from the top: Trinidad, Costa Rica, Costa Rica, Mexico and Texas. Source: From Gilbert, 1975.

A striking feature of the *Passiflora* genus is the diversity of leaf shapes, many resembling those of unrelated rain forest species. Since female butterflies appear to select their oviposition sites visually, this diversity may be due to adaptive evolution to make the plants less conspicuous. The greatest contrasts in leaf shape can often be seen among the *Passiflora* species of *one* locality (Figure 11.1). Gilbert suggests that this is because, as potential host species, they are less susceptible to discovery by heliconiids other than those which usually feed on them, if their leaves are as different as possible from each other. Thus selection has favoured the diversification of leaf appearance. Furthermore, the possible range of distinctive leaf shapes appears to limit the number of *Passiflora* species which can survive in one locality, so that there are usually fewer than ten in one area.

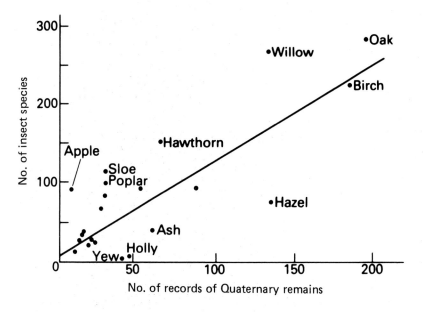

Figure 11.2: The Relationship Between the Number of Species Associated with a Tree and its Abundance in Recent Geological Times. Source: After Southwood, 1961.

A further example comes from UK trees and the insect fauna they support; Southwood (1961) demonstrated a wide variation in the number of herbivorous species supported by the trees and related this variation to the trees' ecological 'conspicuousness' in evolutionary time.

Those trees, such as oak (*Quercus*), which have been present in Britain for a long time and have also been abundant during that period, have acquired a larger insect fauna than rarer, historically less-abundant trees (Figure 11.2). It is postulated that a greater number of opportunities have existed for potential colonisers to adapt to a tree if that tree has been abundant in time and space. The coevolutionary stage of the process would have been reached if we could show that the trees have changed as a result of their being colonised. This was assessed by Wratten, Goddard and Edwards (1982) who used Cates and Orians' method of assessing the palatability of a range of tree species. They showed that

trees currently bearing a large number of herbivores are *less* palatable to a
general feeder than those bearing a few grazing species (Figure 11.3). It is
tempting to conclude that, even among the late-successional trees, commitment
to defence will be related to need; trees like oak we assume to have had a greater
history of defoliation and death of some trees than have species like lime (*Tilia*).
Their increased unpalatability does not necessarily deter the species already
adapted to them but may reduce the chances of being colonised by new species
of potentially damaging grazers. A tree's successional position may be a further
important variable in this process. Not all trees are end-of-succession forest
species; some, like birch (*Betula*) and willow (*Salix*) for instance, are mid-
successional over much of Europe. They are therefore less conspicuous in time
and space and there is recent evidence that they produce defensive compounds
as a within season response to grazing damage (Edwards and Wratten, 1982). This
flexibility makes sense for a species which may suffer more sporadic damage than
does a forest tree; defence is metabolically costly so producing it as required (as
rapidly as six hours following damage in the case of birch) is efficient.

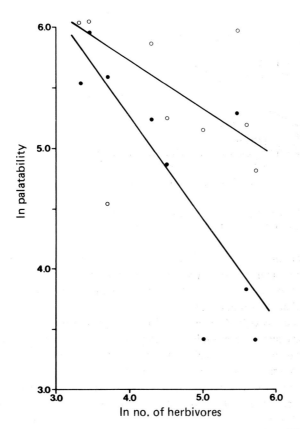

Figure 11.3: The relationship between ln palatability and ln number of Herbivorous Insects
Associated with British Trees. ● = autumn data; ○ = spring data

We should be wary, however, of assuming that all herbivore grazing is as obviously detrimental as it at first appears; there are possibilities that the grazed plant may benefit in subtle ways from some grazing pressure (see Section 11.4).

11.2 Larger Herbivores

Although most of the published work on plant-herbivore interactions has been concerned with phytophagous insects, larger herbivores – such as grazing or browsing ungulates – also have a profound effect upon their forage species; and once again there is a two-way interaction.

Grazing or browsing by a large herbivore can lead to extensive defoliation of the plant which is fed upon: clearly a powerful selection pressure as we have already noted. Under pressure from such large grazers many grasses and other 'preferred' forage species have developed a prostrate growth form for survival – as rosettes, or tillered grasses. All have developed some resistance to being grazed, with enhanced powers of regeneration (it is no coincidence that the most graze-tolerant plants of all, the grasses, grow from the base: there is no growing tip to be removed by the herbivore), and increased reliance on vegetative methods of reproduction. These are rather passive methods enabling the plant to tolerate animal attack, but in the same way that plants may produce active physical or chemical defences against insect attack, so too, plants subject to 'predation' by larger herbivores may develop positive defences. Many palatable species develop spines, thorns or prickles on stem or leaf. Certain species have even developed chemical defences.

Sweet vernal grass (*Anthoxanthum odoratum*) contains a chemical compound, coumarin, which directly reduces voluntary food intake by grazing herbivores. It has been shown to reduce food intake by over 15 per cent in sheep, and when ingested so interferes with the digestive process that it reduces the digestibility of the grass by 32 per cent. The nitrogen-rich clovers and trefoils, so widely sown by farmers amongst their pasture to increase both nitrogen balance of the soil and nutrient content of the animals' feed produce synthetic oestrogens that affect animal fertility. If taken in quantity they suppress fertility, induce abortion and depress lactation (e.g. Bennetts *et al.*, 1946).

The interaction between large herbivores and vegetation is a complex one – and not entirely one-sided. There are indeed a number of plant species which can exist only by virtue of being grazed (see, for example McNaughton, 1976). Such species, themselves resistant to grazing, persist in areas of heavy grazing where vegetative growth of other species (which would otherwise out-compete) is suppressed by herbivores (e.g. Harper, 1977; Whittaker, 1979).

Such observations make it clear that the presence of large herbivores such as the various ungulate species may have a marked effect on the structure of the whole vegetational community in which they occur. Through their grazing activity they may eliminate graze-sensitive species; through grazing they may reduce vegetative growth of more resistant species, and thus, reducing competition, permit the establishment of yet other species which would otherwise not persist — a situation between herbivore and vegetation analagous to that observed by Paine (1966) in the predator-prey community (see page 151). Thus grazing by rabbits on chalk grassland is widely claimed to maintain the diversity of the vegetation and indeed to have created one of the most diverse and species-rich of all natural floras. Where grazing is heavy, as in the New Forest in the UK the whole community becomes 'coloured' by the effects of that grazing. Woodlands display a curious age-structure and species composition — with understorey throughout dominated, unusually, by holly (*Ilex aquifolium*): the only woody species able to withstand the constant browsing. The curious age-structure of the trees themselves was noted by Peterken and Tubbs (1965) who showed that the woodlands were composed essentially of trees of only three age-classes (germinating between 1663-1763, 1860-1910 and 1930-45). Most regeneration is prevented by heavy animal feeding pressure. Only when numbers of grazing animals were reduced could seedlings become established: the regeneration of 1860-1910 followed the killing of most of the Forest deer after the Crown relinquished its hunting rights in 1851, that of 1930-45 coincided with a lowered population of domestic stock (cattle and ponies) due to low market prices in the recession.

Grasslands and heathlands of the New Forest are also predominantly composed of graze-resistant species, and in these communities it becomes clear that the effects of herbivores upon vegetational composition and structure are not restricted to the effects of direct feeding. Both trampling and the effects of both urination and defaecation can also markedly influence community structure. Gorse establishes in the nutrient-rich patches of latrine areas around the edges of favoured grasslands. Animal urine rapidly kills heather and certain acid grassland species, while the rich nitrogen deposit encourages the growth of sweet grasses and annuals (Putman *et al.*, 1982a).

Grazing by large herbivores may thus affect the structure and species composition of the vegetational communities on which they rely in exactly the same way as predators may influence the diversity of their prey community (page 268). And there are other parallels too. As predators may increase or decrease the productivity of their prey populations, so large herbivores may alter the productivity of their forage species. Moderate levels of grazing may act to increase vegetation productivity — removing senescent material, stimulating regeneration; excessive grazing reduces the leaf area index of the forage plant below that level at which it can maintain efficient photosynthesis: productivity falls.

The effects of browsing and grazing upon the forage plant are reviewed by van de Veen (1979). In the case of browsing he details three effects. (a) Removal of buds and leaves may result in immediate regrowth, thus prolonging the availability of higher quality forage. (b) At least in Northern and temperate areas shoot biting tends to stimulate shoot productivity; it has been suggested that for many hardwood species a 50 per cent browse utilisation would result in maximum sustained yield of browse (Krefting *et al.*, 1966). For willows it has been shown that browsing may easily double next year's forage supply (Wolff, 1978) and for heather a 60 per cent browse utilisation has been recommended (Grant and Hunter, 1966). (c) If browsing results in 'hedging', the period over which browse will be available is greatly extended without ultimate deleterious effects on the trees. In semi-arid and arid areas grass feeding will result in a decrease of available forage and intensive grazing will also decrease productivity in the next year and ultimately a desert will be created (Stoddart *et al.*, 1975). But in temperate areas intensive grazing may greatly enhance productivity as well as quality of the grass sward (Hodgson, 1974) and the same holds for dwarfshrubs communities (Gimingham, 1972).

In terms of coevolution between herbivore and plant in this context, it has even been demonstrated that the saliva of, at least some herbivore species actually contains plant growth-stimulating substances (e.g. Dyer, 1980; Detling *et al.*, 1980).

The relationship between large herbivore and vegetation is, however, rather different from that between insect and plant. It is suggested that the difference in the relationship is that, in general terms, most larger herbivores are rather more polyphagous in habit. Plants under any form of grazing pressure — whether by insect or mammal — will be under selection to produce defences: but the animal has alternatives in its 'choice' of reaction to such defence. It can either specialise on one or two particular plant species and develop specific mechanisms to overcome its particular defences or it may adapt so that it becomes polyphagous: feeding on such a variety of foodstuffs that it ingests relatively small quantities of any one toxic compound. Janzen (1975) suggests that the choice of strategy depends on animal size. If an animal is small relative to the food source (as is an insect) (and perhaps has a relatively shorter generation time) then the strategy may be to specialise in one particular foodplant, and overcome its defences. If one is large with respect to the food source, is it 'evolutionarily more sensible' to become polyphagous. And it is perhaps this difference in the type of adaptation to plant defences between large and small herbivores, which results in the rather different form of the herbivore-plant interaction: with that of the larger, polyphagous herbivore more closely resembling that of the polyphagous predators of Chapter 9.

11.3 Interaction of Plant-herbivore Populations

All such coevolutionary effects as we have described above, result from, indeed can only be mediated through, the shorter-term trophic interaction of herbivore and plant. It is the effect of one population upon the population of the other which provides the selection pressure shaping this array of adaptation. The trophic relationship between herbivore and plant is an interaction very similar to that between predator and prey; the dynamics of interacting populations of predator and prey were examined in detail in Chapter 9; plant herbivore systems can be analysed in very much the same way — although the field has been less extensively researched.

By way of illustration let us offer a simple analysis, first presented by Caughley (1976), of the interaction of a grazing ungulate and its forage resources — later extending this analysis to a third trophic level by incorporating into it the effects of predation upon the herbivore.

Plants are usually limited by water, light, and carbon dioxide — resources that are renewed at a rate relatively independent of population density. For the purposes of deriving a simple model for population growth in plants, Caughley requires that the most limited of resources comes in at a rate g per unit area, and that each unit of plant biomass must take up this resource at a rate b, if it is to maintain itself and replace itself in the next generation. A biomass of vegetation V, on the same unit area, must thus use a proportion of the resource bV/g for replacement, leaving a surplus $1\text{-}bV$ which can be utilised for growth. Population growth within the vegetation can thus be represented

$$\frac{dV}{dt} = r_1 (1 - \frac{bV}{g}) \tag{11.1}$$

where r_1 is the intrinsic rate of increase of the plants. The population's rate of growth dV/dt is close to the intrinsic rate of increase when vegetation density is low, but is reduced progressively as b increases. V finally stabilises at a maximum where $bV = g$ and we can thus define a maximum plant density $K = g/b$. In this form, our equation for plant population growth becomes

$$\frac{dV}{dt} = r_1 (1 - \frac{V}{K}) \tag{11.2}$$

— the familiar logistic curve introduced on page 130 and page 155. Substitution of K as a constant is only possible when rate of renewal of the limiting resource (g) is totally independent of the population density — a property which, as Caughley (1976) points out, is with few exceptions, peculiar to the first trophic level.

We may now introduce a population of herbivores into our simple system. We now require two equations: one to describe population growth within the vegetation, one for herbivore population changes. The first equation will be that just derived for rate of increase of vegetation, modified by a term expressing the degree to which this is depressed by herbivores (H), to become

$$\frac{dV}{dt} = r_1 (1 - \frac{V}{K}) - c_1 H (1 - e^{-d_1} V)/V \tag{11.3}$$

In this equation H represents the number of biomass of herbivores, c_1 is the maximum rate of food intake per herbivore unit (i.e. when food is superabundant). As V decreases, the animal can not take in a satiating diet and c_1 is reduced

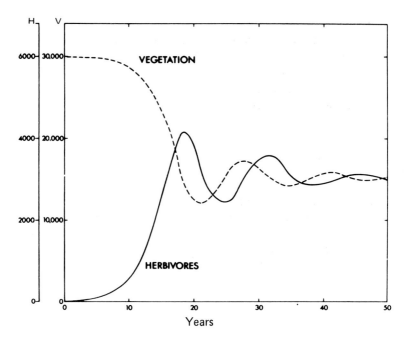

Figure 11.4: Modelled Growth of a Population of Herbivores and the Trend of Biomass for the Plant Population on Which it Feeds. Source: From Caughley, 1976.

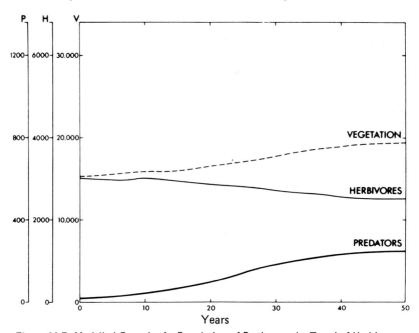

Figure 11.5: Modelled Growth of a Population of Predators, the Trend of Herbivores on Which it Feeds, and the Trend of Plants Sustaining the Herbivores. Source: From Caughley, 1976.

at a rate $(1 - e^{-d_1}V)$ which approaches zero as V approaches zero. The constant d_1 determines the rate of fall and is a function of the grazing efficiency of the herbivore. The whole term $(c_1 (1 - e^{-d_1}V))$ thus becomes the herbivore equivalent of the predatory functional response (page 229), in showing the response of its rate of intake to availability of food.

Our expression for the herbivore populations rate of growth may be derived as

$$\frac{dH}{dt} = -a_2 + c_2 (1 - e^{-d_2}V) \tag{11.4}$$

where a_2 is the rate of decrease per head in the absence of food, and c_2 is the rate at which this decline is ameliorated when food is abundant; the term $c_2 (1 - e^{-d_2}V)$ models the herbivore's numerical response: its reaction to density of food in terms of survival and reproduction.

Figure 11.4 shows these two equations, for vegetation and herbivore, in action. Both populations stabilise – with densities at equilibrium

$$V^X = \frac{1}{d_2} \log_e \left(\frac{c_2}{c_2 - a_2}\right) \tag{11.5}$$

$$\frac{H^X = V^X r_1 \left(1 - \frac{V^X}{K}\right)}{c_1 (1 - e^{-d_1}V^X)} \tag{11.6}$$

The population traces of Figure 11.4 mimic remarkably well the processes which may be observed during a real ungulate eruption; the model reproduces the pattern of growth we see in the field.

This model allows us to draw a number of conclusions about the workings of a vegetation-herbivore system 'which are neither trivial nor immediately obvious' (Caughley, 1976). For example, neither the intrinsic rate of increase of the forage (r_1) nor its equilibrium biomass when ungrazed (K) has any bearing on its biomass at equilibrium with grazing pressure (V^X). In Caughley's terms: should a mutation sweep through the plant population so that its intrinsic rate of increase is doubled, the new equilibrium is still V^X (but $2H^X$). Further, should the herbivores be replaced by a strain that eats twice as much, but is in all other respects unchanged, the vegetative equilibrium is still the same – but the equilibrium biomass of herbivores is halved. In general a change in any parameter of plant growth *or grazing pressure* leaves V^X unchanged. Only those changes which affect the intrinsic rate of increase of the herbivore $(c_2 - a_2)$ will affect V^X.

Caughley extends his analysis of the plant-herbivore system a stage further by introducing a predator upon the herbivore. Our vegetation equation remains unchanged

$$\frac{dV}{dt} = r_1 (1 - \frac{V}{k}) - c_1 H (1 - e^{-d_1} V)/V \tag{11.3}$$

but the model for herbivore growth is influenced by an additional element to represent the effects of predation

$$\frac{dH}{dt} = a_2 + c_2 (1 - e^{-d_2} V) - fP (1 - e^{-d_3} H)/H \tag{11.7}$$

and we introduce a new equation for the predator

$$\frac{dP}{dt} = - a_3 + c_3 (1 - e^{-d_4} H) \tag{11.8}$$

(symbols preserving the senses defined earlier)

The entire system modelled by these equations may either have a stable equilibrium, or it may oscillate, according to the values of the various constants. Figure 11.5 (from Caughley, 1976) models the outcome of introducing predators to the populations diagrammed in Figure 11.4.

The analysis presented here is a simple one which makes a number of, perhaps unrealistic, assumptions. The model assumes

(a) a single plant species, or a set of species which have identical growth functions and are equally grazed; no differences between plant parts;

(b) a single herbivore species;

(c) grazing is on green vegetation in the growing season; plant growth is continuous and the environmental factors affecting it are constant;

(d) herbivore requirements and consumption functions and the environmental and physiological factors affecting them are constant in time.

The predictions of the model are thus applicable without reservations only to such constant-growth, constant-requirements, single-herbivore systems – for instance, a uniform area of a perennial grass (or evergreen shrub) in a tropical (or warm irrigated) environment. However, from current experience with such general models there is good reason to believe that the main qualitative results of the model will not be highly sensitive to moderate deviations from these assumptions. They may hold also for a much wider range of grazing systems for which these assumptions are true only as very rough approximations (Noy-Meir, 1975).

Analyses of this sort – which as we may see, parallel very closely those described in Chapter 9 for population interactions of predators and prey – are explored more fully by Noy-Meir (1975) and Caughley (1976). Perhaps the biggest drawback of all these analyses is their restriction to what are essentially two-species systems. Clearly most plant-herbivore relationships – even more so perhaps than those involving predators and their prey – are multispecies systems, with whole assemblies of plant species and many, often polyphagous herbivores. The implications for population growth and dynamics remain to be considered in detail. The implications for coevolution are reserved for further discussion in Section 11.5.

11.4 Coevolution to Mutualism

With a continuing arms race of adaptation and counteradaptation between coevolving species one might expect some sort of escalation: an ever-increasing spiral of counteradaptation. Selection however continues to act and is far more likely to favour a stable outcome, a diminishing spiral, where each partner in the relationship adapts in such a way as to minimise countermeasures from the other; maximising its own return from the relationship while minimising damage of disturbance to its partner.

We see this quite clearly in parasitic relationships. A poorly adapted parasite risks rejection by the host or alternatively, may itself kill its host; in either case the parasite itself may die. By contrast a well-adapted parasite is hardly registered by the host: it inflicts minimal damage, thereby ensuring tolerance by the host, and, through the host's continued well-being, its own persistence. Coevolution in parasitic relationships of this sort certainly leads to an equilibrium of minimum disturbance with at worst, the parasite siphoning off a little of the hosts' assimilated foodstuffs, or side-tracking some of its food intake while this is still in the gut.

At its extreme the relationship may even develop into a form of mutualism where both partners benefit from the interaction. That such relationships may develop from initially parasitic interactions is quite clear. The Zoochlorella of green hydra (*Hydra viridis*) that give it its green colour and enable it to photosynthesise in conditions of poor food availability were clearly once facultative parasitic protozoa. Lichens have developed as an extraordinarily successful lifeform from an initially parasitic relationship between fungi and algae. Even the mitochondria of eukaryote cells are generally thought to have originated as parasitic, later mutualistic bacteria — and we must presume also, that the symbiotic bacteria of the ungulate rumen — which confer upon it the ability to digest cellulose, and indeed alter its entire digestive physiology — were also once present in purely parasitic role.

Here then we are suggesting that the selection pressures acting on the partners of a coevolutionary relationship will tend to direct adaptation and counteradaptation towards an ultimate relationship of minimum disturbance: each adaptation 'calculated' to provoke minimum counteradaptation from the partner. And such balanced coevolution may not just be restricted to the parasite-host relationships just considered. Certain of the relationships described in Section 11.2 between large herbivores and their forage plants could be considered mutualistic, while in some instances even the battle between phytophagous insects and their food plants (Section 11.1) has been claimed no battle. Owen and Wiegert (1976) suggest that in general, consumers may benefit their host plant, that the anti-herbivore defences developed by the plant are not 'intended' to stop all insect attack, but to restrict it to certain specialised species and to control their numbers. Owen and Wiegert give a number of examples in support of this claim. The production of sugary honeydew by aphids and other Homoptera is a striking ecological phenomenon; estimates have been made that more than 1 $kgm^{-2}a^{-1}$ may fall to the ground in some ecosystems. Owen and Wiegert (1976) suggested that this production, although a drain on the plant's energy supply, may ultimately benefit the plant by acting as an energy source for nitrogen-fixing bacteria in the soil. In nitrogen-poor environments, this trade of carbohydrate for nitrogen could be important. However, this addition of sugar to the soil would probably benefit most micro-organisms, not only the nitrogen-fixers. The fact that more than 40 per cent of the excreted sugar is melizitose in some aphid species, although this sugar is not present in the plant, suggests that this sugar may selectively stimulate the nitrogen fixers. It would thereby stimulate the plant's growth and, subsequently, that of later aphid generations.

Owen and Wiegert (1976) showed that the fall of leaves of holly (*Ilex*) damaged by leaf miners extended throughout the year while that of undamaged ones occurred mainly from June to October. They suggested that this was a mechanism whereby the plant could return nitrogen to the soil over a longer period, to its own benefit in nitrogen-poor environments. Although these ideas are rather tenuous (they are certainly actively debated), there is no doubt that a number of mutualistic relationships *can* be observed between insects and plants – which must be presumed to have evolved from interactions originally antagonistic.

Perhaps the most elegant example comes from the classic studies of Janzen (1966, 1967) on the mutualism between the swollen-thorn acacia and the acacia ant in South America. Ant plants are found throughout the tropics. These are plants which have developed some form of mutualistic relationship with, usually one specific species, of ant. In most cases the ants live in hollow stems or thorns and are dependent upon the plant for shelter and food. Feeding may be direct (as in Janzen's neotropical acacias where the ants feed from extra floral nectaries, and special lipid-rich food bodies (Beltian bodies) produced in modified leaflet tips) or may be indirect: in the African *Barteria* (Passifloraceae) the ants raise scale insects in special galleries on the insides of hollow stems, and harvest nectar and offspring from the scale insects as food. In return for food and shelter, the ants generally protect the plant from herbivores and vines, by attacking fiercely any foreign body that touches the plant.

Janzen (1966, 1967) describes in detail this obligate mutualism for one species of plant in particular: the Bullshorn Acacia (*Acacia cornigera*) and its ant *Pseudomyrmex ferruginea* in neotropical South America. Swollen-thorn acacias of this type are characterised by enlarged stipular thorns, within which colonies of the ant may become established, enlarged foliar nectaries, and Beltian bodies to offer food to the ant tenants, and year round leaf production and maintenance – even in areas with a distinct dry season. Ant colonies established in the hollow thorns protect the acacia from herbivores, attacking any other insects encountered on the plant. In addition they also attack any foreign plant which touches the acacia's foliage or grows within an area of up to 150 cm in diameter below the acacia. A mature acacia with its full complement of ants thus grows in its own cylindrical space free of other plants. The reduction in herbivore attack achieved by the ants – and the competitor-free space around it permits the acacia to grow very rapidly and establish itself within the canopy. New acacia seedlings or root suckers grow only relatively slowly until colonised by ants; once the ants are established, development is much more rapid. Janzen suggests that trees without *Pseudomyrmex* would never reach maturity.

There is a further twist to the tale. One of the most significant ecological pressures in these tropical areas is periodic fire. Periodic flash fires spread through the area killing all in their path. Yet to some degree the ant-acacia is protected even from this: by clearing a space free of vegetation around their tree, the ants produce a natural fire-break. Fires *may* kill the acacia – and the ant colony; they may scorch and kill the acacia but not the ants. But in many cases they kill neither. Which outcome results depends on the structure of the immediately surrounding vegetation, which in turn depends upon how much of it has been killed by the ants (Janzen, 1967).

In a more recent study, Pickett and Clark (1979) have shown similar mutualism between a succulent *Opuntia acanthocarpa* (Cactaceae) and another ant *Crematogaster opuntiae*. In this case the ant does not actually live within the plant but is attracted to feed there upon extrafloral nectar. The nectaries are particularly abundant in areas of new reproductive and vegetative growth; the ants, which are aggressive and efficient defenders of the plant against cactus-feeding insects, thus congregate where they are needed most: at growing points. This mutualism has again developed to a striking degree. Field observations suggest that the cactus nectar is the major and perhaps the only food source for the ant; rarely were *C. opuntiae* seen foraging outside the cactus canopy. With the nectar the main food source for the ant, it is perhaps not entirely coincidental that it is peculiarly nutritive. Most nectar is a solution of sugars with low concentrations of aminoacids – a solution to attract as a 'sweetmeat', and act merely as a feeding stimulus. In support of its ant, however *Opuntia acanthocarpa* produces from its extrafloral nectaries a nectar which contains a far wider range of aminoacids than has been reported in floral or extrafloral nectar of other plants. Further, the concentration of aminoacids is extremely high, suggesting that the nectar has been developed as a nutritional food source for the ant.

Relationships between ants and ant-plants offer interesting parallels to plant chemical defences. In effect the ants are analagous to general purpose chemical defences: an analogy as apparent to the plant as to an outside observer – for it has now been shown that the ant-acacia lacks the chemical defences common among non-ant acacias, using its ants as a substitute. Janzen (1975) argues that it is most likely that as the ant-acacia interaction evolved, the chemical defences of the tree were selected against, since they were at a cost that did not repay itself in lowered herbivory. The equivalence of ant and chemical defence is even exact to the level that, just as there are herbivores which breach the chemical defences of other plants, ant acacias have certain herbivorous insect species which have evolved behavioural mechanisms for avoiding the ants (Janzen, 1975).

Not all coevolutionary relationships are necessarily trophic in nature, nor, even initially damaging. Coevolutionary interaction may occur for example because a plant is fussy about what pollinates it, or what disperses its fruit. We have already noted for example in Chapter 2 that certain species of plants are adapted for pollination only by certain species of rodents (Wiens and Rourke, 1978), whilst there are numerous, and bizarre, obligate mutualisms between plants and their insect pollinators (e.g. Gilbert, 1975, Grant Watson, undated).

11.5 Coadapted Systems

So far, our discussions have been restricted to coevolutionary relationships essentially between two partners — a situation little more realistic than the obviously erroneous one of treating each organism in isolation. In practice organisms do not just exist in a diad of this sort, but as part of a multispecies community. The individual insects of Southwood's trees do not just interact in isolation with the tree itself — but interact with all the other insect species feeding on that same tree. Coevolution does not just exist between pairs of species, but between all the members of an ecological community.

Lions in the Serengeti National Park of Tanzania coevolve with wildebeest but they also influence, and are influenced by ten other prey species. Further, the lions are not the only predators feeding upon those eleven prey species; those same species are also preyed upon by cheetah, leopard, hyaena, wild dog — and interact with these as well. Each prey species interacts with each predator; but, in addition all the predator species themselves influence and interact with each other, too — in competition, and there is a competitive interaction in addition within the prey species complex itself. Here is a true community — and the complexity of coactive effects is colossal. Predators adapt to their individual prey; prey to predators. Predators adopt different hunting styles (Kruuk and Turner, 1967) and prey selection (in terms of species age and sex: Schaller, 1972) to minimise competition between themselves; prey species too adapt their ecologies in competition (Lamprey, 1963; Leuthold, 1978). The whole 'species set' interacts in coevolutionary adaptation. And just the same complex coevolution can be seen within the context of Bell's (1970) grazing succession amongst the herbivores of those same short grass plains of Tanzania: zebra, wildebeest, topi and Thomson's gazelle follow each other across the plains in strict succession, each feeding upon different parts of the sward, and preparing the vegetation for the next in sequence (above page 112). Here again each herbivore interacts directly with the vegetation and each interacts with the other animals in the succession. The whole complicated end result of the sequential grazing system illustrates the complexity of coadaptive responses which have brought it about.

In examples like this, it almost seems as if, indeed, the system itself is evolving. Though we know that natural selection acts almost invariably at the level of the individual, or groups of related individuals, and that populations, much less communities of interacting populations, *cannot* be regarded as units subject to Darwinian evolution, because the interactions between the organisms of a community comprise some of the most powerful selection pressures upon those organisms, coevolution produces almost an evolution of ecosystems. And, while we have considered its effects here, largely in terms of effects upon the individual species within the community, coevolutionary pressures of this sort clearly also play a large part in determining the structure and diversity of those communities as well (May, 1978).

12 Species Diversity

12.1 Diversity as a Descriptor of Ecological Communities

One important dimension of the ecological community to which we have not yet paid explicit attention is that of *diversity*. Yet diversity of the community, both diversity of the species within it and of their relationships to each other is a crucial facet of the system and one with many implications for community function and stability.

Species diversity is a measure of the variety of different animal and plant species of a community. It is in effect an attempt to describe the number of species within that community and their relative abundances, and thus has two basic components. The actual total number of species contained within a community (S, *species richness* of the community) is a simple statistic which clearly goes a long way to establishing at least the first element of this diversity. However, it is in practice virtually impossible to account for all the individual organisms contained within a community and to be sure that one has accurately determined the total number of species represented. As a result, diversity is usually assessed on a sample of individuals drawn from the community; species number recorded must thus be considered in relation to the number of individuals sampled, and is thus a function of the number of species expressed within a certain sample of individuals. In this sense of the term, then, a community in which, amongst a sample of 1000 individuals, only 20 species are represented, will be considered less diverse than one in which a sample of 1000 individuals contains 100 species (or one in which a sample of 200 individuals contains 20 species); in the latter case, more species are represented per unit individual.

There is however, as noted, a second component to this concept of diversity, derived from the *spread of individuals between the species*. That is, of two communities each of which shows five species within a sample of 100 individuals, one in which the distribution of individuals between the species is 20:20:20:20:20 is more diverse than one in which the individuals are spread as 80:5:5:5:5 since in this latter case, the community is dominated by one single species, and is thus less diverse than the one with a more *equitable* spread of species. This equitability is in effect a function of the 'species-abundance distribution' displayed by the community (as discussed in Section 2.11) and describes the evenness with which individuals are 'spread' between the available species.

These separate components of diversity are independent of each other and may be affected differently by different factors; each in its own right is a very important aspect of the community.

12.2 Measures of Diversity

There are a number of indices of species diversity which may be calculated for a community, or for a sample representative of that community. The earliest of these concerned themselves merely with the first element of diversity — with simple relationships between the number of species represented in a given sample of individuals. Such simple 'species number indices' were constructed as D (diversity index) = S/N, or in a more sophisticated form — to take into account the necessary errors of working with samples rather than with entire communities —

$$D = \frac{S - 1}{\log N} \qquad (12.1)$$

Such measures however, do not include any element to incorporate differences in equitability into the index. This equitability, as we have noted is really a description of the species-abundance distribution of the community (Section 2.11). Species-abundance distributions such as that in Figure 12.1, effectively plot the number of species within a community containing a particular number of individuals. If we look at Figure 12.1, it is clear that the variability/equitability in distribution of individuals between species may be described by the mathematical variance of that distribution; this variance may thus itself be used as an index of equitability.

In fact the mathematically minded will realise that both species richness *and* equitability are different moments of the species-abundance distribution of the community. Recognition of this fact provided the starting point for the development of a number of rather different indices, which attempt to combine both elements of diversity into a single measure, and which have been based in large part on a mathematical description of such species-abundance curves. Approximation of a variety of mathematical series to observed or hypothesised species abundance distributions prompted the derivation of diversity indices based upon logarithmic series, power series, exponential expansions etc. etc. Most of these indices were in fact based upon the initial assumption that species abundance distributions were essentially logarithmic (as suggested by Williams (1964) for moth communities at Rothamsted — page 60 here). Diversity (α) is calculated from the equation

$$S = \alpha \log_e (1 + N/\alpha) \qquad\qquad (12.2)$$

(Fisher, Corbet and Williams (1943) — with S and N once again respectively species number and number of individuals.) Later recognition that the shape of the frequency distribution alters markedly in different communities (Figure 2.11) places such diversity indices on a somewhat shakier foundation (although for want of better alternatives they are still in regular use).

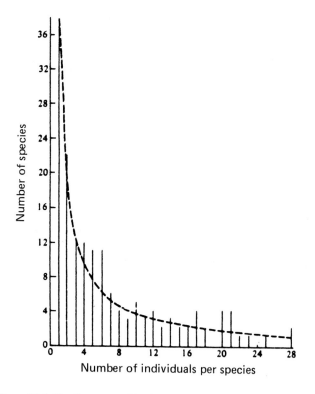

Figure 12.1: The Frequency Distribution of the Abundance of Various Species of Moths in a Light Trap at Rothamsted Experimental Station in England (1935). The expected frequency distribution from a logarithmic series is shown as a dotted line. Source: From Williams, 1964.

Yet another index in common usage in studies of community diversity — which again takes account of distribution of individuals between species as well as species number is the index of Shannon and Weaver (1949).

$$H = - \Sigma \mid \frac{ni}{N} \log \frac{(ni)}{N} \mid \qquad\qquad (12.3)$$

where ni = number of individuals in each species and N = the total number
of individuals.

This statistic may also be written as

$$H = - \Sigma \mid p_i \log (p_i) \mid \qquad (12.4)$$

where $p_i = \dfrac{ni}{N}$ = relative abundance of the ith species.

Developed from rather different premises, this index is calculated from
information theory on the basis that the more species there are in a
community, relative to the number of individuals (taking into account
distribution of individuals between those species) the more complex it is,
and thus the more 'information' the community contains — or needs to
describe it. A similar index, describing the 'concentration' of species
within a community, is calculated as

$$C = \overset{S}{\underset{i}{\Sigma}} pi^2 \qquad (12.5)$$

with diversity, D, as 1/C (Simpson, 1949).

All these later indices compound in one way or another the separate
contributions to diversity of species number and 'spread' of individuals
into a single blanket measure. However it is often useful to be able to
distinguish the two elements — to assess separately differences between
communities in one particular parameter alone: species number or
equitability. Such distinction is possible with those measures based upon
the species-abundance distribution by separation of the original species
number ratio of early diversity analysis, and some function of the variance
of the species-abundance distribution itself; for use with Shannon's index
of diversity, Lloyd and Ghelardi (1964) have derived a separate index of
equitability by relation of observed number of species S, to a hypothetical
species number, S^1, to be expected from the Shannon index if all species
were equally distributed (i.e. if equitability was unity). In this case

$$E = \frac{S^1}{S} \qquad (12.6)$$

Lloyd and Ghelardi's original paper (1964) provides a table for of S' for
any given value of H.

Many other indices of diversity can be calculated, based upon different
formulae and different assumptions and different relative emphases of
species number and equitability elements, and any prospective research
worker would be readily forgiven if he found himself somewhat bemused
by the plethora of such indices. In practice, despite the differences of their
formulation, and the different biological assumptions upon which they are

based all these various indices (except that of Williams, 1964) may in fact be shown to be closely related to each other.

12.3 Resolution of Chaos in Diversity Indices

In an extremely exciting paper – which has been curiously overlooked by many later workers – Hill (1973) demonstrates that many of the common indices (species number, Simpson's index – even Shannon's formula) belong to the same family of mathematical relations – that each is, in effect a different power of the same relation. Defining Na (his index) as

$$Na = \frac{p_1^a + p_2^a + p_3^a \ldots p_n^a}{p_1 + p_2 + p_3 \ldots p_n} \text{ to the power } \frac{1}{1-a} \tag{12.7}$$

where $p_1, p_2 \ldots p_n$ denote the proportional abundances of the n species in a sample, Hill shows by substitution, that N_0 is n – the total number of species in the sample; N_1 approaches the exponent of Shannon's index H (as exp. $- \Sigma \ p_i \ 1_n(p_i)$) while N_2 is the reciprocal of Simpson's index (which, argues Hill, should in any case be used in reciprocal form).

Hill notes that a diversity number is a measure of the *effective* number of species present in a community and that the indices of his family differ only in regard to the different weightings given to rare or abundant species. Figuratively speaking the diversity indices are measures of how many species are present when the sample is examined to a certain depth amongst its rarities. If we examine superficially (by using N_2) we shall see only the more abundant species. If we look deeply (e.g. by using N_0) we shall see all the species present. He summarises his indices as

$N_{-\infty}$ = reciprocal of the proportional abundance of the rarest species
N_0 = total number of species present
N_1 = exp(H)
N_2 = reciprocal of Simpson's index; i.e. $1/(p_1^2 + p_2^2 + \ldots + p_n^2)$
$N_{+\infty}$ = reciprocal of the proportional abundance of the commonest species.

He also offers a new measure of 'even-ness'. Noting that the Lloyd and Ghelandi index is in effect a comparison between the diversity as measured by species number and diversity as measured by some other statistic, he shows that, in a more general case, the same 'even-ness' may be derived in his family of indices by the simple ratio of one index to another. (Note, however, that ratios of indices $> N_0$ are shown to be less dependent on sample size than those which have species number (N_0) as a denominator and Alatalo (1981) has suggested a more refined measure of even-ness as

$$\frac{(N_2 - 1)}{(N_1 - 1)} \hspace{6cm} (12.8)$$

It is not the intention of this chapter to offer a comprehensive review of all the various indices of diversity available. We aim merely to introduce the basic concepts involved, and we must leave the matter here. Fuller review of the different indices, their merits and demerits, and the circumstances in which they should be applied may be found in Whittaker (1965), Pielou (1969), Hill (1973), Routledge (1979) or Hopkins (1983).

12.4 The S Component of Diversity: Why Are There So Many Kinds of Organisms?

12.4a Homage to Santa Rosalia

Although species diversity is accepted as one of the most important parameters/ descriptors of an ecological community, we know remarkably little about the factors promoting such diversity or establishing a particular level of diversity in any one community. What sets the diversity of a given system – a diversity which will have a profound influence on the community's operation and stability? Why are there so many different sorts of organisms in the first place? In 1957 G.E. Hutchinson delivered as a presidential address to the Ecological Society, an explorative discussion which he entitled 'Homage to Santa Rosalia: or why are there so many kinds of animals?'

Clearly, the actual number of species in existence in our world is a complex function of local evolutionary selection pressures and adaptive radiations coupled with biogeographical divisions and barriers. The total diversity of any geographic region – or of a single taxonomic group within that region – is a function of local selection pressures and radiations, of niche availability and needs; because biogeographical barriers and limits to free movement prevent the distribution of any one organism to fill all instances of the niches worldwide for which it may be adapted, such radiations and developments can be repeated and replicated in each separate biogeographical 'continent'. Global 'diversity' – in this case species number – is thus far greater even than the number of potential niches, for equivalent niches in different regions are occupied by different 'local' organisms.

12.4b Saturation, and the Red Queen Hypothesis

However, any particular environment must have a limit to the number of potential niches offered. For a time, new species can be accommodated by progressive 'niche-narrowing' of existing and established members of the community, but there must be a lower limit for each niche-band below which further restriction is impossible if the niche is to remain viable. Other new species can be accommodated if the establishment of yet another new organism creates in itself a new secondary niche: the establishment within a community of a new plant species may open up a new habitat or microenvironment for a number of additional dependent species. But, sooner or later the environment becomes saturated, and no new species can be accommodated. Only when the environment changes markedly or some new biological development opens up new environments or new potential niches (as the amniote egg permitted colonisation of the land by the early vertebrates, or the development of homoiothermy permitted wider colonisation by birds and mammals) can the radiation recommence.

Thus we may see increases in the global diversity of animal and plant species as a series of discrete stepwise expansions. Gould (1979) has suggested that such expansions are likely to be sigmoid: pointing out that the process of speciation in a new environment is exactly analogous to that of population growth in virgin resources (p. 129). Thus increase in the number of species is slow at first (rates of speciation are as fast as they will ever be, but the founder species are few in number, so that development of new species from them is slow). As the numbers increase, in geometric progression, we enter an explosive phase of development — but this increase cannot continue indefinitely and as the environment nears saturation, the rate of increase declines, and the 'population' levels out. Such a sigmoid pattern may be repeated over and over; as — for whatever reason — a new environment becomes available for colonisation, so the process is regenerated (Figure 12.2).

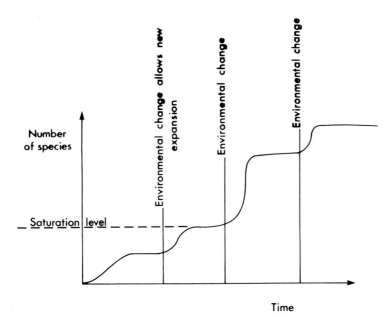

Figure 12.2: Increase in the Total Number of Species in an Area with Time, Through Evolutionary Development. A series of sigmoid curves may be generated as, after each 'explosion' in response to environmental change, species number approach saturation.

But as at each stage the existing environment approaches saturation, so the number of species must tend towards an upper limit (until a major new environment is opened). With no new niches to be exploited, a new species can only evolve at the expense of the extinction of an old one. As the environment approaches saturation so it becomes progressively more and more difficult to establish a new species; species number and diversity become a constant. As the Red Queen advised Alice in *Alice Through the Looking-Glass*: 'From here on it takes all the running you can do just to stay in the same place' (Carroll, 1886). Such an ecological phenomenon has been formalised by Van Valen (1976) as 'The Red Queen Hypothesis' (e.g. Van Valen, 1977; Stenseth, 1979).

This same pattern of rapid growth to saturation, and maintained equilibrium between establishment and extinction is as true for the establishment of organisms within a 'present day' community by colonisation, as it is for the development of species by evolution on a global scale. The development of diversity within any particular community is ultimately arrested when the system is saturated. From then on the level is maintained by successive, balanced colonisations and extinctions. We might conclude that the richness of any particular community is thus defined by those factors which set the different saturation levels. While this is, in some part true — in practice it is not quite so simple.

12.5 Colonisation, Extinction and Island Biogeography

The very development of a community (as we established in Chapter 4) is a function of continued colonisation and extinction — generally with rate of colonisation exceeding that of extinction. If, for purely mechanistic reasons, rates of colonisation and extinctions came into equilibrium during this development, then community diversity, as species number, would be fixed even though this might be below a level at which the community would theoretically be saturated. Thus, if for some reason rate of extinction in a particular community was particularly high, or rate of colonisation (in terms of species number or number of individuals) peculiarly low the community might become balanced with a relatively lower diversity irrespective of its theoretical saturation point.

In an analysis of the way in which rates of colonisation and extinction might influence the species number of islands, MacArthur and Wilson (1967) came to the conclusion that both colonisation and extinction rates were functions of the number of species already on the island (Chapter 4). The rate of immigration of new species to the island decreases as the number of species on the island increases (as more and more of the potential mainland colonists are found on the island, fewer of the new arrivals constitute new species); the rate of extinction likewise increases as with the number of species on the island (Chapter 4 and Figure 4.2). In addition, rates of colonisation were shown to be a function of the distance of the island from the nearest mainland (or 'seeding' point from which migrants might reach the island), while rates of extinction were a function of island size. With the relative number of species on a given island expressed as an equilibrium between the processes of colonisation and extinction, species number can be shown to differ on islands of different size, and distance from the mainland, as varying factors affect rates of colonisation and extinction and produce different equilibria (Figure 12.3). Such considerations may also affect the equilibrium species number expressed by *any* community. Although it may not represent a physical island surrounded by seas, even a mainland ecosystem is in effect an ecological 'island' – in that it represents a particular and discrete area of its type, surrounded by a 'sea' of systems of different type (Chapter 4). It, too, may be defined in terms of its area, or distance from the nearest 'mainland' of a similar system. Thus for these mainland communities, too, rates of colonisation and extinction may vary with 'island size' or 'distance from the mainland' – and different communities may express a different equilibrium species number even though their saturation levels might otherwise be the same; that is to say, diversity in any community may be determined by this relative balance of colonisation and extinction.

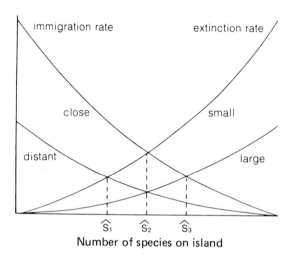

Figure 12.3: Relative Number of Species on Small, Distant Islands (S_1) and Large, Close Islands (S_3) as Predicted by the MacArthur-Wilson Equilibrium Model. The number of species on small, close islands and large, distant islands (S_2) is intermediate. Source: After MacArthur and Wilson, 1967.

MacArthur and Wilson's initial theory is greatly over simplified: as noted by Simberloff (1978b) and Gilbert (1980) it is essentially stochastic and takes no account of species interactions. In consequence it is applicable only for those species which can be considered to be non-interactive. In addition MacArthur and Wilson make the further assumptions (i) that immigration is independent of island size, although in 1963 they recognised that this might not be so and (ii) that while immigration rates decrease with increasing distance from the mainland, extinction rates remain the same. Yet it has been suggested by Osman (1977) that larger islands will receive more colonisers because they present a larger 'catchment surface' to the migrant species. Brown and Kodric-Brown (1977) propose that extinction rates will be unlikely to be independent of mainland distance. The arrival of individuals belonging to species already represented on the island will reduce rates of extinction. This 'rescue effect' will be lower for more distant islands because of the reduced rate of arrival of individuals – and thus extinction rates will increase with increasing distance. Further modifications are suggested by Hopkins (1983). Nonetheless, such modifications merely affect the shape of colonisation and extinction curves and do not affect significantly the basic conclusions (except for the fact that the *relative turnover rates* on large and small islands are seen to be reversed; such reversal is more compatible with such observational data as is available than are the conclusions drawn from the original model of MacArthur and Wilson which paradoxically predicts faster species turnover rates on larger islands). Williamson (1981) however concludes that 'the MacArthur and Wilson theory is true, to the extent that there is an equilibrium between immigration and extinction, but is essentially trivial.' He argues that the major factors influencing species abundance relationships are biotic factors such as heterogeneity of the environment.

Perhaps the major factor influencing species richness within a community then, is still its number of potential niches – its 'saturation level'. But what determines this saturation point?

12.6 Saturation point

Ultimately what must provide the upper limit to the species richness of a community is the number of different niches which can be 'worked' within that community. Saturation level, as we noted in Section 12.4 is determined by the availability of resources and their 'divisibility' – the extent to which those resources can be partitioned.

Brown (1981) concludes that any general theory for diversity must thus contain two kinds of constraints – capacity rules and allocation rules. Capacity rules define the physical characteristics of environments which determine their capacity to support life. Allocation rules determine how the available resources are apportioned amongst species so that each may obtain a sufficient share of the resources to allow it to maintain a population and persist in the community. Such allocation rules are clearly a function of limits to niche design and niche overlap (Chapter 5). What determines environmental capacity is more difficult to define; nor do we actually have any real evidence that natural communities ever *are* saturated so that this theoretical saturation point ever has any real value in practice.

Indeed there is good evidence to the contrary. In an analysis of the community of herbivores on bracken (*Pteridium aquilinum*) in England and North America, Lawton (1982) offers clear evidence that the community of such herbivores in the New World is far from saturated: by comparison with bracken communities of the north of England, which support some 27 species of phytophagous insects (and a further eight possible or occasional members of the community) bracken stands in New Mexico and Arizona have only seven species. Further, Lawton defines the various possible niches which may be identified within such a community. Animals may be either *chewers* (which live externally and bite large pieces out of the plant), *suckers* (which puncture individual cells or the vascular system), *miners* (which live inside the tissues), or *gall-formers*. These various organisms may feed on the pinnae ('leaves'), the rachis (main stem), costae (main stalks of the pinnae) or costules (main veins of the pinnae), or any combination of these. If a two-way table is created to define the potential niches available, and occupation of these various niches scored for both American and English communities (Table 12.1) it becomes *very* clear that many niches in the New World bracken community are left vacant. Yet these niches must be *potentially* tenable since they are occupied by members of the English community: one is forced to the conclusion that these are truly vacant niches and the community is not fully saturated. Reviewing the evidence available from studies of other phytophagous insect systems Lawton suggests that in general such communities appear to be saturated with species very rarely.

Table 12.1: Feeding Sites and Feeding Methods for the Herbivorous Insects Attacking Bracken in the North of England (a) and in New Mexico (b) in the Open and in Woodland. Feeding sites of species exploiting more than one part of the frond are joined by lines.

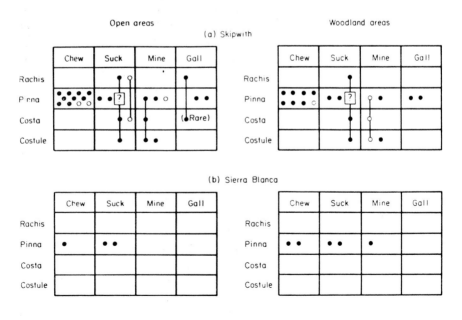

Source: From Lawton, 1982.

12.7 Equitability

Most of this chapter so far has been concerned with only one aspect of diversity — attempting to understand what determines species richness. Perhaps equally important however in consideration of diversity is an analysis of the distribution of individuals between the species of the community. Such distribution is clearly a further result of the niche relationships established in Brown's 'allocation' of resources. For the interactions determining resource division within the community will affect not just the number of species supported, but in addition their relative population sizes. (The inherent 'divisibility' of the resources will determine potential *species number*, while the biotic interactions which effect that division in competition or predation will themselves determine population sizes).

In Chapter 2 we introduced the various observed forms of species-abundance distributions in a search for basic ground rules controlling the structuring of communities. Thus we noted that simple communities, or species assemblages characteristically displayed a logarithmic or so-called broken-stick distribution — attributing such pattern to the fact that such communities are predominantly ordered by some single dominating factor. (Where some major resource dominates the community structure and is roughly evenly divided between its constituent species we observe a broken-stick distribution of species abundances; where such a resource is unevenly allocated between the species, a log series distribution is more characteristic.) We noted also that more complex communities — shaped by a variety of different biotic processes — characteristically showed a log-normal distribution of species abundance. May (1975b) suggests this may not necessarily reflect any intrinsic characteristic of the community itself, pointing out that such log-normal distribution is the necessary mathematical consequence of a multi-variable complex of this sort. Sugihara (1980) however, notes that the log-normal curves characteristic of species-abundance patterns of natural communities conform more closely to pure canonical distributions (where the location of the peak of the abundance distribution in total number of individuals, i.e. the distribution of $N \times S(N)$, coincides precisely with the position of the most abundant species ($S(N)$ max)) than can be explained by mathematical generalities alone. Sugihara thus concludes that such log-normal patterns are indeed a function of specific characteristics of the community (Sugihara, 1980; May, 1981).

None of these analyses actually tells us *why* any community shows a particular pattern of species-abundances, a particular equitability. Stenseth (1979) distinguishes log-normal curves as being characteristic of communities in stable environments, and logarithmic curves as typical of unstable systems (page 64) but in reality we know little of the factors controlling species-abundance distributions and thus little of the factors affecting the equitability component of diversity.

12.8   Factors Promoting Species Diversity

We can explain very little of what may affect species abundance distributions within communities. And, while we can speculate to some degree on what factors may set upper limits to the potential species richness of any community, we have no evidence that species number within a community ever actually reaches those limits. What, then, *does* set the limits to diversity in any given system?

Let us start with a few simple observations: it is noticeable, comparing otherwise equivalent communities, that there appears to be a gradual increase in species diversity from the arctic to the equator. Further, within any one geographical zone, we may record an increase in species diversity through ecological succession. As we have already noted, species number on islands or in isolated systems increases as a function of 'island size'. Finally diversity appears to increase even within established, or climax communities, over time. Based upon such observations, a number of theories have been put forward over the years in attempts to account for differences in diversity.

12.8a   The Time Hypothesis

The earliest of such hypotheses merely postulated that diversity is a simple function of time: all communities tend to diversify with time, therefore older communities are more diverse than young ones. Such a concept embraces increases in diversity due to immigration to an area, due to specialisation of the existing species, due to evolution of new species: all of which processes may contribute in time to a greater diversity within the community. It accounts for increased diversity during succession and also for differences between tropical, temperate and tundra communities – if we may assume that tropical ecosystems are considerably older than these others. But it offers no real explanation for the mechanism: it is an observation rather than a true explanation – and in truth is also really a statement of the obvious. Although overall diversity can change through changes in equitability within the community, species richness can only increase through immigration to an area or through evolution of new species; and both *do* take time. The hypothesis does not explain what *allows* the community to become more diverse in the first place.

12.8b   Spatial Complexity

An alternative hypothesis claims that diversity may be explained in terms of environmental complexity: the more complex and heterogeneous an environment, the more niches it may offer and the more complex and diverse will be its fauna and flora. This is, in a sense an analogue of our earlier ideas of communities becoming filled in response to the abundance of available niches, becoming saturated when all available niches are occupied. And as we have already noted we have no evidence that natural communities ever are completely saturated;

thus spatial complexity may never in practice be limiting. Further it is something of a circular argument, for an increase in species richness may in itself *cause* a further increase in spatial complexity as animals and plants provide habitats for others. Nonetheless, environmental complexity of this sort may indeed be at least a contributing factor in establishing levels of species diversity and there is some evidence for this. Such an effect could for example explain the increased diversity observed with increased island size: that a bigger island is likely to have greater environmental complexity and heterogeneity and will therefore support a greater number of species (although as we note above, MacArthur and Wilson (1967) reach the same conclusion without invoking any intermediate stage). Further the number of species in a freshwater lake increases dramatically from open water to the edge: as the structural complexity from submerged and emergent vegetation also increases. (But once again, the two may not be cause and effect, but coincidental effects of some other common cause).

12.8c Productivity Hypothesis

It has been suggested by other authors that diversity may be a function of community productivity (e.g. Brown, 1973). There is no doubt that very frequently, if we observe within a community, or between communities, an increase in species diversity, it is associated with an increase in primary productivity (e.g. Brown and Davidson, 1977; Cody, 1974). But equally such a relationship is not invariable: Yount (1956) studied the relationship between species number and system productivity in a cold spring (Silver Springs, Florida). His results show clearly (Figure 12.4) that in this situation, low productivity sites have a higher species diversity than high productivity sites. The relationship is thus not a simple one and perhaps the most elegant example of how an increase in productivity may be accompanied either by an increase or a decrease in diversity comes from the work of Abramsky (1978) on small mammal communities. The abundance of food in a natural 1-ha plot of shortgrass prairie in Central Plains, Colorado was artificially manipulated by adding alfalfa pellets and whole oats on a regular schedule. The small mammal species naturally inhabiting the manipulated food plot did not respond in their density to the supplemented food. However, a new specialised seed-eating species, *Dipodomys ordii*, invaded the food plot and persisted in relatively high density. As a result of this colonisation, species diversity was significantly higher on the food plot relative to the unmanipulated control plot. In a second experiment, application of water and nitrogen to two 1-ha plots of shortgrass prairie resulted in increased productivity. However, the increased production of this latter treatment was associated with vegetation growth and thus major changes in habitat structural characteristics relative to the control treatment. Two new species, *Microtus ochrogaster* and *Reithrodontomys megalotis*, colonised the nitrogen + water treatment, but other small mammal species 'resident' to the shortgrass prairie largely avoided this treatment. As a result of this manipulation, species diversity was significantly lower than the species diversity on the control treatment (Abramsky, 1978).

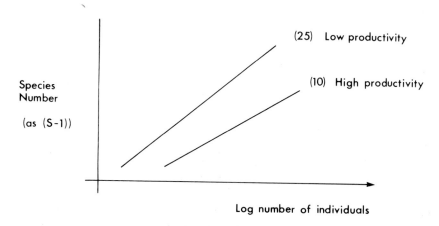

Figure 12.4: Species Diversity as a Function of Productivity. Data from Silver Springs, Florida show that diversity of high productivity sites may be lower than that of low productivity areas. Source: After Yount, 1956.

This difference in response to increased production was suggested by MacArthur (1972) to be dependent on the *pattern* of the increase in production. Thus increase in the abundance of all resources, brought about by an overall increase in production might indeed lead to an increase in diversity; by contrast, increase in only a small part of the total resource spectrum was suggested to lead to a decrease in diversity because the community would become dominated by those species which are competitively superior in exploiting these augmented resources.

12.8d The Theory of Slobodkin and Sanders

In 1968 Sanders suggested that, both in a geographical gradation from the arctic to the tropics and in a successional series, purely physicochemical, abiotic, parameters of the environment become less and less important in determining community-structure (and the presence or absence of individual species), while the role of biotic interactions increases markedly. Sanders claimed that this increase in the importance of biotic interactions would promote diversity.

Predation, as we already know, *can* indeed maintain diversity within a community (Paine, 1966, 1969, Chapter 9). Interspecific competition promotes specialisation and thus increases potential diversity by narrowing niches. But predation can equally eliminate prey species; competition can lead to exclusion. Further, reduced to its essentials the theory suggests that more diverse communities, with biotic interactions playing a more major role in their structure and operation, are more diverse; as we have noted for other hypotheses, it is something of a circular argument. Unless we can explain *why* the communities of pioneer successional stages, or arctic environments are more 'physically shaped' and less biotically ordered than those of tropical or climax communities, we are forced to conclude that 'more diverse communities, because they have more species, will be more powerfully influenced by biotic interactions than are less diverse communities, therefore they are more diverse.' Such justification was, however, presented by Slobodkin and Sanders (1969).

By combination environments may be				With, as example
F	C	and	∴ P	Tropical rain forest coral reef
F	V	but	P	Temperate climate woodland with regular seasonal fluctuations
F	V	but	UP	
S		and	∴ P	Hot springs
S	V	but	P	Arctic tundra
S	V	and	UP	Chemically-polluted stream
				Boulder beach

Figure 12.5: Slobodkin and Sanders Claim That any Environment may be Classified by Three Parameters, in Terms of Severity, Constancy and Predictability. Environments may therefore be Favourable (F) or Severe (S); Constant (C) or Variable (V); Predictable (P) or Unpredictable (UP).

In its final expression this theory suggests that diversity is a function of the severity, stability and predictability of the environment in which a community becomes established. Slobodkin and Sanders claimed that any environment could be defined in terms of these three variables. Thus environments can be for example favourable, constant and therefore predictable, favourable, variable but predictable, severe, constant and therefore predictable, severe, variable and unpredictable. Figure 12.5 expands these potential combinations and offers examples of the different categories.

Clearly no organism can colonise an environment which is severe, variable and unpredictable, but as severity decreases or predictability (or constancy) increases, so possibilities for colonisation increase. Species diversity may be observed to increase along these same gradients (Figure 12.6).

Thus even a severe environment may be colonised if it is constant, or at least predictable, by physiological specialisation. (Such specialisation, incidentally, is always at the expense of lability: various species of *Tilapia* fish can survive in the hot water springs of Eastern Africa, at temperatures of 42°C, but cannot tolerate a change in that temperature of more than 0.5°C; the 'tolerance' and 'lethal' limits (Chapter 1) cluster close to the optimum temperature. This is a trade-off any species must accept if it specialises towards environmental extremes: hence the need for constant or predictable conditions if the environment is severe.) As severity decreases, the 'specialisation' required to colonise environments declines, and more and more species can utilise any particular environment.

Ultimately, in favourable and predictable environments a whole host of organisms may survive (and, as severity decreases, so, too, predictability in fact becomes less important). Thus, diversity is seen to increase as environments become more favourable and/or more predictable. And in regard to our earlier considerations the more severe environments are surely those where abiotic considerations must have the dominant influence on the communities organisms; in the more benign environments, biotic interactions will play a major role. Movement from the tundra to the tropics, or from a pioneer to climax sere will indeed be a movement from environmental severity to an environment which is more favourable and more stable. (NB – while our treatment here is descriptive, various attempts have been made in determining precise mathematical relationships between species diversity and environmental severity.)

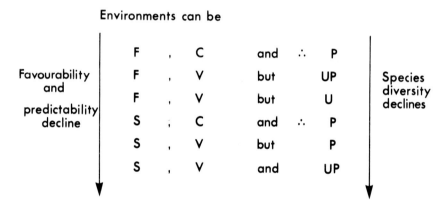

Figure 12.6: Changes in Diversity with Environmental Severity, Constancy and Predictability. Source: After Slobodkin and Sanders, 1969.

12.9 Theories of Diversity

All these various theories are compatible and it is probable that diversity in any instance is a function of many factors in combination, of which these may be only a few. It may be noted however that this last hypothesis carries within it an ultimate explanation for some of those already considered in these pages, and may incorporate their observations in its fundamental conclusions. Thus increased benignity, predictability or stability of an environment will probably be accompanied by an increase in productivity, spatial heterogeneity etc. Further, we may relate these conclusions back to our earlier and more fundamental concepts – that diversity in more general terms must ultimately be limited by environmental capacity and the way the available resources are allocated: that environmental capacity and divisibility of resources may determine species number, while species abundance distributions are also a function of 'allocation rules' within the community (Brown, 1981). Thus spatial complexity can be seen to affect both capacity and potential divisibility of resources; productivity clearly is a major factor determining environmental capacity. Brown himself sees environmental capacity/saturation level as determined primarily by availability of energy and by variability and predictability of resources (Brown, 1973, 1981) – correlates perhaps of the benignity and predictability of Slobodkin and Sanders. Can we perhaps, after all answer the unresolved question 'what determines saturation point?' – as productivity and predictability?

But this is only a beginning – and diversity is an extremely complex function, of all these factors, of balanced rates of colonisation and extinction, and of niche design-rules. There is no simple answer.

13 Stability

One of the most important features of any ecological system, whether population or community, is its inherent stability — or lack of it! It is also — because of its overwhelming implications in both theoretical and applied studies of natural and manipulated systems — one of the ecological parameters around which there has been the most discussion and controversy, about which most has been written. Clearly we cannot attempt in this chapter to cover the vast literature in tremendous detail. Rather, we review the major points of discussion; such summary is however, indubitably biased by our own opinions, and for a fuller treatment the reader is referred to Usher and Williamson (1974) and Van Dobben and Lowe-McConnell (1975).

13.1 Definitions

Ecological stability is variously defined, but may be summarised as the dynamic equilibrium of population, community, or ecosystem size and structure. MacArthur (1955) defined stability as 'the ability of both populations and communities to withstand environmental perturbation, to accommodate change'. More formally it is the extent to which variation in some characteristic of an ecological system is less than the variation in the environmental variables which affect that characteristic (MacArthur, 1957). But from the outset we should emphasise that stability is not a simple character but in fact a multiplicity of distinct attributes. Much of the confusion engendered in the past in discussions about stability in the literature has come from failure to compare like with like, from confusion of these different forms of stability.

Orians (1975) has identified a number of different elements which may be recognised within the overall concept of stability; his rather complex list of 'stability functions' (after Lewontin, 1969) may be resolved into three distinct and quite different 'types' of stability and a number of attributes which may be associated with these (Table 13.1).

The three basic types of stability itself (perhaps better regarded merely as different *facets* of the same phenomenon) may be considered as constancy, resilience and inertia. By *constancy*, we refer to a lack of change in some parameter of a system — such as number of species, taxonomic composition, life form structure of a community, *size* of a population, or feature of the physical environment. *Resilience* may be considered as the ability of a system to recover and continue functioning after disturbance even though it may have changed its form. Thus a community may be described as resilient if, during or after disturbance — even though its constancy of species structure may have changed markedly, it is able to continue to operate as a viable system. Finally, *inertia* is the ability of a system to withstand, or resist such perturbation in the first place. (Strictly speaking, it is this inertia stability to which MacArthur's 1957 definition refers.) *Attributes* of such stability functions are Orians' Persistence, Elasticity and Amplitude. Thus *Elasticity* is a measure of the speed with which the system returns to its former state following a perturbation; this elasticity equates to what Pimm and Lawton (1980) mean by their use of the term resilience (here defined differently), and with what is measured by their 'return time'. The *Amplitude* of a system defines the area over which it is stable; a system has high amplitude if it can be considerably displaced from its previous state, and still return to it. Where a system can return to its previous state following *any* perturbation, however large, it is deemed to be globally stable. *Persistence* refers to the survival time of a system or some component within it. (In this sense one population might be considered more stable than another if its mean 'time to extinction' were larger (Roff, 1974). Finally, Orians' concepts of cyclical or trajectory stability merely provide further descriptors of the stability observed, in relation to whether the system tends towards an end-point of a stable point, or to stable 'limit cycles' (May, 1972, 1976).

Table 13.1: Stability Functions. Concepts and terms used in discussion of ecological stability.

Constancy	: a lack of change in some parameter of a system.
Inertia	: the ability of a system to resist external perturbations.
Resilience	: the ability to continue functioning after perturbation.
Persistence	: the survival time of a system or some component of it.
Elasticity	: the speed with which the system returns to its former state following a perturbation.
Amplitude	: the area over which a system is stable.
Cyclical stability	: the property of a system to oscillate around some central point or zone.
Trajectory stability	: the property of a system to move towards some final end point or zone despite differences in starting point.

Orians himself notes: 'This listing of the meanings attached to the concept of stability is not intended as a classification system because the terms are not comparable. Constancy and persistence are descriptive terms implying nothing about underlying dynamics. Cyclic and trajectory stability have measures of inertia, elasticity and amplitude associated with them, etc. The separation of concepts is presented only to illustrate the many meanings of stability, the existence of which presumably reflects a need for a variety of notions relating to fluctuations.' (Orians, 1975) (see the text).

Source: Adapted from Orians, 1975, after Lewontin, 1969.

As noted earlier it is extremely important to recognise and distinguish these different and independent facets and attributes of stability, for the constancy, resilience and inertia of any ecological system respond differently to different pressures, and as we have already commented, much of the confusion engendered in the past has come from failure to recognise as distinct these different facets of stability.

13.2 Stability of Single Species Populations

While the major part of this chapter will be devoted to considerations of community stability, we will, for simplicity, develop our argument by reviewing *briefly* the factors affecting stability of single species populations (Chapters 7, 8).

Models of population growth of single species populations can be created for populations with continuous growth (overlapping generations) and for those with discrete growth (distinct and separate generations). Such models (a family of differential and difference equations) may be refined to model quite closely, what happens in natural populations (Chapter 7). Study of the way such populations behave has suggested that there is a range of possible outcomes to population growth, depending on the relative reproductive rate and size of the founder population (May, 1976). Populations may show stable equilibrium points (that is, come to equilibrium at a fixed and stable level), they may show stable cycles or 'limit cycles' between constant and defined limits, or they may behave unpredictably and irregularly in a chaotic way (May, 1976 and Figure 13.1). By calculating the appropriate parameters we can predict which behaviour will result in given circumstances. Such outcomes are also demonstrated by Figure 13.2, upon which have been superimposed the positions of 28 populations of insects. (Solid circles are derived from field populations, open circles from populations in laboratory culture.) We may note that there is a tendency for laboratory populations to exhibit cyclic or chaotic behaviour, while natural populations tend to display stable point equilibrium. Indeed these results may be interpreted as indicating a tendency for *natural* populations to exhibit stable point equilibrium, despite the range of possible behaviours they could exhibit.

This fact that laboratory populations, maintained in artificial, predator- and competitor-free environments, tend towards cyclic or chaotic behaviour also suggests that the stable point equilibrium of natural populations is not intrinsic to the population itself, but is conferred upon it by interactions with other populations.

13.3 Stability of Two or Three Species Systems

We have already discussed the kinds of interactions which may occur between populations in interspecific competition (Chapters 6 and 8) or predation (Chapter

9), sufficient to demonstrate that interspecific competition at least is usually a destructive and destabilising influence on a population. Predation and parasitism, while they may be shown to be potentially stabilising under certain circumstances, are equally likely to be destabilising if the population parameters of predator and prey/parasite and host are out of balance, or if efficiency of predation or parasitism is too high (Table 9.3). May (1976b) concludes that predator-prey systems tend to be in tension between the stabilising prey density-dependence, and the often destabilising predator functional and numerical responses. According to the relative values of these in mathematical models of single predator-single prey systems, we can in theory derive damped oscillations or stable limit cycles in both prey and predation populations.

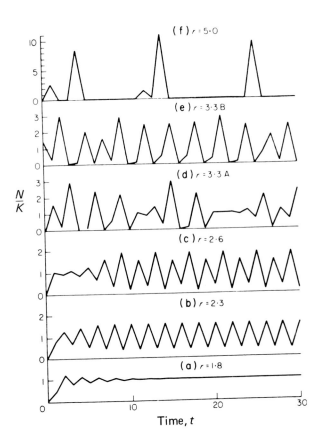

Figure 13.1: The Spectrum of Dynamical Behaviour of the Population Density, N_t/K, as a Function of Time, t, as Described by Difference Equations for Various Values of r. Specifically: (a) $r = 1.8$, stable equilibrium point; (b) $r = 2.3$, stable 2-point cycle; (c) $r = 2.6$, stable 4-point cycle; (d to f) in the chaotic regime, where the detailed character of the solution depends on the initial population value, with (d) $r = 3.3$ ($N_0/K = 0.075$), (e) $r = 3.3$ ($N_0/K = 1.5$), (f) $r = 5.0$ ($N_0/K = 0.02$). Source: From May, 1976b.

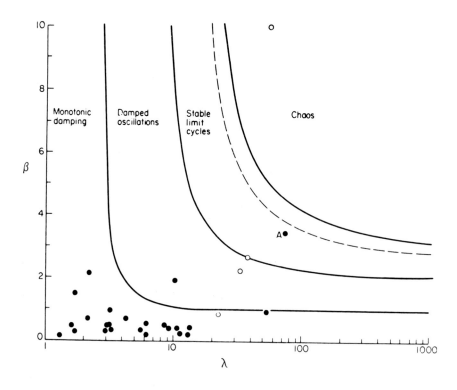

Figure 13.2: Dynamical Behaviour of Different Equations Describing Population Growth. The solid curves separate the regions of monotonic and oscillatory damping to a stable point, stable limit cycles, and chaos; the broken line indicates where 2-point cycles give way to higher order cycles. The solid circles come from the analyses of life table data on field populations, and the open circles from laboratory populations. Source: From Hassell *et al.*, 1976, after May, 1976b.

Table 13.2: Impressionistic Summary of Population Dynamics of 8 Natural Prey-Predator Systems.

Prey-predator	Geographical location	Apparent dynamical behaviour
sparrow — hawk	Europe	equilibrium point
muskrat — mink	central North America	equilibrium point
hare — lynx	boreal North America	cycles
mule deer — mountain lion	Rocky Mountains	equilibrium point
white-tailed deer — wolf	Ontario	equilibrium point
moose — wolf	Isle Royale	equilibrium point
caribou — wolf	Alaska	equilibrium point
white sheep — wolf	Alaska	equilibrium point

Source: After May, 1976b.

Both damped oscillations about an equilibrium point, and stable cycles *may* be observed in natural single predator-single prey systems (Table 13.2). However, single predator-single prey systems are relatively rare in nature and in any case the relevant parameters in *most* natural populations are not 'adjusted for balance', but are such that predation would normally be destabilising and lead to extinction of both predator and prey. We established in Chapter 9, however, that we could restabilise such simple systems if we introduced some complexity – either in terms of physical complexity (refuges for prey: Huffaker's orange mite experiment) or in biotic complexity, such as spatial heterogeneity amongst prey or the development of multi-predator-multi-prey systems. Such deliberations suggest that much of the stability of any population, or interacting set of populations – or indeed the community as a whole (being made up of a set of interacting populations) – is conferred by the complexity of the system itself.

13.4 Community Stability

While the stability of communities is in part conferred upon them by the stability of their component species populations, there are other characteristics of the community which may show additional stability in their own right. Thus there may be a stability in the actual *structure* of the community (Chapter 2), a constancy of trophic form, of foodweb design, a resilience in the mode and balance of the community's operation.

It is, however, important to distinguish between stability of an individual community and apparent constancy of design of communities in general. The most quoted example of stability of trophic structure within a community comes from the work of Simberloff and Wilson (1969) who eliminated the fauna from several very small mangrove islets in the Florida Keys, and then monitored their recolonisation by terrestrial arthropods. In all cases the total number of species on an island returned to around its original value, although the species constituting the total were usually altogether different. Heatwole and Levins (1972) reanalysed these data in terms of trophic organisation, listing for each island the number of species in each of the trophic categories: herbivores, scavengers, detritus feeders, wood borers, ants, predators and parasites (Table 2.2 above). Their results showed that, in terms of trophic structure, the communities appear to display striking stability and constancy. On the other hand, in terms of the detailed taxonomic composition of the community of arthropod species on a particular island, there is great variability. These results are often quoted in illustration of community stability. The rapid recovery of the communities to a working state demonstrates resilience, while the trophic similarity of the new communities to the old despite differences in actual taxonomic composition seem good evidence for postulating constancy of trophic form, inconstancy of species composition. Yet to view the results in this way is, strictly, to misinterpret them. Strictly, the data quoted by Heatwole and Levins show the re-establishment of a new community after total defaunation of an area, *not* recovery of a disturbed system. The fact that this new community shares many features of the old (in species number and trophic design) is not evidence of any resilience in the old community — which is now after all, destroyed. Rather, the results are better considered as pointing towards some fixity of community design under given circumstances (page 56): that there is, so to speak, some optimal way of dividing up the jobs within a community, some design and structure that utilises the resources most efficiently. A comparison as such, between an old community — now destroyed — and a new one established in its place has little bearing on the stability of the initial community.

In practice, evidence for the actual resilience or constancy of *an existing community* can only come from studies where the community has been disturbed, but not entirely destroyed. The fact that the secondary succession in ecological communities generated after some perturbation eventually returns the community to its previous climax state (even if through different developmental pathways) *may* be adduced as evidence for the resilience and constancy of such climax communities while a more specific example may be drawn from the recent analysis by Johns (1983) of faunal communities in montane forest in West Malaysia before, during and after logging operations.

Johns conducted a census of vertebrate communities in hill forest areas of West Malaysia — surveying areas of primary forest before logging operations commenced, during logging and up to six years after logging was completed. Logging was selective, for main timber species only, and at comparatively low density: extraction rate for the study area was at an average of 17 trees per hectare. Surveys of birds, and most mammals within the area may be assumed to be complete (although Chiroptera and Muridae were probably inadequately sampled in logged forest); data for amphibians and reptiles are less detailed.

Immediate effects of logging are clear: numbers of species of both mammals and birds declined dramatically (Table 13.3). Such changes in species number did not purely result from loss of existing species due to disturbance. The entire faunal composition altered dramatically: many species were lost but other new species arrived within the logged areas. The 20 species of mammal of 1-2 year old logged forest include only 15 of the original 45 species of primary forest; at least five are new to the community. Even within the existing species there were marked changes in dominance. Diversity of bird faunas fell (Table 13.3), not just because of a decrease in species number but due also to a shift in equitability within the community.

Timber extraction, even at relatively low intensity, has a tremendous impact upon the vegetational structure of the community — opening glades and trackways — and there is correspondingly a marked change in faunal composition. Certain species are favoured such as elephants and sambhur deer (*Elephas maximus*; *Cervus unicolor*) which travel long distances along logging roads and browse on colonising vegetation in more open areas; others (such as sunbears (*Helarctos malayanus*); dhole (*Cuon alpinus*)) seem to avoid logged forest. Amongst reptiles and amphibians, too, there are clear changes in faunal composition. Many species of amphibians typical of primary forest (such as *Megophrys monticola*, most *Microhyla* species and many ranids) are absent or rare in disturbed forest. Other species normally present at very low density increase rapidly in numbers following logging, while several species that were abundant in logged forest and along logging roads were never seen in primary forest. Amongst reptiles cobras (*Naja naja* and *Ophiophagus hannah*) appeared to be commoner in logged forest. Monitor lizards (*Varanus* spp.) were also frequently seen in logged forest but rarely encountered in undisturbed areas. That changes of faunal composition are primarily due to the vegetational changes of logging is most clearly reflected amongst the birds in an altered pattern of foraging strategies shown by birds of logged and unlogged forest. It is immediately apparent to the observer that the logged forest community is completely dominated by aerial gleaners (swifts, tree swifts and swallows) which in primary forest of course forage entirely above the canopy. However, even if these are excluded as an artefact of sampling there remains still a major shift towards fruit-eating — presumably due to the loss of much understorey vegetation and its replacement with rapidly-fruiting colonisers (Table 13.4).

Table 13.3: Species Number and Diversity of Birds and Mammals of Hill Forest in Sungei Tekam, West Malaysia.

	Number of mammal species	Number of bird species	Diversity of birds (Shannon-Weiner index)	Equitability
PRIMARY FOREST	45	90	5.37	0.796
LOGGED FOREST				
Just logged	35	44	4.21	0.781
1-2 years old	20	96	5.22	0.792
3-4 years old	31	78	5.54	0.881
5-6 years old	28	80	5.13	0.812

Data for mammals are for total sightings; figures for birds are corrected to a common time base of an eight-day sample period.

Source: After Johns, 1981.

After 5-6 years the forest has started to recover. Table 13.4 shows considerable displacement of the trophic structure of the forest birds during logging. Table 13.5, this time considering the resident mammal fauna, shows that after 5-6 years the trophic structure of the community has returned towards its original state. Species composition has changed (as in Heatwole and Levins mangrove islets), as has species number: amongst 28 species of mammals only 24 remain of the original 45 species of primary forest, and of those the relative abundances of six have changed completely — but trophic composition (Table 13.5) has shown at least some recovery towards its original state.

Johns himself notes however:

It should be stressed that this study was located in an area where logging was at a comparatively low density. Logging at higher intensities may be expected to have a proportionally much greater effect on community survival. Many species are capable of recolonising old logged forest, but this can only occur if an area of primary forest exists as a reservoir, from which colonisers can reenter the logged areas. Success of recolonisation for some species thus depends on the presence of areas of undisturbed or already recolonised logged forest. Where extensive areas are completely given over to logging there is likely to be a large and permanent reduction of species diversity.

Table 13.4: Differences in Foraging Strategies of Bird Species in Primary and Logged Forest in West Malaysia. Figures as total numbers and percentage of total sample.

| | Before logging | | After logging | |
| | | % | | % |
Foraging technique	n	total	n	total
Terrestrial frugivore	15	1.1	13	1.0
Undergrowth frugivore	32	2.4	4	0.3
Canopy frugivore	331	25.3	517	38.6
Undergrowth insectivore/frugivore	93	7.1	63	4.7
Canopy insectivore/frugivore	171	13.1	207	15.3
Insectivore/nectarivore	36	2.8	28	2.1
Bark-gleaning insectivore	80	6.1	43	3.2
Terrestrial insectivore	49	3.7	25	1.9
Foliage-gleaning insectivore	366	28.0	274	20.4
Flycatching insectivore	105	8.0	111	8.3
Piscivore	5	0.4	0	0
Raptor	26	2.0	56	4.2
N	1309		1341	

Source: After Johns, 1983.

One of the most controversial questions surrounding such discussions of community stability is this one of whether or not such stability is achieved determininistically or whether it arises so to speak, merely by its own definition: that if a community is 'unstable' it will continue to change until it achieves some constancy. Do communities develop along some deliberate and predetermined pathway towards a stable state — or is it just that unstable communities must by definition change? The observed convergence in ecological successions towards a limited range of end-communities argues cogently for some such 'predetermined' development towards a stable situation. The most striking feature of Heatwole and Levins' analysis of community development in defaunated mangrove islets — and the strongest intuitive support for the implied stability — is the speed with which the system recovers to its 'former state' (within approximately one year). Heatwole and Levins claim this as clear evidence for a deterministic, directional recovery (evidence in turn for a dynamic stability of the system as a whole). However, in his own reanalysis of the original data on which this is based, Simberloff (1978b) suggests that such a conclusion is premature: that both degree and speed of recovery are functions of the small systems involved, and that it is not yet possible to say how much of the structuring of communities is deterministic, and how much the same result could be achieved purely stochastically. Such reservation is applicable to many other examples. The apparent 'determinism' in succession may be explained in terms of the statistical properties of successional series as Markov chains (Horn, 1976, and above, page 103), while the fact that the developmental pathways of secondary succession are not constant and do not reflect the original seral stages of primary succession is surely strong evidence that such development is *not* deterministic — that the stability of ecological communities is developed not because they 'seek' such stability, but because only the stable associations will persist.

Table 13.5: Trophic Strategies of Mammals of West Malaysian Hill-forest Before Logging and After Five to Six Years Recovery Following Logging.
Numbers of species in each trophic category are accompanied, in brackets, by corresponding percentages of total sample.

Trophic categories	Primary forest		Logged forest after 5-6 years	
Terrestrial folivore	(12)	5	7	(25)
Arboreal frugivore/folivore	(5)	2	2	(7)
Terrestrial frugivore/folivore	(5)	2	1	(4)
Arboreal frugivore	(19)	8	6	(21)
Terrestrial frugivore	(2)	1	0	(0)
Arboreal insectivore/frugivore	(21)	9	6	(21)
Terrestrial insectivore/frugivore	(10)	4	2	(7)
Arboreal insectivore	(2)	1	0	(0)
Terrestrial insectivore	(5)	2	1	(4)
Predators	(19)	8	3	(11)
		42	28	

(Excluding Chiroptera, Muridae)
Source: After Johns, 1983.

13.5 Diversity and Stability

Throughout our deliberations we have been suggesting (Sections 13.2 and 13.3) that complexity enhances stability. In terms of community stability, too, it has been observed that the most 'stable' communities (tropical communities by comparison with temperate or arctic ones, late successional communities by comparison with pioneer stages; the measure of stability is not defined) tend to have greater complexity of structure, greater species diversity.

This has developed into a widely accepted axiom: that diversity begets stability. Intuitively it is sound: irregularities within a complex system are easily compensated for by minor adjustments elsewhere. If some species begins to increase markedly in numbers, or indeed decline, such a change in equilibrium is buffered by the complexity of the system: a variety of predators may switch their attention to a newly abundant prey, or conversely, switch to alternative prey if one particular species becomes uncommon (Chapter 9). The deviation is rapidly corrected. Equally, if a species in a complex food web actually declines to extinction, the operation of the community is not rocked upon its foundations; minor adjustments in the roles, or relative abundances of other related species can absorb the 'perturbation' by redirecting the community relationships.

Nor is the link between complexity and stability built merely upon such supposition. Such a relationship has been established experimentally. Pimentel (1961) compared the levels of several insect populations on a group of collard plants (*Brassica oleracea*) set out in a field which had been uncultivated for 15 years, and on another group planted in a single species stand. In the single species plantings, the abundances of several species of insects − particularly aphids and flea-beetles − reached outbreak proportions; the same species were kept under control in the collard plants in the old field. Odum, Barrett and Pulliam (in Odum, 1971) conducted a similar experiment on arthropod populations in a field planted with millet in one season and left to develop a natural vegetational community in a second season. Each group of arthropods sampled was represented by more species during the second year than the first. Numbers of individuals of predatory and parasitic species were also higher during the second year, but numbers within herbivore populations were lower.

Such evidence has led to the widespread supposition that it is species diversity that promotes stability. Regrettably, just because something is intuitively sound does not necessarily mean it is correct. Is the relationship one of cause or effect? Is stability the result of diversity − or is it that diversity is the consequence of the stability? We have already noted (Section 12.8) that environmental constancy or predictability permits the development of greater species diversity. Further, one of the most consistent effects of environmental *disturbance* is to increase the variation in the relative abundances of species within a community. [Patrick (1963, 1975), in a study of diatoms in clear and in polluted streams, has shown clearly that polluted streams support fewer species, and also display a far less even distribution of individuals between the species; in polluted streams a few species are extremely abundant. The log-normal distribution of relative abundances tends to be lower and broader for polluted streams than for clear streams, indicating greater variation in the size of populations (Figure 13.3).] May (1973) has lent further support to this thesis that species diversity within a community is a function of environmental stability rather than the other way around, concluding from theoretical studies of the behaviour of both communities and populations that in fact, as a system becomes more complex it becomes more fragile; that it is the simplest systems that are the most robust, and that thus, any correlation between species diversity and stability results because complex communities can only be supported by very stable environments

(May, 1976). May argues that as a mathematical generality, increasing complexity makes for dynamical fragility rather than robustness. This is *not* to say that, in nature, complex ecosystems need appear less stable than simple ones. A complex system in an environment characterised by a low level of random fluctuation and a simple system in an environment characterised by a high level of random fluctuation can well be equally likely to persist, each having the dynamical stability properties appropriate to its environment. In a predictable environment, the system need only cope with relatively small perturbations, and can therefore achieve this fragile complexity, yet persist. Conversely, in an unpredictable environment, there is need for the stable region of parameter space to be extensive, with the implication that the system must be relatively simple. In brief, a predictable ('stable') environment may permit a relatively complex and delicately balanced ecosystem to exist; an unpredictable ('unstable') environment is more likely to demand a structurally simple, robust ecosystem. Such conclusions, so much at variance with intuition — or indeed evidence from observation of most natural communities — have become known in the literature as May's Paradox.

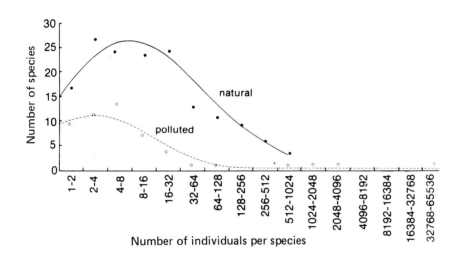

Figure 13.3: Log-normal Distribution of Relative Abundances of Diatom Species in a Natural Stream Community and in a River That is Subjected to Pollution. Source: After Patrick, 1963.

13.6 May's Paradox

May's work suggests that the more complex a system, the *less* stable it becomes — that the apparent stability is only conferred upon it by virtue of the fact that it is founded upon a stable environment and has nothing to do with the intrinsic properties of the system itself. But closer examination reveals that May is referring primarily to stability of species composition and population sizes — that in fact he is referring to 'constancy'. To some extent, May's Paradox is resolved in the fact that while his conclusions thus relate to community constancy, our discussions above are, implicitly, concerned with 'resilience'. If we thus distinguish the separate facets of stability, May's conclusions no longer conflict with intuition. The more complex a system is, the less likely it is indeed to be able to maintain constancy of species composition, specific population sizes, etc. etc. By contrast the more complex it is, the more likely it is to be able to survive, to continue to *function* even after perturbation, because 'damage' to one part of the system can be compensated for by adjustments in other parts of the system.

But even this intuitive reaction, that complex systems are more likely to show resilience than simple ones, because of a supposedly greater ability to compensate may in itself be over-simplistic. One can draw a simple food web and, by studying it, can easily appreciate that if one pathway is blocked, material cannot be moved by another route; conversely in a complex web, there are many possible routes to the same destination. But this does not take into account the relative importance of the different pathways (page 54) or how likely they are to become blocked. P.J. Hopkins (unpublished), offers an analogy with the arteries and capillaries of animal circulatory systems. The few links of simple communities are major through-routes; such arteries are robust, and although blockage of one energy exchange pathway would have serious repercussions for the community as a whole, such blockage is in fact highly unlikely. In a more complex system however the small links, like capillaries, are in fact much more *easily* blocked, but are inconsequential in terms of continued community function. Hopkins concludes that in practice, even on intuitive grounds it is not easy to predict any necessary correlation between diversity and apparent stability in that under appropriate conditions even relatively simple communities may indeed show considerable stability as May suggests.

Community resilience is in fact better regarded not as a function of *species* diversity *per se*, but as a function of the diversity of energy exchange pathways within the community. For stability in this sense results, as we have discussed, from an ability to accommodate a blockage in the system by 'shunting' material through other routes. Such recasting of the relationship as one between stability (resilience) and diversity of energy exchange pathways, also allows us to appreciate that *species* diversity and stability need *not* necessarily be connected (providing yet further explanation for May's Paradox even in relation to community resilience). For, in theory, greater complexity of exchange pathways can be achieved in two ways: (i) by an increase in the number of species in a community; (ii) by an increased tendency to generalised feeding by existing species.

In other words such diversity of energy exchange may be achieved by an increase in the number of specialist species within the community, or by generalisation of existing species. In practice it is more likely to be achieved by an increase in species number (at least in stable environments!) because interspecific competition will tend to promote specialisation. Further, increased diversity by diversification of relationships of existing species may be in itself destabilising. (Pimm and Lawton (1978) show, for example, that omnivory tends to destabilise a food web.) As a result, in general, increased stability *will* be accompanied by increased species diversity. But this is not a necessary relationship, and in other circumstances stability might equally be achieved by a community of generalists of low diversity.

Such a scheme also fits well with Odum's (1975) idea that community structure (and associated with that diversity) reflects the nature of energy transfer within a community. He holds that optimum diversity is a function of the quantity and quality of energy flow: that low diversity may be optimum in ecosystems strongly subsidised by external inputs of energy, while a higher diversity may be optimal in systems which are dependent on internal production of transfer of energy. Odum claims that a species matrix adapts to the strength and variety of energy input, and the resource flows coupled with it (cf May's abundance and predictability of resources (above, page 349)). When one or a few sources of high utility energy are available and are coupled with high nutrient surplus low diversity has advantages: a concentrated and specialised structure is more efficient at exploiting the 'bonanza' than is a dispersed structure; likewise where energy is limiting or of low utility, then a higher diversity appears to be optimum to produce a steady state. Thus, again, any relationship between diversity and stability will be purely secondary: 'Quite stable systems can have either a low or high diversity, depending on the energy-forcing function' (Odum, 1975). Such an idea is clearly compatible with our earlier discussions on the importance of energy-exchange pathways and also provides a more formal resolution of May's Paradox. From Odum's hypothesis of diversity adjusted to the quantity and quality of energy flow we can devise two systems. One with limited energy input − self-contained − best suited by high diversity. The other, heavily subsidised by auxillary energy inputs, best suited by a concentrated and specialised structure of low diversity. He writes 'high energy low diversity systems can be quite stable if the input subsidies are regular or continue at the same level over long periods of time − it is perhaps in such systems that high diversity is *destabilising* as May's theoretical models show' (Odum, 1975). (Odum's considerations of the design of communities best adapted for stability in relation to quantity and quality of energy input, clearly also have important implications in regard to our 'rules for community organisation' in Chapter 2 (page 56) as offering additional design constraints.)

De Angelis (1980) follows this argument one stage further, demonstrating, in a theoretical treatment of community dynamics that stability increases as the *flux* of energy increases: that is to say stability of the community increases as energy flow per unit total energy in the system increases (De Angelis, 1980).

13.7 Stability and Food Web Design

Such conclusions beg certain questions about the role of food web design in community stability. The relationship between stability and the structure of food webs has been investigated more formally by Lawton and Pimm (Pimm, 1979a, 1980; Pimm and Lawton, 1977, 1978).

Paine (1980) in a review of this work notes that there are a number of important properties of food webs that must be taken into account. These are: the number of links involved in any food chain; the degree of omnivory within the community; the number of relationships in which any one species may be involved (its connectedness, or connectance (May, 1973a)) within the community as a whole, etc. etc. (He also notes that there are links in community webs other than between predators and prey: that is there may be cross-links between trophic levels or within them due to competition or other interactive relationships which should be taken into account.) Another important consideration is the relative strengths of the various linkages. For the strength or importance of any relationship cannot be assumed to be equivalent for all web members. A consumer will be a strong interactor if, in its absence, pronounced changes ensue (MacArthur, 1972) while removal of a weakly interacting species will yield little or no change.

In their work on food web design, based on mathematical models, but supported by back-reference to natural systems, Pimm and Lawton suggest that – as we have already noted – certain trophic relationships within a community tend to be destabilising: that is, in terms of persistence, communities become progressively less stable as the degree of omnivory within them increases (Pimm and Lawton, 1978). They predict that food webs with high numbers of omnivores will thus be rare in the real world: a prediction supported by data on real communities assembled by Cohen (1978). Further, different relationships clearly have a different relative importance in maintaining community structure and function depending on their 'linkage strength', thus it is not merely the shape of the web that will affect its stability, but the distribution of stronger or weaker linkages. Pimm (1980) notes more generally that:

> On the assumption that systems of interacting species, when perturbed from equilibrium, should return to equilibrium quickly, one can predict four properties of food webs: (1) food chains should be short, (2) species feeding on more than one trophic level (omnivores) should be rare, (3) those species that do feed on more than one trophic level should do so by feeding on species in adjacent trophic levels, and (4) host-parasitoid systems are likely to be exceptions to (1)-(3) when interaction coefficients permit greater trophic complexity.

By generating random, model food webs (with many features identical to webs described from a variety of marine, freshwater, and terrestrial systems), Pimm was able to test these predictions by generating expected values for the number of trophic levels and the degree of omnivory within webs and comparing these with real world webs. Real world webs were shown to have fewer trophic levels, less omnivory, and very few omnivores feeding on non-adjacent trophic levels than random webs. Pimm concludes 'The confirmation of all these predictions from stability analyses suggests that system stability places necessary, though not sufficient, limitations on the possible shapes of food webs.

Such work on the properties of food webs and their significance for stability is still in its infancy and much remains to be done. Paine (1980) concludes that:

> Food webs, along with their associated 'cross-links' (relationships within trophic levels) provide a realistic framework for understanding complex highly interactive multi-species relationships. (But) I believe that the next generation of models must be more sensitive to interaction strength, less so to trophic complexity.

13.8 The Energetics of Stable Systems

While the stability of communities may be explained in terms of the complexity, or form, of energy-exchange pathways within a community, community energetics can also be invoked as another method of studying community stability: in terms of energy value. In this regard the community is seen to be at its most stable (in terms of constancy or inertia) when community productivity equals community respiratory loss: (CP = CR; NCP = 0) i.e. when there is no opportunity for further accumulation of biomass, and thus the community has no capacity for change.

In *succession*, as one particular community develops, if CP > CR, biomass is accumulating and community structure is changing. By the same token, the relative stabilities of different communities — in terms of structural change — may be evaluated in terms of this balance of CP and CR. Whether or not we are talking in terms of *successional* change the relative balance of CP/CR is a measure of the structural constancy of a community. Thus Odum (1959) suggests that while CP/CR = 1 signifies a stable system, situations where CP/CR > 1 (autotrophy) are characteristic of a state where the community is still developing; and where CP/CR < 1 (hypertrophy), the community is unstable and degenerate. Odum gives examples of the various different communities as in Figure 13.4.

While, empirically it is certainly true that for a stable system CP/CR = 1 — and that when CP/CR is greater or less than unity the community is changing and thus must be seen as relatively less stable, evaluation of community stability in this way is somewhat problematical. For, in the same way that May (1973 *et seq.*) finds that constancy is not a function of increased community complexity, so it is difficult to see how mature communities, energetically speaking, should show such stability. By definition, when CP/CR = 1 the community supports the maximum possible biomass. Yet the more biotically complex a system, the more biota are supported by a limited nutrient pool; the more of the community's available nutrients are bound up in living tissue (Odum, 1959 and Table 4.1). The nutrient cycle becomes extremely 'tight' with no stockpile in the soil to even out irregularities of nutrient return; as a result, the system becomes more at risk to bottlenecks in nutrient recycling (Section 3.7) and is thus, potentially more fragile. In energetic terms, the most complex systems with high biomass of organic matter, show maximum deviation from entropy. According to the purely physical laws of thermodynamics, such systems must be highly unstable.

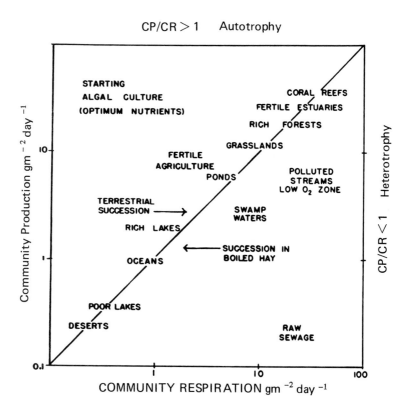

Figure 13.4: Position of Various Types of Communities in a Classification Scheme Based on Community Metabolism. Photosynthetic production (CP) exceeds the community respiratory consumption (CR) in the upper left side of the diagram (P/R is greater than 1; autotrophic types). In the lower left side of the diagram the activities of the respiratory processes exceed photosynthesis. Such communities are importing organic matter or living off previous storage. (P/R is less than 1; heterotrophic type.) Source: From Odum, 1959.

Such arguments return us to the complex issues of whether or not complexity does promote stability (Sections 13.5 and 13.6) incidentally, complicating the issue still further by suggesting that complexity may not only beget fragility in terms of community *constancy*, but may indeed be associated with loss of both resilience and inertia, too. (Certainly, with the physical necessity that systems seek to maximise entropy, the inertia of energetically complex systems must be in some doubt.) In so doing, they also cast considerable doubt on the use of measures of CP/CR in determining ecosystem stability.

13.9 Causes for Stability

An understanding of the various factors underlying or affecting the stability of populations, communities or ecosystems is fundamental to our understanding of the ecological world. More importantly such understanding of the uneasy balance of a stable system is indispensible if we are to undertake to manipulate natural systems for our own ends — without destroying them. Pimentel's experiments comparing monocultures and complex systems in 'farming' of the brassica *B. oleracea* (Section 13.5) emphasise the significance of such understanding for agriculture. But: the stability of such systems is a complex function and the factors underpinning such stability are incompletely resolved. What is the role of foodweb-design — of the diversity of energy exchange pathways? Is species diversity important? Where does the physical stability of the environment come in? It must be a compounding factor, as May suggests, but is it the overriding one? Is community structure irrelevant; and does environmental stability merely promote community complexity and community stability *in despite* of each other? Smith (1972) suggests that trophic pathways do contribute little to ecosystem stability, and that the answers lie in spatial patterning within the environment. Environmental stability must (and does) affect the type of community developed. But does that community add nothing of its own to the stability of its ecosystem? Indeed is the fact that the majority of natural communities are stable merely an accidental artefact due to the fact that unstable systems have not persisted?

References

Abrahamson, W.G. & Gadgil, M. (1973) 'Growth form and reproductive effort in Goldenrods (*Solidago*: Compositae)', *Amer. Nat.*, *107*, 651-61

Abrams, P. (1980) 'Some comments on measuring niche overlap', *Ecology*, *61*, 44-9

Abramsky, Z. (1978) 'Small mammal community ecology: changes in species diversity in response to manipulated productivity', *Oecologia*, *34*, 113-24

Alatalo, R.V. (1981) 'Problems in the measurement of evenness in ecology', *Oikos*, *37*, 119-204

Anderson, R.M. & May, R.M. (1978) 'Regulation and stability of host-parasite population interactions. I. Regulatory processes', *J. Anim. Ecol.*, *47*, 219-49

Andrewartha, H.G. (1959) 'Self-regulatory mechanisms in animal populations', *Austr. J. Sci.*, *22*, 200-5

Andrewartha, H.G. (1970) *Introduction to the Study of Animal Populations*, 2nd edn. Methuen, p. 283

Andrewartha, H.G. & Birch, L.C. (1954) *The Distribution and Abundance of Animals*, Chicago University Press

Auer, C. (1968) 'Erste Ergebnisse einfacher stochastischer Modelluntersuchungen über die Ursachen der Populationsbewegung des grauen Lärchenwicklers *Zeiraphera diniana*, Gn. (= *Z. griseana Hb.*) im Oberengadin, 1949/66', *Z. angew. Ent.*, *62*, 202-35

Ayala, F. (1970) 'Competition, Coexistence and Evolution', in *Essays in Evolution and Genetics* (eds. M.K. Hecht & W.C. Steere), Appleton-Century-Crofts, New York

Bakker, K., Bagchee, S.N., Van Zwet, W.R. & Meelis, E. (1967) 'Host discrimination in *Pseudeucoila bochei* (Hymenoptera: Cynipidae)', *Ent. exp. appl.*, *10*, 295-311

Barbour, M.G., Burk, J.H. & Pitts, W.D. (1980) *Terrestrial Plant Ecology*, Benjamin/Cummings, 604 pp

Barlow, N.D. & Dixon, A.F.G. (1980) *Simulation of Lime Aphid Population Dynamics*, PUDOC (Centre for Agricultural Publishing and Documentation), Wageningen

Bartholomew, G.A. (1968 & 1982) 'Body temperature and energy metabolism', in *Animal Physiology: Principles and Applications*, 1st & 4th edns (ed. Gordon M.S.) Collier Macmillan, London

Beauchamp, R.S.A. & Ullyett, P. (1932) 'Competitive relationships between certain species of fresh-water triclads', *J.Ecol.*, *20*, 200-8

Beaver, R.A. (1977) 'Non-equilibrium 'island' communities: diptera breeding in dead snails', *J. Anim. Ecol.*, *46*, 783-98

Beddington, J.R. (1974) 'Age structure, sex ratio and population density in the harvesting of natural animal populations', *J. appl. Ecol.*, *11*, 915-24

Beddington, J.R. (1978) 'On the risks associated with different harvesting strategies', *Int. Whaling Comm. Report*, *28*, 165-7

Beddington, J.R., Free, C.A. & Lawton, J.H. (1975) 'Dynamic complexity in predator-prey models framed in different equations', *Nature*, *255*, 58-60

Beddington, J.R., Free, C.A. & Lawton, J.H. (1976) 'Concepts of stability and resilience in predator-prey models', *J. Anim. Ecol.*, *45*, 791-816

Beddington, J.R., Free, C.A. & Lawton, J.H. (1978) 'Characteristics of successful natural enemies in models of biological control of insect pests', *Nature*, *273*, 513-9

Beddington, J.R., Hassell, M.P. & Lawton, J.H. (1976) 'The components of arthropod predation II The predator rate of increase', *J. Anim. Ecol.*, *45*, 165-85

Bell, R.H.V. (1970) 'The use of the herb layer by grazing ungulates in the Serengeti', in *Animal Populations in Relation to Their Food Supply* (ed. Watson, A.) *B.E.S. Symposium*, *10*, 111-24

Bennetts, H.W., Underwood, E.J., Shier, F.L. (1946), 'A specific breeding problem of sheep on subterranean clover pastures in Western Australia', *Austr. Vet. J.*, *22*, 2-

Benson, T.F. (1973) 'Population dynamics of cabbage root fly in Canada and England', *J. appl. Ecol.*, *10*, 437-46

Billings, W.D., Godfrey, P.J., Chabot, B.F. & Bourque, D.P. (1971) 'Metabolic acclimation to temperature in arctic and alpine ecotypes of *Oxyria digyna*', *Arctic and Alpine Research*, *3*, 277-89

Birch, L.C. (1948) 'The intrinsic rate of natural increase of an insect population', *J. Anim. Ecol.*, *17*, 15-26

Birch N. & Wratten, S.D. (1983) 'Patterns of aphid resistance in the genus *Vicia*', *Ann. appl. Biol.* (in press)

Blank, T.H., Southwood, T.R.E. & Cross, D.J. (1967) 'The ecology of the partridge. I. Outline of population processes with particular reference to chick mortality and nest density', *J. Anim. Ecol.*, *36*, 549-56

Blest, A.D. (1963) 'Relations between moths and predators', *Nature*, *197*, 1046-7

Bobek, B. (1969) 'Survival, turnover and production of small rodents in a beech forest', *Acta theriologica*, *14*, 191-210

Bonner, J.T. (1965) *Size and Cycle: an Essay on the Structure of Biology*. Princeton University Press

Botkin, D.B., Janak, J.F. & Wallis, J.R. (1972) 'Some ecological consequences of a computer model of forest growth', *J. Ecol.*, *60*, 849-72

Brenchley, W.E. (1918) 'Buried weed seeds', *J. agric. Sci.*, *9*, 1-31

Brown, J.H. (1973) 'Species diversity of seed-eating desert rodents in sand-dune habitats', *Ecology*, *54*, 775-87

Brown, J.H. (1981) 'Two decades of homage to Santa Rosalia: toward a general theory of diversity', *Amer. Zool.*, *21*, 877-88

Brown, J.H. & Davidson, D.W. (1977) 'Competition between seed-eating rodents and ants in desert ecosystems', *Science*, *196*, 800-82

Brown, J.H. & Kodric-Brown, A. (1977) 'Turnover rates in insular biogeography: effect of immigration on extinction', *Ecology*, *58*, 445-9

Brown, J.L. (1975) *The Evolution of Behaviour*, Norton: New York

Brown, W.L. & Wilson, E.O. (1956) 'Character displacement', *Syst. Zool.*, 5, 49-64

Burnett, T. (1964) 'Host larval mortality in an experimental host-parasite population', *Can. J. Zool.*, 42, 745-65

Caraco, T. & Wolf, L.L. (1975) 'Ecological determinants of group-size in foraging lions', *Amer. Nat.*, 109, 343-52

Carl, E.A. (1971) 'Population control in arctic ground squirrels', *Ecology*, 52, 395-413

Carroll, L. (Dodgson, C.L.) (1886) *Alice Through the Looking-Glass*

Cates, R.G. & Orians, G.H. (1975) 'Successional status and the palatability of plants to generalised herbivores', *Ecology*, 56, 410-8

Caughley, G. (1976) *Analysis of Vertebrate Populations*, Wiley, New York

Caughley, G. & Birch, L.C. (1971) 'Rate of increase', *J. Wildl. Magmt.*, 35, 658-63

Caughley, G. & Lawton, J.H. (1981) 'Plant-Herbivore Systems', in *Theoretical Ecology: Principles and Applications* (ed. R.M. May), Blackwell, pp. 132-66

Chaetum, E.L. & Severinghaus, C.W. (1950) 'Variations in fertility of white-tailed deer related to range conditions', *Trans. N. Amer. Wildl. Conf.*, 15, 170-89

Chambers, R.J., Sunderland, K.D., Stacey, D.L. & Wyatt, I.J. (1982) 'A survey of cereal aphids and their natural enemies in winter wheat in 1980', in *Natural Enemies and Insect Pest Dynamics* (ed. S.D. Wratten), *Proc. Ann. appl. Biol.*, 101, 175-8

Charnov, E.L., Orians, G.H. & Hyatt, K. (1976) 'The ecological implications of resource depression', *Amer. Nat.*, 110, 247-59

Chiang, H.C. & Hodson, A.C. (1950) 'An analytical study of population growth in *Drosophila melanogaster*', *Ecol. Monogr.*, 20, 173-206

Chitty, D. (1960) 'Population processes in the vole and their relevance to general theory', *Can. J. Zool.*, 38, 99-113

Christian, J.J. (1950) 'The adreno-pituitary system and population cycles in animals', *J. Mammal.*, 31, 247-57

Christian, J.J. (1959) 'The roles of endocrine and behavioural factors in the growth of mammalian populations', in *Symposium on Comparative Endocrinology* (ed. A. Gorbman)

Christian, J.J. & Davis, D.E. (1964) 'Endocrines, behaviour and populations', *Science*, 146, 1550-60

Clements, F.E. (1916) 'Plant Succession; an analysis of the development of vegetation', *Carnegie Institution of Washington Publication*, 242, 1-512

Cloudesley-Thompson, J.L. & Chadwick, M.J. (1964) *Life in Deserts*, Foulis, London

Clutton Brock, T.H. (1974) 'Primate social organisation and ecology', *Nature*, 250, 539-42

Clutton Brock, T.H. & Harvey, P.H. (1978) 'Cooperation and disruption', in *Readings in Sociobiology* (eds T.H. Clutton Brock & P.H. Harvey), W.H. Freeman, pp. 135-41

Cock, M.J.W. (1978) 'The assessment of preference', *J. Anim. Ecol.*, 47, 805-16

Cody, M.L. (1971) 'Finch flocks in the Mohave Desert', *Theoret. Pop. Biol.*, 2, 142-58

Cody, M.L. (1974) *Competition and the Structure of Bird Communities*, Princeton University Press

Cody, M.L. (1975) 'Towards a theory of continental species diversities' in *Ecology and Evolution of Communities* (eds M.L. Cody, J.M. Diamond) Harvard University Press, pp. 214-57

Coe, M.J. (1969) 'Microclimate and animal life in the equatorial mountains', *Zool. Africana*, 4, 101-28

Cogger, H.G. (1974) 'Thermal relations of the Mallee dragon: *Amphibolurus fordi* (Lacertilia: Agamidae)', *Austr. J. Zool.*, 22, 219-39

Cohen, J.E. (1966) *A Model of Simple Competition*, Harvard University Press

Cohen, J.E. (1968) 'Alternative derivation of a species-abundance relation', *Amer. Nat.*, 102, 165-72

Cohen, J.E. (1978) *Food Webs and Niche Space*, Princeton University Press

Cole, L.C. (1951) 'Population cycles and random oscillations', *J. Wildl. Mgmt.*, 15, 233-52

Collier, B.D., Cox, G.W., Johnson, A.W. and Miller, P.C. (1973) *Dynamic Ecology*, Prentice-Hall

Colwell, R.K. & Futuyma, D.J. (1971) 'On the measurement of niche breadth and overlap', *Ecology*, 52, 567-76

Connell, J.H. (1961) 'Effects of competition, predation by *Thais lapillus* and other factors on natural populations of barnacles', *Ecol. Monogr.*, 31, 61-104

Connor, E.F. & Simberloff, D. (1979) 'The assembly of species communities: chance or competition?' *Ecology*, 60, 1132-40

Cooke, G.D., Beyers, R.J. & Odum, E.P. (1969) 'The case for the multispecies ecological system, with special reference to succession and stability', *NASA report SP – 165*, pp. 129-39

Cowan, I. McT. (1950) 'Some vital statistics of big game on overstocked mountain range', *Trans. N. Amer. Wildl. Conf.*, 15, 581-8

Cowie, R.J. (1977) 'Optimal foraging in Great Tits (*Parus major*)', *Nature*, 268, 137-9

Crook, J.H. & Gartlan, J.S. (1966) 'Evolution of primate societies', *Nature*, 210, 1200-3

Crowell, K.L. & Pimm, S.L. (1976) 'Competition and niche shifts of mice introduced onto small islands', *Oikos*, 27, 251-8

Darlington, H.T. & Steinbauer, G.P. (1961) 'The eighty-year period for Dr Beal's seed viability experiment', *Amer. J. Bot.*, 48, 321-5

Davidson, J. (1938) 'On the growth of the sheep population in Tasmania', *Trans. Roy. Soc. S. Australia*, 62, 342-6

Davidson, J. & Andrewartha, H.G. (1948) 'Annual trends in a natural population of *Thrips imaginis* (Thysanoptera), *J. Anim. Ecol.*, 17, 133-99

Davies, N.B. (1976) 'Food, flocking and territorial behaviour of the pied wagtail (*Motacilla alba yarelli*) in winter', *J. Anim. Ecol.*, 45, 235-54

Davies, N.B. (1977) 'Prey selection and the search strategy of the spotted flycatcher (*Muscicapa striata*), a field study on optimal foraging', *Anim. Beh.*, 25, 1016-33

Davies, N.B. & Houston, A.E. (1981) 'Owners and satellites: the economics of territory defence in the pied wagtail *Motacilla alba*', *J. Anim. Ecol.*, 50, 157-80

Davis, D.E. (1951) 'The relationship between level of population and pregnancy of Norway rats', *Ecology*, 32, 459-61

Dawkins, M. (1971) 'Perceptual changes in chicks, another look at the "search image" concept', *Anim. Beh., 19*, 566-74

Dawkins, R. (1976) *The Selfish Gene*, Oxford University Press

Dawkins, R. & Carlisle, T.R. (1976) 'Parental investment and mate-desertion, a fallacy', *Nature, 262*, 131-3

De Angelis, D.L. (1980) 'Energy flow, nutrient cycling and ecosystem resilience', *Ecology, 61*, 764-71

De Angelis, D.L., Gardner, R.H., Mankin, J.B., Post, W.M. & Carney, J.H. (1978) 'Energy flow and the number of trophic levels in ecological communities', *Nature, 273*, 406-7

De Bach, P. (1964) *Biological Control of Insect Pests and Weeds*, Chapman and Hall, London

Deevey, E.S. (1947) 'Life tables for natural populations of animals', *Q. Rev. Biol., 22*, 283-314

Dempster, J.P. (1971) 'The population ecology of the cinnabar moth, *Tyria jacobaeae* L. (Lepidoptera: Arctiidae)', *Oecologia, 7*, 26-67

Dempster, J.P. (1975) *Animal Population Ecology*, Academic Press, London

Dempster, J.P. & Pollard, E. (1981) 'Fluctuations in resource availability and insect populations', *Oecologia, 50*, 412-6

Detling, J.K., Dyer, M.I., Procter-Gregg, C. & Winn, D.T. (1980) 'Plant-herbivore interactions: Examination of potential effects of bison saliva on regrowth of *Bouteloua gracilis* (H.B.K.) Lag.', *Oecologia, 45*, 26-31

DeVita, J. (1979) 'Niche separation and the broken-stick model', *Amer. Nat., 114*, 171-8

Diamond, J.M. (1973) 'Distributional ecology of New Guinea birds', *Science, 179*, 759-69

Diamond, J.M. (1975) 'Assembly of species communities', in *Ecology and Evolution of Communities* (eds M.L. Cody & J.M. Diamond), Harvard University Press, pp. 342-444

Dixon, A.F.G. (1959) 'An experimental study of the searching behaviour of the predatory coccinellid beetle *Adalia decempunctata* (L.)', *J. Anim. Ecol., 28*, 259-81

Dixon, A.F.G. (1970a) 'Stabilization of aphid populations by an aphid induced plant factor', *Nature, 227*, 1368-9

Dixon, A.F.G. (1970b) 'Factors limiting the effectiveness of the coccinellid beetle *Adalia bipunctata* (L.) as a predator of the sycamore aphid, *Drepanosiphum platanoidis* (Schr.)', *J. Anim. Ecol., 39*, 739-51

Dixon, A.F.R. & McKay, S. (1970) 'Aggregation in the sycamore aphid, *Drepanosiphum platanoidis* (Schr.) (Hemiptera: Aphididae) and its relevance to the regulation of population growth', *J. Anim. Ecol., 39*, 439-54

Dixon, A.F.G. & Wratten, S.D. (1971) 'Laboratory studies on aggregation, size and fecundity in the black bean aphid, *Aphis fabae*, Scop', *Bull. ent. Res., 61*, 97-111

Dyer, M.I. (1980) 'Mammalian epidermal growth factor promotes plant growth', *Proc. Nat. Acad. Sci. USA, 77*, 4836-7

Edwards, C.A., Sunderland, K.D. & George, K.S. (1979) 'Studies on polyphagous predators of cereal aphids', *J. appl. Ecol., 16*, 811-24

Edwards, P.J. & Wratten, S.D. (1980) 'Ecology of Insect-Plant Interactions', *Studies in Biology, 121*, Edward Arnold, London

Edwards, P.J. & Wratten, S.D. (1982) 'Wound-induced changes in palatability in

Birch (*Betula pubescens* Ehrh. ssp. *pubescens*)', *Am. Nat.*, *120*, 816-18

Edwards, P.J. & Hollis, S. (1983) 'The distribution of excreta on New Forest grassland used by cattle, ponies and deer', *J. appl. Ecol.*, *19*, 953-64

Einarsen, A.S. (1945) 'Some factors affecting ring-necked pheasant population density', *Murrelet*, *26*, 39-44

Ellenberg, H. (1958) 'Boden reation (einschlieplich Kalfrage)', in *Handbuch der Pflanzenphysiologie*, Vol. 4 (ed. W. Ruhland), Springer-Verlag, pp. 638-708

Elliott, H. (1971) 'Wandering Albatross', in *Birds of the World* (ed. M. Gooders), IPC, London, pp. 29-31

Elliott, J.M. (1977) 'Some Methods for the Statistical Analysis of Samples of Benthic Invertebrates', 2nd ed, *Freshwater Biological Association Scientific Publication* No. *25*

Elton, C.S. (1927) *Animal Ecology*, Methuen

Elton, C.S. & Nicholson, M. (1942) 'The ten-year cycle in numbers of the Lynx in Canada', *J. Anim. Ecol.*, *11*, 215-44

Embree, D.G. (1966) 'The role of introduced parasites in the control of the winter moth in Nova Scotia', *Can. Ent.*, *98*, 1159-68

Emlen, J.M. (1973) *Ecology: an evolutionary approach*, Addison Wesley: New York

Engelmann, M.D. (1966) 'Energetics, terrestrial field studies and animal productivity', *Adv. Ecol. Res.*, *3*, 73-115

Errington, P.L. (1945) 'Some contributions of a fifteen-year local study of the Northern Bobwhite to a knowledge of population phenomena', *Ecol. Monogr.*, *15*, 1-34

Evans, H.F. (1973) *A study of the predatory habits of Anthocoris species* (*Hemiptera: Heteroptera*), unpublished D.Phil. Thesis, University of Oxford

Feinsinger, P. (1976) 'Organisation of a tropical guild of nectarivorous birds', *Ecol. Monogr.*, *46*, 257-91

Feinsinger, P., Spears, E.E. & Poole, R.W. (1981) 'A simple measure of niche breadth', *Ecology*, *62*, 27-32

Fenchel, T. (1975) 'Character displacement and coexistence in mud snails (Hydrobiidae)', *Oecologia*, *20*, 19-32

Fenchel, T. & Christiansen, F.B. (1976) *Theories of Biological Communities*, Springer-Verlag, New York

Fenner, M. (1978a) 'Susceptibility to shade in seedlings of colonising and closed turf species', *New Phytol.*, *81*, 739-44

Fenner, M. (1978b) 'A comparison of the abilities of colonisers and closed-turf species to establish from seed in artificial swards', *J. Ecol.*, *66*, 953-63

Fenner, M. (1980) 'The inhibition of germination of *Bidens pilosa* seeds by leaf canopy shade in some natural vegetation types' *and* 'The induction of a light requirement in *Bidens pilosa* seeds by leaf canopy shade', *New Phytol.*, *84*, 95-101; 103-6

Fisher, R.A. Corbet, A.S. & Williams, C.B. (1943) 'The relation between the number of species and the number of individuals in a random sample of an animal population', *J. Anim. Ecol.*, *12*, 42-58

Forcier, L.K. (1975) 'Reproductive strategies and the co-occurrence of climax tree species', *Science*, *189*, 808-10

Frank, P.W., Boll, C.D. & Kelly, R.W. (1957) 'Vital statistics of laboratory cultures of *Daphnia pulex* as related to density', *Physiol. Zool.*, *30*, 287-305

Freeland, W.J. (1974) 'Vole cycles: another hypothesis', *Amer. Nat.*, *108*,

238-45

Frith, H.J. (1959) 'Incubator Birds', *Sci. Amer.*, *CCI*, 52-8

Fry, F.E.J., Brett, J.R. & Clawson, G.H. (1942) 'Lethal limits of temperature for young goldfish', *Rev. Can. Biol.*, *1*, 50-6

Fuentes, E.R. (1976) 'Ecological convergence of lizard communities in Chile and California', *Ecology*, *57*, 3-17

Fujii, K. (1968) 'Studies on interspecies competition between the azuki bean weevil and the southern cowpea weevil, III, Some characteristics of strains of two species', *Res. Popul. Ecol.*, *10*, 87-98

Fuller, M.E. (1934) 'The insect inhabitants of carrion: a study in animal ecology', *CSIRA Bull.*, *82*, 5-62

Gause, G.F. (1934) *The struggle for existence*, Williams & Wilkins, Baltimore, reprinted 1964, Hafner, New York

Georghiou, G.P. & Taylor, C.E. (1977) 'Pest resistance as an evolutionary phenomenon', *Proc. XV Int. Cong. Ent. Washington D.C.*, 759-85

Gibson, C.W.D. (1977) 'Biology of Heteroptera associated with Tor Grass: *Brachypodium pinnatum*', unpublished D.Phil. thesis, Oxford University

Gilbert, L.E. (1975) 'Ecological consequences of a coevolved mutualism between butterflies and plants', in *Coevolution of Animals and Plants* (eds L.E. Gilbert and P.H. Raven), University of Texas Press

Gilbert, F.S. (1980) 'The equilibrium theory of island biogeography: fact or fiction?' *J. Biogeography*, *7*, 209-35

Gilbert, N. & Gutierrez, A.P. (1973) 'A plant-aphid-parasite relationship', *J. Anim. Ecol.*, *42*, 323-40

Gilbert, N., Gutierrez, A.P., Frazer, B.D. & Jones, R.E. (1977) *Ecological Relationships*, W.H. Freeman

Gill, F.B. & Wolf, L.L. (1975) 'Economics of feeding territoriality in the golden-winged sunbird', *Ecology*, *56*, 333-45

Gimingham, C.H. (1972) *Ecology of Heatherlands*, Chapman & Hall, London

Gladfelter, W.B., Ogden, J.C. & Gladfelter, E.H. (1980) 'Similarity and diversity among coral reef fish communities: a comparison between tropical western Atlantic (Virgin Islands) and tropical central Pacific (Marshall Islands) patch reefs', *Ecology*, *61*, 1156-68

Glen, D.M. (1973) 'The food requirements of *Blepharidopterus angulatus* (Heteroptera: Miridae) as a predator of the lime aphid, *Eucallipterus tiliae*', *Ent. exp. appl.*, *16*, 255-67

Goodall, D.W. (1973) 'Sample similarity and species correlation', in *Handbook of Vegetation Science. V.* (ed. R.H. Whittaker), Junk, The Hague, pp. 107-56

Goodman, D. (1975) 'The theory of diversity-stability relationships in ecology', *Q. Rev. Biol.*, *50*, 237-66

Goss-Custard, J.D. (1977a) 'Optimal foraging and the size selection of worms by the redshank *Tringa totanus*', *Animal Behaviour*, *25*, 10-29

Goss-Custard, J.D. (1977b) 'Predator responses and prey mortality in the red-shank *Tringa totanus* (L) and a preferred prey *Corophium volutator* (Pallas)', *J. Anim. Ecol.*, *46*, 21-36

Gould, S.J. (1979) 'Is the Cambrian Explosion a Sigmoid Fraud?' Ch. 15 in Gould, S.J. *Ever Since Darwin: Reflections in Natural History*, Norton: New York

Graham, J.B. (1972) 'Low temperature acclimation and the seasonal temperature sensitivity of some tropical marine fishes', *Physiol. Zool.*, *45*, 1-13

Grant, P.R. (1968) 'Bill size, body size and the ecological adaptations of bird species to the competitive situations on islands', *Syst. Zool.*, *17*, 319-33

Grant, P.R. (1972) 'Convergent and divergent character displacement', *Biol. J. Linn. Soc.*, *4*, 39-68

Grant, S.A. & Hunter, R.F. (1966) 'The effects of frequency and season of clipping on the morphology, productivity and chemical composition of *Calluna vulgaris*', *New Phytol.*, *65*, 125-33

Grant Watson, E.L. (undated) *Enigmas of Natural History*, The Cresset Press Ltd., London

Greene, H.W. (1977) 'The aardwolf as a hyaena mimic: an open question', *Anim. Behav.*, *25*, 245

Greenstone, M.H. (1979) 'Spider feeding behaviour optimises dietary essential amino acid composition', *Nature*, *282*, 501-3

Griffiths, E. (1983) 'The carabid *Agonum dorsale* as an aphid predator in cereals', Unpublished Ph.D. thesis, University of Southampton

Grubb, P.J. & Suter, M.B. (1970) 'Acidification of soil by *Calluna* and *Ulex* and the significance for conservation', in *The Scientific Management of Animal and Plant Communities for Conservation*, (eds. E. Duffey & A.S. Watt), Blackwell, Oxford

Gulland, J.A. (1962) 'The application of mathematical models to fish populations', in *The Exploitation of Natural Animal Populations* (eds. E.D. Le Cren & M.W. Holdgate), Wiley, New York, pp. 204-18

Gutierrez, A.P., Havenstein, D.E., Nix, H.A. & Moore, P.A. (1974) 'The ecology of *Aphis craccivora* Koch and subterranean clover stunt virus in South East Australia II. A model of cowpea aphid populations in temperate pastures', *J. appl. Ecol.*, *11*, 1-20

Haefner, J.W. (1981) 'Avian community assembly rules: the foliage-gleaning guild', *Oecologia*, *50*, 131-42

Haefner, P.A. (1970) 'The effect of low dissolved oxygen concentrations on temperature-salinity tolerance of the sand shrimp *Crangon septemspinosa*', *Physiol. Zool.*, *43*, 30-7

Hairston, N.G. (1980) 'The experimental test of an analysis of field distributions: competition in terrestrial salamanders', *Ecology*, *61*, 817-26

Hamilton, W.D. (1963) 'The evolution of altruistic behaviour', *Amer. Nat.*, *97*, 354-6

Hamilton, W.D. (1971) 'Geometry for the selfish herd', *J. theoret. Biol.*, *31*, 295-311

Hanski, I. (1981) 'Coexistence of competitors in patchy environments with and without predators', *Oikos*, *37*, 306-12

Harborne, J.B. (1977) *Introduction to Ecological Biochemistry*, Academic Press, London

Harcourt, D.G. (1971) 'Population dynamics of *Leptinotarsa decemlineata* (Say) in eastern Ontario. III. Major population processes', *Can. Ent.*, *103*, 1049-61

Harper, J.L. (1964) 'The individual in the population', *J. Ecology*, *52*, (Suppl.) 149-58

Harper, J.L. (1977) *Population Biology of Plants*, Academic Press, London

Harper, J.L. & White, J. (1971) 'The dynamics of plant populations', in *Dynamics of Populations*, (eds P.J. Boer & G.R. Gradwell), Pudoc: Wageningen, pp. 41-63

Hassell, M.P. (1968) 'The behavioural response of a tachinid fly (*Cyzenis albicans* (Fall.)) to its host, the winter moth (*Operophtera brumata* (L))' *J. Anim. Ecol.*, *37*, 627-39

Hassell, M.P. (1975) 'Density dependence in single-species populations', *J. Anim. Ecol.*, *44*, 283-95

Hassell, M.P. (1976) *The Dynamics of Competition and Predation*, Edward Arnold, London

Hassell, M.P. (1977) 'Some practical implications of recent theoretical studies in host-parasitoid interactions', *Proc. XV. Int. Congr. Ent.*, 608-16

Hassell, M.P. (1978) *The Dynamics of Arthropod Predator-Prey Systems*, Princeton University Press

Hassell, M.P. & Comins, H.N. (1978) 'Sigmoid functional responses and population stability', *Theor. Pop. Biol.*, *9*, 202-21

Hassell, M.P. & Huffaker, C.B. (1969) 'Regulation processes and population cyclicity in laboratory populations of *Anagasta kühniella* (Zeller) (Lepidoptera: Phycitidae) III The development of population models', *Res. Popul. Ecol.*, *11*, 186-210

Hassell, M.P., Lawton, J.H. & May, R.M. (1976) 'Patterns of dynamical behaviour in single-species populations', *J. Anim. Ecol.*, *45*, 471-86

Hassell, M.P., Lawton, J.H. & Beddington, J.R. (1977) 'Sigmoid functional responses by invertebrate predators and parasitoids', *J. Anim. Ecol.*, *46*, 249-62

Hassell, M.P. & May, R.M. (1973) 'Stability in insect host-parasite models', *J. Anim. Ecol.*, *42*, 693-736

Hassell, M.P., & May, R.M. (1974) 'Aggregation in predators and insect parasites and its effects on stability', *J. Anim. Ecol.*, *43*, 567-94

Hassell, M.P. & Varley, G.C. (1969) 'New inductive population model for insect parasites and its bearing on biological control', *Nature*, *223*, 1133-6

Hatton, H. (1938) 'Essais de bionomie explicative sur quelques especes intercotidales d'algues et d'animaux', *Ann. Inst. Oceanogr.*, *17*, 241-348

Heatwole, H. & Levins, R. (1972) 'Trophic structure, stability and faunal change during recolonisation', *Ecology*, *53*, 531-4

Heron, A.C. (1972) 'Population ecology of a colonising species: the pelagic tunicate: *Thalia democratica*', *Oecologia*, *10*, 269-93; 294-312

Hill, M.O. (1973) 'Diversity and evenness: a unifying notation and its consequences', *Ecology*, *54*, 427-32

Hinde, R.A. (1974) *Biological Bases of Human Behaviour*, McGraw-Hill, New York

Hodgson, J. (1974) 'The effect of the grazing animal on herbage quality and utilisation', *Växtodling*, *28*, 74-80

Holling, C.S. (1959) 'Some characteristics of simple types of predation and parasitism', *Can. Ent.*, *91*, 385-98

Holling, C.S. (1961) 'Principles of insect predation', *Ann. Ent. Rev.*, *6*, 163-82

Holling, C.S. (1964) 'The analysis of complex population processes', *Can. Ent.*, *96*, 335-47

Holling, C.S. (1965) 'The functional response of predators to prey density and its role in mimicry and population regulation', *Mem. Ent. Soc. Canada*, *45*, 1-60

Holmes, R.T., Bonney, R.E. & Pacala, S.W. (1979) 'Guild structure of the Hubbard Brook bird community: a multivariate approach', *Ecology*, *60*, 512-20

Holt, R.D. (1977) 'Predation, apparent competition and the structure of prey communities', *Theoret. Pop. Biol.*, *12*, 197-229

Hopkins, P.J. (1983) *Invertebrate Diversity and Composition on Fragmented Heathland*, unpublished Ph.D. Thesis, University of Southampton

Horn, H.S. (1968) 'The adaptive significance of colonial nesting in the Brewer's Blackbird *Euphagus cyanocephalus*', *Ecology*, *49*, 682-94

Horn, H.S. (1975) 'Markovian properties of forest succession', in *Ecology and Evolution of Communities* (eds M.L. Cody & J.M. Diamond), Harvard University Press, pp. 196-211

Horn, H.S. (1976) 'Succession', in *Theoretical Ecology: Principles and Applications*, 1st edn (ed. R.M. May), Blackwell, Oxford

Horn, H.S. (1978) 'Optimal tactics of reproduction and life history', in *Behavioural Ecology: an Evolutionary Approach* (eds J.R. Krebs & N.B. Davies), Blackwell, Oxford, pp. 411-29

Horn, H.S. & May, R.M. (1977) 'Limits to similarity among coexisting competitors', *Nature*, *270*, 660-1

Horwood, M.T. & Masters, E.H. (1970) *Sika Deer*, published British Deer Society, 30 pp

Hudson, J.W. (1962) 'The role of water in the biology of the antelope ground squirrel, *Citellus tereticandus*', *University of California Publ. Zool.*, *64*, 1-56

Huey, R.B. & Pianka, E.R. (1974) 'Ecological character displacement in a lizard', *Amer. Zool.*, *14*, 1127-36

Huey, R.B., Pianka, E.R., Egan, M.E., Coons, L.W. (1974) 'Ecological shifts in sympatry: Kalahari fossorial lizards (*Typhlosaurus*)', *Ecology*, *55*, 304-16

Huey, R.B. & Pianka, E.R. (1981) 'Ecological consequences of foraging mode', *Ecology*, *62*, 991-9

Huffaker, C.B. (1958) 'Experimental studies on predation: dispersion factors and predator-prey oscillations', *Hilgardia*, *27*, 343-83

Hughes, R. & De Gilbert, N. (1968) 'A model of an aphid population – a general statement', *J. Anim. Ecol.*, *37*, 553-63

Humphreys, W.F. (1979) 'Production and respiration in animal populations', *J. Anim. Ecol.*, *48*, 427-53

Hurlbert, S.H. (1978) 'The measurement of niche overlap and some derivatives', *Ecology*, *59*, 67-77

Hutchinson, G.E. (1953) 'The Concept of Pattern in Ecology', *Proc. Acad. Nat. Sci. Philadelphia*, *105*, 1-12

Hutchinson, G.E. (1957) 'Concluding remarks', *Cold Spring Harbor Symp. Quant. Biol.*, *22*, 415-27

Hutchinson, G.E. (1959) 'Homage to Santa Rosalia, or why are there so many kinds of animals?', *Amer. Nat.*, *93*, 145-59

Hutchinson, G.E. (1961) 'The paradox of the plankton', *Amer. Nat.*, *95*, 137-45

Janzen, D.H. (1966) 'Coevolution of mutualism between ants and acacias in Central America', *Evolution*, *20*, 249-75

Janzen, D.H. (1967) 'Fire, vegetation structure and ant x acacia interaction in Central America', *Ecology*, *48*, 26-35

Janzen, D.H. (1975) *Ecology of Plants in the Tropics; Studies in Biology*, *58*, Edward Arnold, London

Jarman, P.J. (1974) 'The social organisation of antelope in relation to their ecology', *Behaviour*, *48*, 215-66

Johns, A.D. (1981) 'Effects of selective logging on animal communities', Unpublished report, Cambridge University

Johns, A.D. (1983) Unpublished Ph.D. thesis, Cambridge University

Johnston, D.W. & Odum, E.P. (1956) 'Breeding bird populations in relation to plant succession on the Piedmont of Georgia', *Ecology*, *37*, 50-62

Jones, J.R.E. (1949) 'A further ecological study of calcareous streams in the "Black Mountain" district of South Wales', *J. Anim. Ecol.*, *18*, 142-59

Kamil, A.C. (1979) 'Systematic foraging for nectar by Amakihi (*Loxops virens*)', *J. Comp. Physiol. Psychol.*, *92*, 288-396

Karr, J.R. & James, J.C. (1975) 'Eco-morphological configurations and convergent evolution in species and communities', in *Ecology and Evolution of Communities* (eds. M.L. Cody and J.M. Diamond), Harvard University Press, pp. 258-91

Kemeny, J.G. & Snell, J.L. (1960) *Finite Markov Chains*, Van Nostrand, New York

Ketchum, B.H., Lillick, J. & Redfield, A.C. (1949) 'The growth and optimum yields of unicellular algae in mass culture', *J. Cell. Comp. Physiol.*, *33*, 267-79

Kirkwood, R.S.M. & Lawton, J.H. (1981) 'Efficiency of biomass transfer and the stability of model food-webs', *J. theor. Biol.*, *93*, 225-37

Kitching, R.L. (1977) 'An ecological study of water-filled tree-holes and their position in the woodland ecosystem', *J. Anim. Ecol.*, *40*, 281-302

Kluyver, H.N. (1951) 'The population ecology of the Great tit *Parus m. major* L.', *Ardea*, *39*, 1-135

Koller, D. (1969) 'The physiology of dormancy and survival of plants in desert environments', *Symp. Soc. Exptl. Biol.*, *23*, 449-69

Krebs, C.J. (1971) 'Genetic and behavioural studies on fluctuating vole populations', in *Dynamics of Populations* (eds. P.J. den Boer and G.R. Gradwell), Pudoc, Wageningen, pp. 243-54

Krebs, C.J., Keller, B.L. & Tamarin, R.H. (1969) '*Microtus* population biology: demographic changes in fluctuating populations of *M. ochrogaster* and *M. pennsylvanicus* in Southern Indiana', *Ecology*, *50*, 587-607

Krebs, J.R. (1970) 'Regulation of numbers in the Great Tit (Aves: Passeriformes)', *J. Zool. Lond.*, *162*, 317-33

Krebs, J.R. (1974) 'Colonial nesting and social feeding as strategies for exploiting food resources in the great blue heron (*Ardea herodias*)', *Behaviour*, *51*, 99-134

Krebs, J.R. (1978) 'Optimal foraging: decision rules for predators', in *Behavioural Ecology: an Evolutionary Approach*, Ch. 2 (eds. J.R. Krebs and N.B. Davies), Blackwell, Oxford

Krebs, J.R. & Cowie, R.J. (1976) 'Foraging strategies in birds', *Ardea*, *64*, 98-116

Krebs, J.R., Erichsen, J.T., Webber, M.I. & Charnov, E.L. (1977) 'Optimal prey selection in the Great Tit (*Parus major*)', *Animal Behaviour*, *25*, 30-8

Krebs, J.R., MacRoberts, M.H. & Cullen, J.M. (1972) 'Flocking and feeding in the great tit *Parus major*, an experimental study', *Ibis*, *114*, 507-30

Krefting, L.W., Stenlund, M.H. & Seemel, R.K. (1966) 'Effect of simulated and natural browsing on mountain maple', *J. Wildl. Mgmt.*, *30*, 481-8

Kruuk, H. & Turner, M. (1967) 'Comparative notes on predation by lion, leopard, cheetah and wild dog in the Serengeti area, E. Africa', *Mammalia*, *31*, 1-27

Lack, D. (1943) 'The age of some more British birds', *Br. Birds.*, *36*, 214-21

Lack, D. (1954) *The Natural Regulation of Animal Numbers*, Oxford University Press

Lamprey, H.F. (1963) 'Ecological separation of the large mammal species in the Tarangire Game Reserve Tanganyika', *E. Afr. Wildl. J.*, *1*, 63-92

Law, R. (1979) 'Ecological determinants in the evolution of life histories', in *Population Dynamics* (eds R.M. Anderson *et al.*), BES Symposium, *20*, 81-104

Lawlor, L.R. (1980) 'Overlap, similarity and competition coefficients', *Ecology*, *61*, 245-51

Lawton, J.H. (1973) 'The energy costs of food-gathering', in *Resources and Populations* (eds. B. Benjamin, P.R. Cox & J. Peel)

Lawton, J.H. (1982) 'Vacant niches and unsaturated communities: a comparison of bracken herbivores at sites on two continents', *J. Anim. Ecol.*, *51*, 573-96

Lawton, J.H., Beddington, J.R. & Bonser, R. (1974) 'Switching in invertebrate predators', in *Ecological Stability* (ed. M.B. Usher and M.H. Williamson), Chapman & Hall, London

Leak, W.B. (1970) 'Successional change in northern hardwoods predicted by birth and death simulation', *Ecology*, *51*, 794-801

Le Cren, E.D. (1973) *The Mathematical Theory of the Dynamics of Biological Populations*, Academic Press, London

Leslie, P.H. (1945) 'On the use of matrices in certain population mathematics', *Biometrika*, *33*, 183-212

Leslie, P.H. (1948) 'Some further notes on the use of matrices in population mathematics', *Biometrika*, *35*, 213-45

Leuthold, W. (1978) 'Ecological separation among browsing ungulates in Tsavo East National Park, Kenya', *Oecologia*, *35*, 241-

Levine, S.H. (1976) 'Competitive interactions in ecosystems', *Amer. Nat.*, *110*, 903-10

Levins, R. (1968) *Evolution in Changing Environments*, Princeton University Press

Lewis, C.J. (1977) 'The economics of pesticide research', in *Origins of Pest, Parasite, Disease and Weed Problems* (eds. J.M. Cherrett & G.R. Sagar), British Ecological Society Symposium No. *18*

Lewontin, R.C. (1969) 'The meaning of stability', *Brookhaven Symposia in Biology*, *22*, 13-24

Lindeman, R. (1942) 'The tropho-dynamic aspect of ecology', *Ecology*, *23*, 399-418

Lloyd, L. (1937) 'Observations on sewage flies; their seasonal incidence and abundance', *J. Inst. Sewage Purification* (1937), 1-16

Lloyd, L. (1943) 'Materials for a study in animal competition II The fauna of sewage beds', *Ann. appl. Biol.*, *30*, 47-60

Lloyd, M. & Ghelardi, R.J. (1964) 'A table for calculating the "equitability" component of species diversity', *J. Anim. Ecol.*, *33*, 217-25

Lotka, A.J. (1925) *Elements of Physical Biology*, Williams & Wilkins: Baltimore

Louw, G.N. & Seely, M.K. (1982) *Ecology of Desert Organisms*, Longman, London & New York, 194 pp

Lüscher, M. (1961) 'Air-conditioned termite nests', *Sci. Amer.*, *CCV*, 138-45

MacArthur, R.H. (1955) 'Fluctuations of animal populations and a measure of community stability', *Ecology*, *36*, 533-6

MacArthur, R.H. (1957) 'On the relative abundance of bird species', *Proc. Nat. Acad. Sci. USA*, *43*, 293-5

MacArthur, R.H. (1960) 'On the relative abundance of species', *Amer. Nat.*, *94*, 25-36

MacArthur, R.H. (1970) 'Species packing and competitive equilibrium for many species', *Theoret. Pop. Biol.*, *1*, 1-11

MacArthur, R.H. (1972) *Geographical Ecology*, Harper & Row, New York

MacArthur, R.H. & Connell, J.H. (1966) *The Biology of Populations*, Wiley, New York

MacArthur, R.H. & Levins, R. (1964) 'Competition, habitat selection and character displacement in a patchy environment', *Proc. Nat. Acad. Sci. USA*, *51*, 1207-10

MacArthur, R.H. & Levins, R. (1967) 'The limiting similarity, convergence and divergence of coexisting species', *Amer. Nat.*, *101*, 377-85

MacArthur, R.H. & Wilson, E.O. (1963) 'An equilibrium theory of insular zoogeography', *Evolution*, *17*, 373-87

MacArthur, R.H. & Wilson, E.O. (1967) *The Theory of Island Biogeography*, Princeton University Press

MacFarland, D.J. (1977) 'Decision-making in animals', *Nature*, *269*, 15-21

Mack, R.N. (1976) 'Survivorship of *Cerastium atrovirens* at Aberffraw, Anglesey', *J. Ecol.*, *64*, 309-12

Malthus, T.R. (1798) *An Essay on the Principle of Population*, (Reprinted by Macmillan, New York)

Mann, J.C.E. (1983) 'The Social Organisation and Ecology of the Japanese Sika deer (Cervus nippon) in Southern England', unpublished Ph.D. thesis, University of Southampton

Mann, K.H. (1957) 'The breeding, growth and age structure of a population of the leech, *Helobdella stagnalis* L.', *J. Anim. Ecol.*, *26*, 171-7

May, R.M. (1972a) 'Will a large complex system be stable?', *Nature*, *238*, 413-4

May, R.M. (1972b) 'Limit cycles in predator-prey communities', *Science*, *177*, 900-2

May, R.M. (1973a) 'Qualitative stability in model ecosystems', *Ecology*, *54*, 638-41

May, R.M. (1973b) *Stability and Complexity in Model Ecosystems*, Princeton University Press

May, R.M. (1974) 'On the theory of niche overlap', *Theor. Pop. Biol.*, *5*, 297-332

May, R.M. (1975a) 'Some notes on estimating the competition matrix α', *Ecology*, *56*, 737-41

May, R.M. (1975b) 'Patterns of Species Abundance and Diversity', Ch. 4 in *Ecology and Evolution of Communities* (eds M.L. Cody and J.M. Diamond), Harvard University Press, pp. 81-120

May, R.M. (1976a) 'Mathematical aspects of the dynamics of animal populations', in *Studies in Mathematical Biology* (ed. S.A. Levin), American Mathematical Society

May, R.M. (1976b) 'Models for single populations' *and* 'Models for two interacting populations', Chs. 2 and 5 in *Theoretical Ecology* (ed. R.M. May), Blackwell, Oxford

May, R.M. (1978) 'The evolution of ecological systems', *Scientific American*, *239* (3), 118-35

May, R.M. (1981) 'Models for single populations', Ch. 2 in *Theoretical Ecology: Principles and Applications* (2nd edn), Blackwell, Oxford

May, R.M. & MacArthur, R.H. (1972) 'Niche overlap as a function of environmental variability', *Proc. Nat. Acad. Sci. USA*, *69*, 1109-13

May, R.M. & Anderson, R.M. (1978) 'Regulation and stability of host-parasite population interactions II. Destabilizing processes', *J. Anim. Ecol.*, *47*, 249-68

Maynard Smith, J. (1964) 'Group selection and kin selection: a rejoinder', *Nature*, *201*, 1145-7

Maynard Smith, J. (1976a) 'Group selection', *Q. Rev. Biol.*, *51*, 297

Maynard Smith, J. (1976b) 'Evolution and the theory of games', *Amer. Sci.*, *64*, 41-5

Maynard Smith, J. (1977) 'Parental investment: a prospective analysis', *Animal Behaviour*, *25*, 1-9

McCleery, R.H. (1978) 'Optimal behaviour sequences and decision-making', Ch. 13 in *Behavioural Ecology: an evolutionary approach* (eds J.R. Krebs & N.B. Davies), Blackwell, Oxford

McGinnis, S.M. & Dickson, L. (1967) 'Thermoregulation in the Desert iguana: *Dipsosaurus dorsalis*', *Science*, *156*, 1757-9

McNaughton, S.J. (1976) 'Grazing as an optimisation process: grass-ungulate relationships in the Serengeti', *Amer. Nat.*, *113*, 691-703

McNeill, S. (1973) 'The dynamics of a population of *Leptoterna dolobrata* (Heteroptera: Miridae) in relation to its food resources', *J. Anim. Ecol.*, *42*, 495-507

McNeill, S. & Lawton, J.H. (1970) 'Annual production and respiration in animal populations', *Nature*, *225*, 472-4

McQueen, D.J. (1969) 'Reduction of zooplankton standing stocks by predaceous *Cyclops bicuspidatus thomasi* in Marion Lake, British Columbia', *J. Fish. Res. Bd. Can.*, *26*, 1605-18

Mech, L.D. (1966) 'The wolves of Isle Royale', *U.S. Natl. Park Service Fauna Series*, No. 7

Mech, L.H. (1972) *The Wolf*, Nat. Hist. Press, New York

Menge, B.A. (1972) 'Competition for food between two intertidal starfish species, and its effect on bodysize and feeding', *Ecology*, *53*, 635-44

Metcalfe, J.R. (1972) 'An analysis of the population dynamics of the Jamaican sugar-cane pest *Saccarosydne saccharivora* (Westw.) (Hom., Delphacidae)', *Bull. ent. Res.*, *62*, 73-85

Miller, C.A. (1963) 'Parasites of the spruce budworm' in *The Dynamics of Epidemic Spruce Budworm Populations* (ed. R.F. Morris), *Mem. ent. Soc. Canada*, *31*, 228-44

Miller, R.S. (1967) 'Pattern and process in competition', *Adv. Ecol. Res.*, *4*, 1-74

Mills, N.J. (1981) 'Satiation and the functional response: a test of a new model', *Ecol. Ent.*, *7*, 305-15

Mills, N.J. (1982) 'Voracity, cannibalism and coccinellid predation', in *Natural Enemies and Insect Pest Dynamics* (ed. S.D. Wratten), *Ann. appl. Biol.*, *101*, 144-8

Milne, A. (1961) 'Definition of competition among animals', in *Mechanisms in Biological Competition* (ed. F.L. Milthorpe), *Symp. Soc. Exp. Biol.*, *15*, 40-61

Monteith, J.L. (1968) 'Analysis of the photosynthesis and respiration of field-crops from vertical fluxes of carbon dioxide', in *Functioning of Terrestrial Ecosystems at the Primary Production Level*, (ed. F.E. Eckardt), *Natural Resources Research*, *5*, 349-58, UNESCO

Morris, R.F. (1959) 'Single factor analysis in population dynamics', *Ecology*, *40*, 580-8

Moran, P.A.P. (1953) 'The statistical analysis of the Canadian lynx cycle. 1: Structure and prediction. 2: Synchronisation and meteorology', *Austr. J. Zool.*, *1*, 163-291, 291-98

Motomura, I. (1932) 'A statistical treatment of associations', *Japan J. Zool.*, *44*, 279-383

Mukerji, M.K. & Le Roux, E.J. (1969) 'A quantitative study of food consumption and growth of *Podisus maculiventris* (Hemiptera: Pentatomidae)', *Can. Ent.*, *101*, 387-403

Murdoch, W.W. (1969) 'Switching in general predators: experiments on predator specificity and stability of prey populations', *Ecol. Monogr.*, *39*, 335-54

Murdoch, W.W. & Oaten, A. (1975) 'Predation and population stability', *Adv. Ecol. Res.*, *9*, 2-131

Murie, A. (1944) 'The Wolves of Mt. McKinley', *Fauna Natn. Pks. U.S.*, No. 5

Neyman, J., Park, T. & Scott, E.L. (1956) 'Struggle for existence. The *Tribolium* model: biological and statistical aspects', *Proc. 3rd Berkeley Symp. on Mathematical Statistics and Probability*, Vol. *IV*, pp. 41-79

Nicholson, A.J. (1933) 'The balance of animal populations', *J. Anim. Ecol.*, *2*, 132-78

Nicholson, A.J. (1950) 'Competition for food amongst *Lucilia cuprina* larvae', *8th Int. Cong. Ent.* (Stokholm), 277-81

Nicholson, A.J. (1954) 'An outline of the dynamics of animal populations', *Austr. J. Zool.*, *2*, 9-65

Nicholson, A.J. (1955) 'Compensating reactions of populations to stresses, and their evolutionary significance', *Austr. J. Zool.*, *2*, 1-8

Nilsson, L. (1969) 'Food consumption of diving ducks wintering at the coast of south Sweden', *Oikos*, *20*, 128-35

Noy-Meir, I. (1975) 'Stability of grazing systems: an application of predator-prey graphs', *J. Ecol.*, *63*, 459-81

Oaten, A. & Murdoch, W.W. (1975) 'Switching, functional response and stability in predator-prey systems', *Amer. Nat.*, *109*, 299-318

Odum, E.P. (1959/1971) *Fundamentals of Ecology*, 2nd edn./3rd edn., W.B. Saunders, New York

Odum, E.P. (1969) 'The strategy of ecosystem development', *Science*, *164*, 262-70

Odum, E.P. (1975) 'Diversity as a function of energy flow', in *Unifying Concepts in Ecology* (eds W.H. Van Dobben & R.H. Lowe-McConnell), Junk, The Hague, pp. 11-14

Odum, E.P., Connell, C.E., Davenport, L.B. (1962) 'Population energy flow of three primary consumer components of old-field ecosystems', *Ecology*, *43*, 88-96

Odum, H.T. (1957) 'Trophic structure and productivity of Silver Springs, Florida', *Ecol. Monogr.*, *27*, 55-112

Orians, G.H. (1969) 'On the evolution of mating systems in birds and mammals', *Amer. Nat.*, *103*, 589-604

Orians, G.H. (1975) 'Diversity, stability and maturity in natural ecosystems', in *Unifying Concepts in Ecology* (eds W.H. Van Dobben & R.H. Lowe-McConnell), Junk, The Hague, pp. 139-50

Orians, G.H. & Horn, H.S. (1969) 'Overlap in foods and foraging of four species of blackbirds in the potholes of Central Washington', *Ecology*, *50*, 930-8

Osman, R.W. (1977) 'The establishment and development of a marine epifaunal community', *Ecol. Monogr.*, *47*, 37-63

Owen, D.F. & Weigert, R.G. (1976) 'Do consumers maximise plant fitness?', *Oikos*, *27*, 488-92

Owen, D.F. & Wiegert, R.G. (1982) 'Grasses and grazers: is there a mutualism?', *Oikos*, *38*, 258-9

Paine, R.T. (1966) 'Food web complexity and species diversity', *Amer. Nat.*, *100*, 65-75

Paine, R.T. (1969) 'A note on trophic complexity and community stability', *Amer. Nat.*, *103*, 91-3

Paine, R.T. (1974) 'Intertidal community structure', *Oecologia*, *15*, 93-120

Paine, R.T. (1980) 'Food webs: linkage, interaction strength and community infra-structure', *J. Anim. Ecol.*, *49*, 667-86

Park, T. (1948) 'Experimental studies of interspecies competition I. Competition between populations of the flour beetles, *Tribolium confusum* (Duval) and *Tribolium castaneum* (Herst.)', *Ecol. Monogr.*, *18*, 267-307

Park, T. (1954) 'Experimental studies of interspecies competition. II Temperature, humidity and competition in two species of *Tribolium*', *Physiol. Zool.*, *27*, 177-238

Park, T. (1962) 'Beetles, competition and populations', *Science*, *138*, 1369-75

Park, T., Leslie, P.H. & Mertz, D.B. (1964) 'Genetic strains and competition in populations of *Tribolium*', *Physiol. Zool.*, *37*, 97-162

Patrick, R. (1963) 'The structures of diatom communities under varying ecological conditions', *Ann. N.Y. Acad. Sci.*, *108*, 353-8

Patrick, R. (1975) 'Stream Communities', in *Ecology and Evolution of Communities* (eds. M.L. Cody and J.M. Diamond), Harvard University Press, pp. 445-59

Patterson, I.J. (1965) 'Timing and spacing of broods in the black-headed gull *Larus ridibundus*', *Ibis*, *107*, 433-59

Pelikan, J. (1970) 'Embryonic resorption in *Microtus arvalis* (Pall.)', *Zool. Listy*, *19*, 93-102

Perrins, C.M. (1971) 'Population studies of the Great Tit, *Parus major*', in *Dynamics of Populations* (eds. P.J. den Boer and G.R. Gradwell), Pudoc, Wageningen, pp. 524-31

Peterken, G.F. & Tubbs, C.F. (1965) 'Woodland regeneration in the New Forest, Hampshire since 1650', *J. appl. Ecol.*, *2*,

Petrusewicz, K. (1967) 'Suggested list of more important concepts in productivity studies, (definitions and symbols)', in *Secondary Productivity in Terrestrial Ecosystems*, Vol. 1 (ed. K. Petrusewicz), Panstwowe Wydawnictwo Naukowe, Warsaw, pp. 51-8

Phillipson, J. (1966) *Ecological Energetics*, Edward Arnold, *Studies in Biology*, *1*

Pianka, E.R. (1969) 'Sympatry of desert lizards (*Ctenotus*) in Western Australia', *Ecology*, *50*, 1012-30

Pianka, E.R. (1970) 'On r- and K-selection', *Am. Nat.*, *104*, 592-7

Pianka, E.R. (1973) 'The structure of lizard communities', *Ann. Rev. Ecol. Syst.*, *4*, 53-74

Pianka, E.R. (1975) 'Niche relations of desert lizards', in *Ecology and Evolution of Communities* (eds M.L. Cody and J.M. Diamond), Harvard University Press, pp. 291-314

Pianka, E.R. (1976 & 1981) 'Competition and Niche Theory', in *Theoretical Ecology: Principles and Applications* (ed. R.M. May), Blackwell, Oxford

Pianka, E.R. (1978) *Evolutionary Ecology*, 2nd edn, Harper & Row, 397 pp

Pianka, E.R. (1980) 'Guild structure in desert lizards', *Oikos*, *35*, 194-201

Pianka, E.R., Huey, R.B. and Lawlor, L.R. (1979) 'Niche segregation in desert lizards', in *Analysis of Ecological Systems* (eds. D.J. Horn, R. Mitchell and G.R. Stairs), Ohio State Univ. Press, pp. 67-115

Pickett, C.H. & Clark, W.D. (1979) 'The function of extra-floral nectaries in *Opuntia acanthocarpa* (Cactaceae)', *Amer. J. Bot.*, *66*, 618-25

Pielou, E.C. (1969) *An Introduction to Mathematical Ecology*, Wiley-Interscience, New York

Pielou, E.C. (1975) *Ecological Diversity*, Wiley-Interscience, New York

Pimentel, D. (1961) 'Species diversity and insect population outbreaks', *Ann. Ent. Soc. Amer.*, *54*, 76-86

Pimentel, D., Feinberg, E.H., Wood, P.W. & Hayes, J.T. (1965) 'Selection, spatial distribution and the coexistence of competing fly species', *Amer. Nat.*, *99*, 97-109

Pimlott, D.H. (1967) 'Wolf predation and ungulate populations', *Amer. Zool.*, *7*, 267-78

Pimm, S.L. (1979a) 'The structure of food webs', *Theor. Pop. Biol.*, *16*, 144-58

Pimm, S.L. (1979b) 'Complexity and Stability: another look at MacArthur's original hypothesis', *Oikos*, *33*, 351-7

Pimm, S.L. (1980) 'Properties of food webs', *Ecology*, *61*, 219-25

Pimm, S.L. (1983) 'Niche relationships of tropical humming-birds', in press

Pimm, S.L. & Lawton, J.H. (1977) 'The number of trophic levels in ecological communities', *Nature*, *268*, 329-31

Pimm, S.L. & Lawton, J.H. (1978) 'On feeding on more than one trophic level', *Nature*, *275*, 542-4

Pimm, S.L. & Lawton, J.H. (1980) 'Are food webs divided into compartments?', *J. Anim. Ecol.*, *49*, 879-98

Pollard, J.H. (1973) *Mathematical Models for the Growth of Human Populations*, Cambridge University Press

Podoler, H. & Rogers, D. (1975) 'A new method for the identification of key factors from life table data', *J. Anim. Ecol.*, *44*, 85-115

Pontin, A.J. (1969) 'Experimental transplanation of nestmounds of *Lasius flavus* (F)', *J. Anim. Ecol.*, *38*, 747-54

Poole, R.W. (1974) *An Introduction to Quantitative Ecology*, McGraw-Hill

Potts, G.R. & Vickerman, G.P. (1974) 'Studies on the cereal ecosystem', *Adv. Ecol. Res.*, *8*, 108-97

Pratt, D.M. (1943) 'Analysis of population development in *Daphnia* at different temperatures', *Biol. Bull.*, *85*, 116-40

Preston, F.W. (1948) 'The commonness and rarity of species', *Ecology*, *29*, 254-83

Prosser, C.L. (1973) *Comparative Animal Physiology*, 3rd edn, W.B. Saunders, Philadelphia

Pulliam, H.R. (1975a) 'Coexistence of sparrows: a test of community theory', *Science*, *189*, 474-6

Pulliam, H.R. (1975b) 'Diet optimization with nutrient constraints', *Amer. Nat.*, *109*, 765-8

Putman, R.J. (1977) 'Dynamics of the blowfly, *Calliphora erythrocephala* within carrion', *J. Anim. Ecol.*, *46*, 853-66

Putman, R.J. (1981) 'Social systems of deer: a speculative review', *Deer*, *5*, 186-8

Putman, R.J. (1983a) 'The geography of animal communities', in *Themes in Biogeography* (ed. J.A. Taylor), Croom Helm, London

Putman, R.J. (1983b) *Carrion and Dung: the Decomposition of Animal Wastes*, Edward Arnold, *Studies in Biology*, *156*

Putman, R.J. (1983c) 'Efficient exploitation of natural populations of Fallow deer: decision rules', in *The Biology of Deer Production*, (eds P.F. Fennessy and K.R. Drew), N.Z. Soc. Animal Production, Dunedin

Putman, R.J., Edwards, P.J., Ekins, J.R. & Pratt, R.M. (1982a) 'Food and feeding behaviour of cattle and ponies in the New Forest: a study of the inter-relationships between the large herbivores of the forest and their vegetational environment', Report HF3/03/127 to Chief Scientists Team, Nature Conservancy Council, Huntingdon, England

Putman, R.J., Pratt, R.M., Ekins, J.R. & Edwards, P.J. (1982b) 'Habitat use and grazing by free-ranging cattle and ponies in the New Forest, Hampshire, and impact upon the forest vegetation', *Proc. III Int. Theriol. Congress*, Helsinki, August 1982

Rappoldt, C. & Hogeweg, P. (1980) 'Niche packing and number of species', *Amer. Nat.*, *116*, 480-92

Reynoldson, T.B. & Bellamy, L.S. (1970) 'The establishment of interspecific competition in field populations with an example of competition in action between *Polycelis nigra* (Mull.) and *P. tenuis* (Ijima) (Turbellaria, Tricladida)', in *Dynamics of Populations*, (eds P.J. den Boer and G.R. Gradwell), Pudoc, Wageningen, pp. 282-97

Reynoldson, T.B. & Davies, R.W. (1970) 'Food niche and coexistence in lake-dwelling triclads', *J. Anim. Ecol.*, *39*, 599-617

Ricklefs, R.E. (1971) *Ecology*, Nelson

Ricklefs, R.E. & Travis, J. (1980) 'A morphological approach to the study of avian community organisation', *Auk.*, *97*, 321-38

Ridley, M. (1978) 'Paternal care', *Animal Behaviour*, *26*, 904-32

Rivard, I. (1962) 'Some effects of prey density on survival, speed of development, and fecundity of the predaceous mite *Melichares dentriticus* (Berl.) (Acarina: Aceosejidae)', *Can. J. Zool.*, *40*, 1233-6

Roff, D.A. (1974) 'Spatial heterogeneity and the persistence of populations', *Oecologia*, *15*, 245-58

Rogers, D.J. (1972) 'Random search and insect population models', *J. Anim.*

Ecol., *41*, 369-83

Rogers, D.J. & Hassell, M.P. (1974) 'General models for insect parasite and predator searching behaviour: interference', *J. Anim. Ecol.*, *43*, 239-53

Root, R.B. (1967) 'The niche exploitation pattern of the blue-grey gnatcatcher', *Ecol. Monogr.*, *37*, 317-50

Roth, V.L. (1981) 'Constancy in the size-ratios of sympatric species', *Amer. Nat.*, *118*, 394-404

Roughgarden, J. (1974) 'Species packing and the competition function with illustrations from coral reef fish', *Theor. Pop. Biol.*, *5*, 163-86

Roughgarden, J. & Feldman, M. (1975) 'Species packing and predation pressure', *Ecology*, *56*, 489-92

Routledge, R.D. (1979) 'Diversity indices: which ones are admissible?', *J. Theor. Biol.*, *76*, 503-15

Royama, T. (1970) 'Factors governing the hunting behaviour and selection of food by the Great Tit (*Parus major* L.)', *J. Anim. Ecol.*, *39*, 619-68

Sanders, H. (1968) 'Marine benthic diversity: a comparative study', *Amer. Nat.*, *102*, 243-82

Sandness, J.N. & McMurtry, J.A. (1970) 'Functional response of three species of Phytoseiidae (Acarina) to prey density', *Can. Ent.*, *102*, 692-704

Sang, J.H. (1950) 'Population growth in *Drosophila* cultures', *Biol. Rev.*, *25*, 188-219

Saunders, P.T. (1978) 'Population dynamics and the length of food chains', *Nature*, *272*, 189-90

Schaller, G. (1972) *The Serengeti Lion: a study of predator-prey relations*, University of Chicago Press

Scheffer, V.B. (1951) 'The rise and fall of a reindeer herd', *Sci. Monthly*, *73*, 356-62

Schimitschek, E. (1931) 'Forstentomologische untersuchungen aus dem Gebiete von Lunz. I. Standortsklima und Kleinklima in ihren Beziehungen zunn Entwicklungsablauf und zur Mortalität von Insekten', *Z. f. angew Entomol.*, *18*, 460-91

Schmidt-Nielsen, B. & Schmidt-Nielsen, K. (1952) 'Water metabolism of desert mammals', *Physiol. Rev.*, *32.*, 135-66

Schoener, T.W. (1965) 'The evolution of bill size differences among sympatric, congeneric species of birds', *Evolution*, *19*, 189-213

Schoener, T.W. (1968) 'The *Anolis* lizards of Bimini: resource partitioning in a complex fauna', *Ecology*, *49*, 704-26

Schoener, T.W. (1969) 'Models of optimal size for solitary predators', *Amer. Nat.*, *103*, 277-313

Schoener, T.W. (1970) 'Nonsynchronous spatial overlap of lizards in patchy habitats', *Ecology*, *51*, 408-18

Schoener, T.W. (1974) 'Resource partitioning in ecological communities', *Science*, *185*, 27-39

Schultz, A.M. (1964) 'The nutrient-recovery hypothesis for arctic microtine cycles', in *Grazing in Terrestrial and Marine Environments* (ed. D.J. Crisp), Blackwell, Oxford, pp. 57-68

Scientific American (ed.) (1970) *The Biosphere*, W.H. Freeman

Severinghaus, C.W. (1951) 'A study of productivity and mortality of corralled deer', *J. Wildl. Magmt.*, *15*, 73-80

Shannon, C.E. & Weaver, W. (1949) *The Mathematical Theory of Communication*, University of Illinois Press, 117 pp.

Shelford, V.E. (1913) 'The reactions of certain animals to gradients of evaporating power and air. A study in experimental ecology', *Biol. Bull.*, *25*, 79-120

Silliman, R.P. & Gutsell, J.S. (1957) 'Response of laboratory fish populations to fishing rates', *Trans 22 N. Am. Wildl. Conf.*, 464-71

Silliman, R.P. & Gutsell, J.S. (1958) 'Experimental exploitation of fish populations', *Fish. Bull. U.S.*, *58*, 214-52

Silvertown, J.W. (1982) *Introduction to Plant Population Ecology*, Longman

Simberloff, D.S. (1976) 'Trophic structure determination and equilibrium in an arthropod community', *Ecology*, *57*, 395-8

Simberloff, D.S. (1978a) in *Diversity of Insect Faunas* (eds L.A. Mound and N. Waloff), *Symp. R. ent. Soc. Lond.*, *9*, 139-53

Simberloff, D.S. (1978b) 'Using island biogeographic distributions to determine if colonisation is stochastic', *Amer. Nat.*, *112*, 713-26

Simberloff, D.S. & Wilson, E.O. (1969) 'Experimental zoogeography of islands: the colonisation of empty islands', *Ecology*, *50*, 278-96

Simpson, E.H. (1949) 'Measurement of diversity', *Nature*, *163*, 688

Slatkin, M. (1980) 'Ecological character displacement', *Ecology*, *61*, 163-77

Slobodkichoff, C.N. and Schulz, W.C. (1980) 'Measures of niche overlap', *Ecology*, *61*, 1051-5

Slobodkin, L.B. (1961) *Growth and Regulation of Animal Populations*, Holt, Rinehart and Winston

Slobodkin, L.B. (1968) 'How to be a predator', *Amer. Zool.*, *8*, 43-57

Slobodkin, L.B. & Richman, S. (1956) 'The effect of removal of fixed percentages of the newborn on size and variability in populations of *Daphnia pulicaria*', *Limnol. Oceanogr.*, *1*, 209-37

Slobodkin, L.B. & Sanders, H.L. (1969) 'On the contribution of environmental predictability to species diversity', in *Brookhaven Symposia in Biologica*, *22*, 82-95

Smith, F.E. (1972) 'Spatial heterogeneity, stability and diversity in ecosystems', in *Growth by Intussusception: Ecological essays in honor of G. Evelyn Hutchinson* (ed. E.S. Deevey), *Trans. Conn. Acad. Arts. Sci.*, *44*, 309-35

Smith, J.N.M. & Sweatman, H.P.A. (1974) 'Food searching behaviour of titmice in patchy environments', *Ecology*, *55*, 1216-32

Solbrig, P.T. & Simpson, B.B. (1974) 'Components of regulation of a population of dandelions in Michigan', *J. Ecol.*, *62*, 473-86

Sømme, L. & Conradi-Larsen, E.M. (1977) 'Cold hardiness of collembolans and oribatid mites from windswept mountain ridges', *Oikos*, *29*, 118-26

Southern, H.N. and Lowe, V.P.W. (1982) 'Predation by Tawny Owls (*Strix aluco*) on Bank Voles (*Clethriononys glaseolus*) and Wood Mice (*Apodemus sylvaticus*)' *J. Zool. London 198*, 83-102

Southwood, T.R.E. (1961) 'The number of species of insect associated with various trees', *J. Anim. Ecol.*, *30*, 1-8

Southwood, T.R.E. (1975) 'The dynamics of insect populations', in *Insects, Science and Society* (ed. D. Pimentel), Academic Press, New York

Southwood, T.R.E. (1977) 'The relevance of population dynamic theory to pest status', in *Origins of Pest, Parasite, Disease and Weed Problems* (eds J.N. Cherrett and G.R. Jagar), Blackwell, Oxford, pp. 35-54

Southwood, T.R.E. (1978) *Ecological Methods*, 2nd edn, Chapman and Hall, London

Southwood, T.R.E. (1979) 'Seventh Bawden Lecture – Pesticide Usage, Prodigal or Precise', *Proc. 1979 Brit. Crop. Prot. Conf. – Pests & Diseases*, 603-19

Southwood, T.R.E. (1976/81) 'Bionomic strategies and population parameters', in *Theoretical Ecology*, 2nd edn, (ed. R.M. May), Blackwell, Oxford, Ch. 3

Southwood, T.R.E. & Comins, H.N. (1976) 'A synoptic population model', *J. Anim. Ecol., 45*, 949-65

Southwood, T.R.E., May, R.M., Hassell, M.P. & Conway, G.R. (1974) 'Ecological strategies and population parameters', *Amer. Nat., 108*, 791-804

Spellerberg, I.F. (1972) 'Temperature tolerances of Southeast Australian reptiles examined in relation to reptile thermoregulatory behaviour and distribution', *Oecologia, 9*, 23-46

Spellerberg, I.F. & Hoffmann, K. (1972) 'Circadian rhythm in lizard critical minimum temperature', *Die Naturwissenschaften, 59*, 517-8

Stenseth, N.C. (1979) 'Where have all the species gone? On the nature of extinction and the Red Queen Hypothesis', *Oikos, 33*, 196-227

Stenseth, N.C. & Hansson, L. (1979) 'Optimal food selection: a graphic model', *Amer. Nat., 113*, 373-89

Stoddart, L.A., Smith, A.D. & Box, T.W. (1975) *Range Management*, 3rd edn, McGraw-Hill, New York

Stubbs, M. (1977) 'Density-dependence in the life-cycles of animals and its importance in K- and r-strategies', *J. Anim. Ecol., 46*, 677-88

Sugihara, G. (1980) 'Minimal Community Structure: an explanation of species-abundance patterns', *Amer. Nat., 116*, 770-87

Sunderland, K.D. & Vickerman, G.P. (1980) 'Aphid feeding by some polyphagous predators in relation to aphid density in cereal fields', *J. appl. Ecol., 17*, 389-96

Svihla, A., Bowman, H.R. & Ritenour, R. (1951) 'Prolongation of clotting time in dormant estivating mammals', *Science, 114*, 298-99

Swift, M.J., Heal, O.W. & Anderson, J.M. (1979) *Decomposition in Terrestrial Ecosystems*, Blackwell, Oxford

Symonides, E. (1979) 'The structure and population dynamics of psammophytes on inland dunes: II Loose-sod populations', *Ekol. Pol., 27*, 191-234

Takahashi, F. (1968) 'Functional response to host density in a parasitic wasp, with reference to population regulation', *Res. Popul. Ecol., 10*, 54-68

Taylor, C.R. (1972) 'The Desert Gazelle: a paradox resolved', *Symp. Zoo. Soc. Lond., 31*, 215-27

Taylor, R.A.J. & Taylor, L.R. (1979) 'A behavioural model for the evolution of spatial dynamics', in *Population Dynamics* (eds R.M. Anderson, B.D. Turner and L.R. Taylor), *BES Symposium, 20*, 1-28

Teal, J.M. (1959) 'Community metabolism in a temperate cold spring', *Ecol. Monogr., 27*, 283-302

Thorman, S. (1982) 'Niche dynamics and resource partitioning in a fish guild inhabiting a shallow estuary on the Swedish West Coast', *Oikos, 39*, 32-9

Thompson, D.J. (1975) 'Towards a predator-prey model incorporating age structure: the effects of predator and prey size on the predation of *Daphnia magna* by *Ischnura elegans*', *J. Anim. Ecol., 44*, 907-16

Toole, E.H. & Brown, E. (1946) 'Final results of the Duvel buried seed experiment', *J. Agric. Research, 72*, 201-10

Toth, R.S. & Chew, R.M. (1972) 'Development and energetics of *Notonecta undulata* during predation on *Culex tarsalis*', *Ann. ent. Soc. Am.*, *65*, 1270-9

Trivers, R.L. (1971) 'The evolution of reciprocal altruism', *Q. Rev. Biol.*, *46*, 35-57

Trivers, R.L. (1972) 'Parental investment and sexual selection', in *Sexual selection and the Descent of Man* (ed. B. Campbell), Aldine, Chicago, reproduced pp. 52-97 in *Readings in Sociobiology*, (eds T.H. Clutton Brock & P. Harvey), W.H. Freeman (1978)

Turnbull, A.L. (1962) 'Quantitative studies of the food of *Linyphia triangularis* Clerck (Aranae: Linyphiidae)', *Can. Ent.*, *94*, 1233-49

Turner, M. & Polis, G.A. (1979) 'Patterns of coexistence in a guild of raptorial spiders', *J. Anim. Ecol.*, *48*, 509-20

Ullyett, G.C. (1949) 'Distribution of progeny by *Cryptus inornatus* Pratt (Hymenoptera: Ichneumonidae)', *Can. Ent.*, *81*, 285-99

Ullyett, G.C. (1950) 'Competition for food and allied phenomena in sheep blowfly populations', *Phil. Trans.*, *234*, 77-175

Usher, M.B. & Williamson, M.H. (eds) (1974) *Ecological Stability*, Chapman and Hall, London

Vandermeer, J.H. (1972) 'Niche theory', *Ann. Rev. Ecol. Syst.*, *3*, 107-32

Vandermeer, J.H. (1980) 'Indirect mutualism: variations on a theme by Stephen Levine', *Amer. Nat.*, *116*, 441-8

Van Dobben, W.H. & Lowe-McConnell, R.H. (eds) (1975) *Unifying Concepts in Ecology*, Junk, The Hague

Van de Veen, H.E. (1979) *Food Selection and Habitat Use in the Red Deer (Cervus elaphus* L.), Ph.D. thesis, Rijksuniversiteit te Groningen

Van Valen, L. (1976) 'The Red Queen lives', *Nature*, *260*, 575

Van Valen, L. (1977) 'The Red Queen', *Amer. Nat.*, *111*, 809-10

Varley, G.C. (1947) 'The natural control of population balance in the Knapweed Gall-fly (*Urophora jaceana*)', *J. Anim. Ecol.*, *16*, 139-87

Varley, G.C. & Gradwell, G.R. (1960) 'Key factors in population studies', *J. Anim. Ecol.*, *29*, 399-401

Varley, G.C. & Gradwell, G.R. (1968) 'Population models for the winter moth', *Symp. R. Ent. Soc. London*, *4*, 132-42

Varley, G.C. & Gradwell, G.R. (1970) 'Recent advances in insect population dynamics', *Ann. Rev. Ent.*, *15*, 1-24

Varley, G.C., Gradwell, G.R. & Hassell, M.P. (1973) *Insect Population Ecology an Analytical Approach*, Blackwell, Oxford

Vaurie, C. (1951) 'Adaptive differences between two sympatric species of nuthatches (*Sitta*)', *Proc. Xth Int. Ornith. Congr.*, 163-6

Vickerman, G.P., Potts, G.R. & Sunderland, K.D. (1983) 'Changes in the insect fauna of cereals, 1970-1980', in prep.

Vincent, T.L. & Anderson, L.R. (1979) 'Return time and vulnerability for a food chain model', *Theor. Pop. Biol.*, *15*, 217-31

Vine, A.A. (1971) 'The risk of visual detection and pursuit by a predator and the selective advantage of flocking behaviour', *J. Theor. Biol.*, *30*, 405-22

Volterra, V. (1926) 'Variations and fluctuations of the number of individuals in animal species living together', *J. Cons. perm. int. Ent. Mer.*, *3*, 3-51, reprinted in Chapman, R.N. (1931) *Animal Ecology*, McGraw-Hill, New York

Waggoners, P.E. & Stephens, G.R. (1970) 'Transition probabilities for a forest', *Nature*, *255*, 1160-1

Walter, H. (1973) *Vegetation of the earth in relation to climate and the eco-physiological conditions*, Springer-Verlag, New York

Ward, P. (1965) 'Feeding ecology of the Black-faced Dioch *Quelea quelea* in Nigeria', *Ibis, 107*, 173-214

Wareing, P.P. (1966) 'Ecological aspects of seed dormancy and germination', in *Reproductive Biology and Taxonomy of Vascular Plants*, (ed. J.G. Hawkes), Pergamon, Oxford

Watkinson, A.R. & Harper, J.L. (1978) 'The demography of a sand dune annual *Vulpia fasiculata*: I The natural regulation of populations', *J. Ecol., 66*, 15-33

Watson, A. (1967) 'Population control by territorial behaviour in red grouse', *Nature, 215*, 1274-5

Watt, A.S. (1947) 'Pattern and process in the plant community', *J. Ecol., 35*, 1-22

Watt, K.E.F. (1955) 'Studies on population productivity. I Three approaches to the optimum yield problem in populations of *Tribolium confusum*', *Ecol. Monogr., 25*, 269-90

Watt, K.E.F. (1962) 'The conceptual formulation and mathematical solution of practical problems in population input-output dynamics', in *The Exploitation of Natural Animal Populations*, (eds E.D. Le Cren & M.W. Holdgate), Wiley, New York

Webb, D.J. (1974) 'The statistics of relative abundance and diversity', *J. Theor. Biol., 43*, 277-92

Werner, E.E. & Hall, D.J. (1974) 'Optimal foraging and the size selection of prey by the Bluegill Sunfish (*Lepomis macrochirus*)', *Ecology, 55*, 1042-52

Westoby, M. (1974) 'An analysis of diet selection by large generalist herbivores', *Amer. Nat., 108*, 290-304

White, T.C.R. (1978) 'The importance of a relative shortage of food in animal ecology', *Oecologia, 33*, 71-86

Whittaker, J.B. (1979) 'Invertebrate grazing, competition and plant dynamics', in *Population Dynamics* (eds R.M. Anderson, B.D. Turner & L.R. Taylor), *BES Symp., 20*, 207-22

Whittaker, R.H. (1965) 'Dominance and diversity in land plant communities', *Science, 147*, 250-60

Whittaker, R.H. (1970/1975) *Communities and Ecosystems*, 1st/2nd edn, Collier Macmillan

Whittaker, R.H. & Likens, G.E. (1973) 'Primary production: the biosphere and Man', *Human Ecol., 1*, 357-69

Wiens, D. & Rourke, J.P. (1978) 'Rodent pollination in Southern African *Protea* spp.', *Nature, 276*, 71-3

Williams, C.B. (1953) 'The relative abundance of different species in a wild animal population', *J. Anim. Ecol., 22*, 14-31

Williams, C.B. (1964) *Patterns in the Balance of Nature*, Academic Press

Williamson, M. (1981) *Island Ecology*, Oxford University Press

Wilson, E.O. (1975) *Sociobiology, the New Synthesis*, Belknap Press, Harvard

Wolff, J.O. (1978) 'Burning and browsing effects on willow growth in interior Alaska', *J. Wildl. Magmt., 42*, 135-40

Wood, D. (1977) *Energetics of the grey squirrel (Sciurns carolinensis)*, unpublished thesis, Oxford University

Woodwell, G.M. & Dykeman, W.R. (1966) 'Respiration in a forest measured by carbon dioxide accumulation during temperature inversions', *Science, 154,* 1031-4

Wratten, S.D. (1973) 'The effectiveness of the coccinellid beetle, *Adalia bipunctata* (L.) as a predator of the lime aphid, *Eucallipterus tiliae* L.', *J. Anim. Ecol., 42,* 785-802

Wratten, S.D. & Fry, G.L.A. (1980) *Field and Laboratory Exercises in Ecology,* Edward Arnold, London

Wratten, S.D., Goddard, P. & Edwards, P.J. (1982) 'British trees and insects: the role of palatability', *Am. Nat., 118,* 916-9

Wratten, S.D. & Pearson, J. (1982) 'Predation of sugar beet aphids in New Zealand', in *Natural Enemies in Insect Pest Dynamics* (ed. S.D. Wratten), *Proc. Ann. appl. Biol., 101,* 178-81

Wratten, S.D. & Watt, A.D. (1984) *Ecological Basis of Pest Control,* George Allen and Unwin

Wright, D.W., Hughes, R.D. & Worrall, J. (1960) 'The effects of certain predators on the numbers of cabbage root fly (*Erioischia brassicae* (Bouché)) and on the subsequent damage caused by the pest', *Ann. appl. Biol., 48,* 756-63

Wylie, H.G. (1965) 'Some factors that reduce the reproductive rate of *Nasonia vitripennis* (Walk.) at high adult population densities', *Can. Ent., 97,* 970-7

Wynne Edwards, V.C. (1962) *Animal Dispersion in Relation to Social Behaviour,* Oliver and Boyd, Edinburgh

Wynne Edwards, V.C. (1963) 'Intergroup selection in the evolution of social systems', *Nature, 200,* 623-6

Yeaton, R.I. (1978) 'A cyclic relationship between *Larrea tridentata* and *Opuntia leptocaulis* in the northern Chihuahuan Desert', *J. Ecol., 66,* 651-6

Yoda, K., Kira, T., Ogawa, H. & Hozumi, H. (1963) 'Self-thinning in overcrowded pure stands under cultivated and natural conditions', *J. Biol. Osoka Univ., 14,* 107-29

Yoshiyama, R.M. & Roughgarden, J. (1977) 'Species packing in two dimensions', *Amer. Nat., 111,* 107-21

Young, A.M. & Muyshondt, A. (1972) 'Biology of *Morpho polyphemus* (Lepidoptera: Morphidae) in El Salvado', *J. N.Y. Ent. Soc., 80,* 18-42

Yount, J.L. (1956) 'Factors that control species numbers in Silver Springs, Florida', *Limnol. Oceanogr., 1,* 286-95

Index

Acknowledgements

We would like to acknowledge with thanks the kindness of many individuals and publishers in allowing us to reproduce figures and tables in this volume. Specifically, we thank:

Dr. H.G. Cogger: Figs 1.14, 1.15;

Dr J.B. Graham for photographs of fish on which is based Fig. 1.10;

Dr. A.D. Johns: Tables 13.3, 13.4, 13.5;

Dr. G.P. Vickerman: Figs 7.46, 9.2, Table 9.1;

Prof. D. Wiens for photograph of *Aethomys namaquensis*: Fig. 2.2;

D.P. Wilson/E. Hosking: Fig. 2.1, photograph of *Calliactis*

Academic Press: Figs 7.19, 7.22, from Dempster, J.P. (1975) *Animal Population Ecology*; Figs 8.5, 8.6, from Harper, J.L. (1977) *Population Biology of Plants*; Figs 9.25, 9.26, 9.27a, from *Advances in Ecological Research 8*, 108-97 (1974);

Edward Arnold Ltd: Figs 8.8, 8.9, 9.13, 9.19; from Hassell, M.P. (1976) *Dynamics of Competition and Predation*;

American Association for the Advancement of Science: Fig. 1.11, from *Science, 164*, 262-70 (1969);

Association of Applied Biologists: Fig. 9.17, from *Annals of Applied Biology 101*, 144-8 (1982);

Blackwell Scientific Publications: Figs 7.2, 7.17, 7.18, 9.4, 9.5, 9.20, 9.22, 9.33, 9.34, from Varley, G.C., Gradwell, G.R. & Hassell, M.P. (1973), *Insect Population Ecology*; Figs 2.8, 5.2, 5.4, 5.6, 5.9, 5.10, 6.6, 6.7, 6.12, 7.6, 10.8, 10.9, 13.1, 13.2, Tables 2.2, 7.1, 13.2, reprinted from May, R.M. (1976/81), *Theoretical Ecology, Principles and Applications*; Figs 10.3, 10.5; reprinted from Krebs, J.R. & Davies, N.B. (1978), *Behavioural Ecology, An Evolutionary Approach*;

Blackwell Scientific Publications and the British Ecological Society: Fig. 9.2, from Southwood, T.R.E. (1977), 'The relevance of population dynamic theory to pest status', in *Origins of Pest, Parasite, Disease and Weed Problems* (eds.) J.M. Cherrett & G.R. Sagar; Figs 2.3, 4.7, 7.10, 7.11, 7.12, 7.13, 7.14, 7.21, 7.23, 7.43, 8.3, 8.4, 8.7, 9.7, 9.8, 9.9, 9.12, 9.14, 9.15, 9.16, 9.21, 9.27(b), 9.28, 9.29, 9.30, 10.1, 10.2, 10.4, 10.5, 11.2, Tables 7.7. 7.8, 7.9, 9.4, 12.1, from *Journal of Animal Ecology*, *Journal of Applied Ecology* or *Journal of Ecology*;

British Crop Protection Council: Fig. 9.1; from Southwood, T.R.E. (1979) 'Seventh Bawden Lecture — Pesticide Usage, Prodigal or Precise', *Proc. 1979 Brit. Crop Prot. Conf. — Pests and Diseases*, 603-19;

British Trust for Ornithology: Fig 7.1 (unpublished data);

Chiron Press: Figs. 6.9, 6.10, 6.11, 6.13, 9.35, Tables 3.4, 4.3, 6.1, reprinted from Ricklefs (1971) *Ecology*;

Collier Macmillan: Fig. 4.6, reprinted from Whittaker, R.H. (1970/1975) *Communities and Ecosystems*;

CSIRO Publications: Figs 1.14, 1.15, from *Austr. J. Zool, 22*, 219-39 (1974);

Ecological Society of America: Fig. 3.3, from *Ecol. Monogr. 27*, 283-302 (1957);

Entomological Society of Canada: Table 7.5, from *Can. Ent. 103*, 1049-61 (1971); Fig. 9.10; from *Mem. Ent. Soc. Canada 45*, 1-60 (1965);

W. Junk: Table 13.1, from Van Dobben, W.H. & Lowe McConnell, R.H. (1975) *Unifying Concepts in Ecology*;

Harper & Row: Fig. 1.4, from Pianka, E.R. (1981) *Evolutionary Ecology*, Fig. 5.4, from MacArthur, R.H. (1972) *Geographical Ecology*;

Longman: Figs 7.3, 7.8, 8.16, from Silvertown, J.R. (1982) *Introduction to Plant Population Ecology*;

Macmillan Journals: Fig. 2.2, *Nature 225*, 472-4 (1970), Fig. 8.10, *Nature 227*, 1368-9 (1970), Fig. 10.4a, *Nature 276*, 71-3 (1978); Prentice Hall International: Fig. 6.8, from Collier *et al.* (1976) *Dynamic Ecology*;

Princeton University Press: Fig. 2.8, from Cody, M.L. (1974); *Competition and the Structure of Bird Communities*; Fig. 4.2, from MacArthur, R.H. & Wilson, E.O. (1967) *The Theory of Island Biogeography*, Figs. 9.11, 9.18, 9.23, 9.24, from Hassell, M.P. (1978) *The Dynamics of Arthropod Predator/Prey Systems*; Fig. 10.8, from Bonner, J.T. (1965) *Size and Cycle: an essay on the structure of biology*;

PUDOC — Centre for Agricultural Publishing & Documentation, Wageningen, Netherlands: Figs. 7.24, 7.42, from Barlow & Dixon (1980) *Simulation of Lime Aphid Population Dynamics*;

Regents of the University of Colorado; Table 1.1, from *Arctic & Alpine Research 3*, 277-89 (1971);

Saunders College Publishing: Figs 6.2, 13.4; Tables 3.1, 3.2, 3.3, 3.5, 4.1, 4.2, from Odum, E.P. (1959) *Fundamentals of Ecology*;

Springer Verlag: Fig. 1.6, from Walter, H. (1973) *Vegetation of the Earth in Relation to Climate and the Eco-physiological Conditions*, Fig. 7.19 from *Oecologia 7*, 26-67 (1971); Figs 7.44, 7.45, from *Oecologia 50*, 412-16 (1981); Fig. 10.9 from *Oecologia 10*, 294-312 (1972), Table 5.1, from *Oecologia 35*, 241 (1978), Table 8.2, from *Oecologia 7*, 26-67 (1971);

US National Academy of Sciences: Fig. 6.8, from *Proc. Acad. Nat. Sci. Philadelphia 105*, 1-12 (1953);

University of California Press: Fig. 1.18, from *Univ. Calif. Publs. Zool. 64*, 1-56 (1962);

University of Chicago Press: Fig. 10.2, from *American Naturalist 93*, 145-59

(1959), Fig. 1.5, from *Physiological Zoology 43*, 30-7 (1970), Fig. 1.10, from *Physiol. Zool. 45*, 1-13 (1972);

University of Texas Press: Fig. 11.1, from Gilbert, L.E. & Raven, P.H. (1975) *Coevolution of Animals and Plants*;

John Wiley & Sons: Figs 5.7, 5.8, from MacArthur, R.H. & Connell, J.H. (1969) *The Biology of Populations*, Figs 11.4, 11.5, from Caughley, G. (1976) *Analysis of Vertebrate Populations*, Fig. 9.35, Table 9.5 both from 'The Application of Mathematical Models of Fish Populations', in *The Exploitation of Natural Animal Populations*, eds. E.D. Le Cren & M.W. Holdgate;

Zoological Society of London: Fig. 1.19, from *Symp. Zoo. Soc. Lond. 31*, 215-27 (1972), Fig. 8.11 from *J. Zool. Lond. 198*, 83-102 (1982)